ALLAN WILLIAMS

OPERATION CROSSBOW

arrow books

Published by Arrow Books 2014

2 4 6 8 10 9 7 5 3 1

First published in Great Britain in 2013 by
Preface Publishing
Random House, 20 Vauxhall Bridge Road,
London SW1V 2SA

www.randomhouse.co.uk

Addresses for companies within The Random House Group Limited can be found at:
www.randomhouse.co.uk/offices.htm

The Random House Group Limited Reg. No. 954009

A CIP catalogue record for this book
is available from the British Library

ISBN 978 0 0995 5733 3

The Random House Group Limited supports the Forest Stewardship
Council® (FSC®), the leading international forest-certification organisation.
Our books carrying the FSC label are printed on FSC®-certified paper. FSC is
the only forest-certification scheme supported by the leading environmental
organisations, including Greenpeace. Our paper procurement policy can
be found at: www.randomhouse.co.uk/environment

Typeset in Dante MT by SX Composing DTP, Rayleigh, Essex
Printed and bound by CPI Group (UK) Ltd, Croydon, CR0 4YY

'Reconnaissance, or observation, can never be superseded; knowledge comes before power; and the air is first of all a place to be seen from'

– Sir Walter A. Raleigh, *The War in the Air, Being the Story of the Part Played in the Great War by the Royal Air Force: Volume One* (1922)

'The military organisation with the best aerial reconnaissance will win the next war'

– *Generaloberst* Werner Freiherr von Fritsch of the German High Command, 1938

To the Photographic Interpreters and Imagery Analysts, whose professionalism, skill and dedication have served to maintain freedom and peace in the world.

Contents

Organisational Structure of Medmenham, 1945

Technical Control Officer
Progress Control
Senior Photographic Intelligence Officer (SPIO)
Intelligence (Ground Intelligence / Ground Liaison)

Sections
A – Naval
B – Army (Army Photographic Interpretation Section)
      B2 – Enemy Coastline; Bodyline / Crossbow
      B3 – Coastal Defences & Siegfried Line
      B4 – USAAF Southern Germany and Austria
      B5 – Far East Sub-Section
      B6 – Underground Factories and Facilities
      B7 – Dutch Coast and Polder flooding
C – Airfields
D – Industry
E – Camouflage
F – Transport and Communications
G – Wireless (subsequently Signals Intelligence)
H – Control Commission
J – Press & Publicity
K – Damage Assessment (previously Bomber Command PI Section)
L – Aircraft and Aircraft Factories
N – Night Photography
P – Plotting
Q – Enemy Decoys
R – Combined Operations
S – Survey Liaison

T – Target Intelligence (post 1942 merged with Ground Intelligence)
V – Model-Making
W – Photogrammetric (Wild Machines) and Drawing
X – Topographic (Navigational Aids and Landmarks)
Z – Second Phase
 Shipping
 Airfields
 Railways
 Balloons
 Coverage

Support Sections

PI School
Cover Search
Photographic
Film Library
Print Library
Map Library

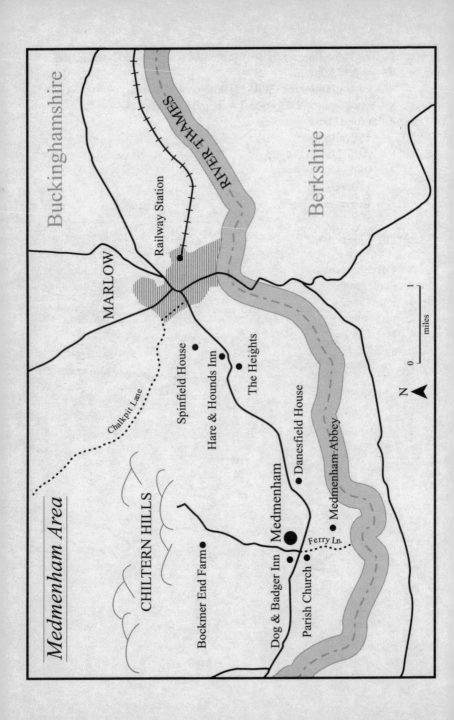

Medmenham Area

Buckinghamshire

RIVER THAMES

Railway Station

MARLOW

Berkshire

Chalkpit Lane

Spinfield House

Hare & Hounds Inn

The Heights

Danesfield House

Medmenham Abbey

CHILTERN HILLS

Bockmer End Farm

Medmenham

Dog & Badger Inn

Ferry Ln.

Parish Church

N

0   1
miles

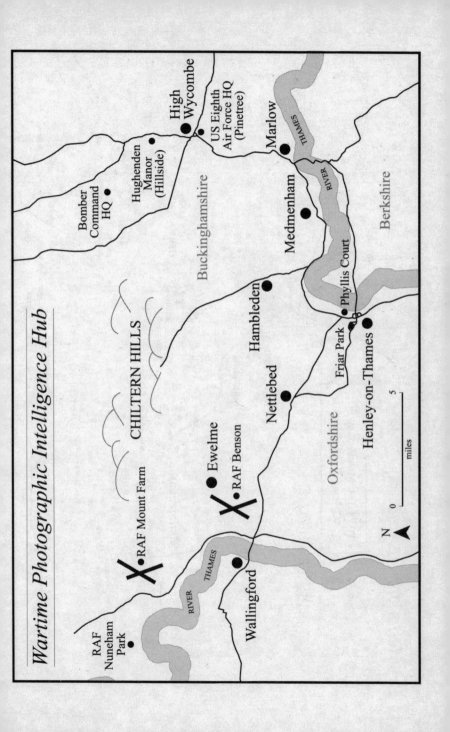

Wartime Photographic Intelligence Hub

RAF Nuneham Park

THAMES
RIVER

Wallingford

RAF Mount Farm

CHILTERN HILLS

Ewelme
RAF Benson

Nettlebed

Hambleden

Oxfordshire

Henley-on-Thames

Friar Park

Phyllis Court

Medmenham

RIVER

THAMES

Marlow

Berkshire

Buckinghamshire

Bomber
Command
HQ

Hughenden
Manor
(Hillside)

High
Wycombe

US Eighth
Air Force HQ
(Pinetree)

N

0   5

miles

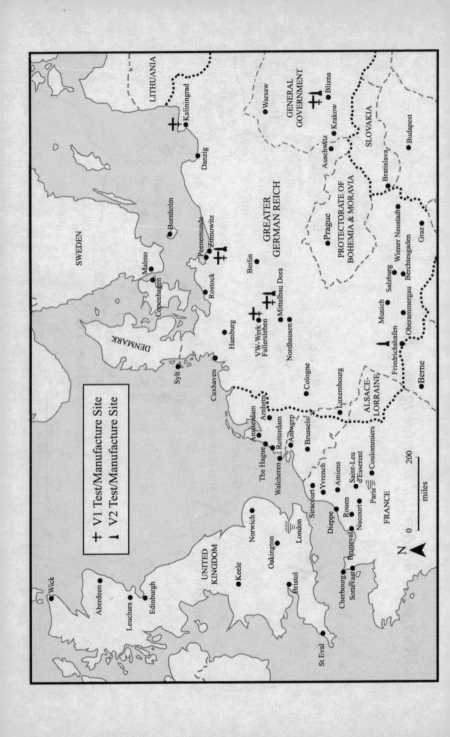

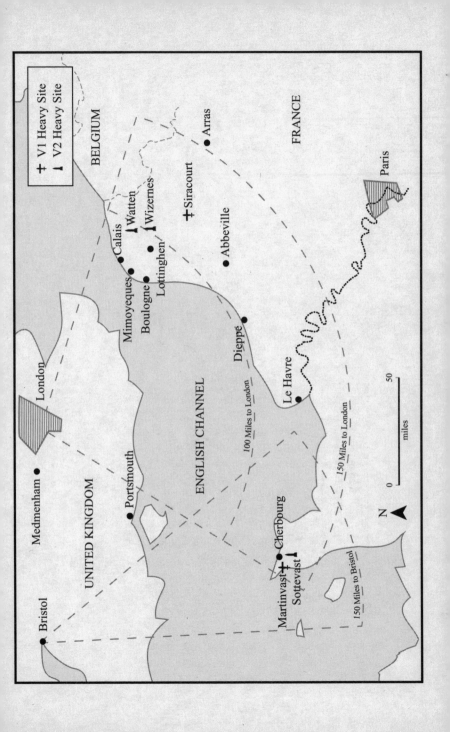

Prologue

On the morning of 6 June 1944, a few hours after the D-Day landings began in Normandy, the German High Command gave immediate orders for the first of their revenge weapons – *Vergeltungswaffe* – to be used against Britain. A week later, sent aloft from a network of launch sites in the Pas-de-Calais, flying bombs appeared in the skies over southern England. They flew northwards at between 2,000 and 3,000 feet – on a course that never appeared to deviate – and the war-hardened British civilians on the ground witnessed strange objects travelling through the air at speeds that appeared to outpace even fighter aircraft.

By the time of this first assault the Germans had been playing a high-stakes game of cat and mouse with Allied photographic intelligence for the previous year and a quarter. For Operation Crossbow, photographic reconnaissance pilots systematically photographed the enemy's top-secret experimental and manufacturing sites across the German Reich, and all of northern France within a 130-mile radius of London, in order that photographic interpreters could pinpoint the network of launch sites. But despite the Allied bomber offensive that destroyed all ninety-six first-generation sites in northern France before D-Day, in the spring of 1944 the Germans began to craftily develop modified emplacements, heavily camouflaged and aimed at specific targets across the English Channel. Whilst the launch sites directed against Plymouth, Bristol and other British towns and cities would be captured or destroyed before the Germans could use them, the new Battle of London had begun.

The weapon concerned was the *Vergeltungswaffen 1* – V1 – a pilotless aircraft armed with a warhead packed with high explosives and propelled by a pulse-jet engine that eerily cut out over the target area after it had flown a predetermined distance. It had been developed by the Luftwaffe under the code name *Cherrystone* and secretly financed by the German Ministry of Aviation who had been using prominent engineering companies to work on a flying bomb for Nazi Germany since the 1930s. Code-named DIVER by British intelligence, and quickly nicknamed the 'doodlebug' by the British public, the flying bomb was a kind of prototype cruise missile and unquestionably a weapon of mass destruction. In the first V1 attack – on 13 June 1944 – two flying bombs landed harmlessly on farmland, another carried on towards London, and a fourth landed in the garden of a large house in Borough Green, Kent. The one that reached London landed on the railway bridge over Grove Road, Bethnal Green, at 4:25 a.m., destroying the tracks and two nearby houses. It also killed six people and injured thirty. The first V1 fatalities in Britain were nineteen-year-old Mrs Ellen Woodcraft and her eight-month-old son Tom, twelve-year-old Lennie Sherman, fifty-five-year-old Dora Cohen, Willie Rogers, aged fifty, and Connie Day, aged thirty-two. It was the moment that the Operation Crossbow team – the photographic interpreters, MI6, members of RAF Fighter and Bomber Commands, all assembled by Churchill to identify and destroy Hitler's secret weapons – had feared. But, as an opening salvo, it was not what the Germans had planned, nor what the British authorities had expected, and it prompted Lord Cherwell – Churchill's friend and scientific adviser who for a long time had doubted the very existence of German secret weapons – to quip that 'the mountain hath groaned and brought forth a mouse'.

The early victims of the V1 could have had little or no sense of what had befallen them. Early survivors and eyewitnesses described a strange sound from the sky, often likened to an old-fashioned motorbike without a silencer. Those who were more observant or more distant saw what they thought was a burning enemy aircraft crossing the sky, with a long flame coming from its tail. People learnt quickly to listen for the trademark noise of the engine: when that stopped there were just a few surreal heart-stopping seconds in which to find some sort of cover that *might* make the difference between life, maiming or death. When 850 kilograms of amatol high explosive exploded on impact, it caused blast damage over an area of around 400–600 yards in each direction and a huge blast wave rippled out from the epicentre. As it did so it left a vacuum, which caused a second rush of air as the vacuum was filled, causing a devastating pushing and pulling effect. At the impact sites there were reports of buildings entirely demolished. Further away from the points of impact blast injuries were inflicted by shards of flying glass. People killed and wounded close to the explosions would be blown apart or would suffer crush injuries from falling masonry. Others lay trapped below collapsed buildings. 'We had to use saucepans and tin baths to pick up the remains,' explained one London firefighter. He had never seen anything like it: 'either on battlefields or during the worst of the previous blitzes.'

A few days after Lord Cherwell's quip – and although many V1s were crashing at the point of launch, or while en route in France or southern England – Luftwaffe Colonel Max Wachtel and Flakregiment 155(W) were bombing London and its suburbs with unremitting frequency. To defend themselves, the British had carefully planned DIVER defence measures. Fighters were the first line of defence, patrolling the Kent and Sussex coasts, backed by a concentrated belt of ack-ack guns in front of a barrage-balloon array that grew thicker as the campaign progressed. With Spitfire, Mustang and Thunderbolt fighters struggling to keep pace with the V1, the pilots of the Tempest V – which could reach speeds of 416 miles per hour at low level – were sent up on anti-doodlebug missions. Reaching their peak in the first week of July 1944, when in seven days more than 800 V1s were launched, the attacks continued until the Allied armies advanced through the Pas-de-Calais. When Paris was liberated on 23 August 1944, and as the British and Canadian armies swept through northern France, Wachtel's men evacuated their modified sites and retreated eastward towards the Netherlands. As a result, there was a rapid drop in the amount of V1 launches. With the number of attacks greatly diminished, the politician and Churchill's son-in-law Duncan Sandys – Chairman of the Crossbow Committee – boldly announced at a press conference on 7 September 1944 that 'except possibly for a few last shots, the Battle of London is over'. By that stage, there had been eighty days of bombardment from over 8,000 flying bombs of which 2,300 made it through to the British capital.

But despite the reassurance by Duncan Sandys, on Friday, 8 September 1944 an even deadlier reprisal weapon struck Staveley

Road, Chiswick: *Vergeltungswaffe 2* – the long-range rocket. At dawn the previous day, V2 rockets had hit Paris, killing six people in the suburb of Maisons-Alfort, but with the Allies advancing so quickly on the ground through Belgium towards Germany, Hitler's rocket soldiers withdrew and joined the batteries at The Hague. Paris might have been saved, but at the expense of London. Since the V2 rocket travelled from the Netherlands at more than 3,000 miles per hour, the three people killed in Staveley Road would have been completely unaware of the thirteen-tonne rocket with its one-tonne payload of high explosive heading towards them. Everyone else in west London, however, would have heard it. Not only the immense impact explosion but also the additional thunderclap as the rocket broke the sound barrier. This incident, and ones that followed, would be met with a conspiracy of silence as the government disguised the first V2 attack as a 'gas-line explosion'. But Londoners naturally became increasingly suspicious when the number of 'gas explosions' escalated steadily. Prime Minister Churchill was forced to inform his public of Hitler's secret weapon that dived without warning from the stratosphere at supersonic speed. For years the V2 had puzzled Britain's best brains – boffins, bureaucrats, spooks, generals, politicians, intelligence-gatherers and intelligence-sifters. Under the code name BIG BEN British intelligence had been investigating the long-range rocket. But opinion had been divided among Britain's top scientists over whether a liquid-fuelled rocket could be developed and, if so, what form it would take. This ensured that, from the outset of their research in February 1943 – initially known as the BODYLINE investigation on the basis that such weapons were just not cricket – the photographic interpreters were given the near-impossible task of finding the unknown.

By late November 1944, Londoners had grown accustomed to the sound of rocket explosions and with fatalities in Forest Hill, Peckham, Welling, Deptford, Aldgate, Bethnal Green, Poplar, Greenwich and Wandsworth, people frequently wondered if they would be next.

As food shortages made rationing and queuing commonplace in wartime Britain, the family chore for schoolboy Robin Hatton on Saturday, 25 November 1944 was predictable. Late in the morning he joined the queue in front of Kennedy's Sausage Shop – one of a popular chain of butchers' shops in south London – at 301 New Cross Road in the London Borough of Lewisham. To keep the family's place in the queue, his sisters Pat and Sylvia changed places with him while he walked 75 yards to the nearest Woolworths to buy some lemonade. On leaving the department store he headed across the road, and on reaching the other side was suddenly hurled through a fence by the explosive force of a rocket blast. When the ensuing chaos died down he looked back across the street to see that Woolworths and the neighbouring Co-op had disappeared. The dead 'were all over the place, in the road, in cars and lorries and buses and trams'. As a result of the trauma he'd suffered Robin briefly lost the ability to speak. He now faced the double distress of being able to see his sisters frantically looking for him across the road but being unable either to call them or reach them through the carnage.

That Saturday lunchtime New Cross Road had been thronging with pre-Christmas shoppers, women, children and servicemen queuing at the tills, their numbers swelled by rumours of the arrival of a consignment of cooking utensils at Woolworths. At 12:26 p.m. a rocket exploded immediately behind the department store, the walls of which bulged outwards slightly before collapsing inwards in a great cloud of dust and smoke. Whilst people inside and passing by stood little chance of survival, many more were killed in the neighbouring Co-op store when it disintegrated around them. With 168 people killed, this was the worst-ever V-weapon attack on Britain and it took the Civil Defence authorities whole days and nights to extricate the injured and dead from the mountains of rubble. An ordinary London shopping street had been turned into a battlefield.

As the Allied soldiers advanced through northern France they discovered NO-BALL sites in a landscape which had been so comprehensively bombed that it frequently resembled a moonscape. Presumably christened by a cricketing fan, NO-BALL targets were those suspected of having any connection with Operation Crossbow. They included enormous bunkers at Mimoyecques, Siracourt, Sottevast, Watten and Wizernes, which had been constructed with one purpose in mind: the bombardment of Britain with long-range rockets, flying bombs and artillery shells. But whilst the V1 might have required a fixed launch site when fired from the ground, the V2 rocket was by comparison highly mobile. This meant that the only defence from this particular

OPERATION CROSSBOW

weapon was being outside its range. This was shown to be the case
for areas around the large target of London during Operation Market
Garden, when Hans Kammler, the SS General then responsible for
the V2, fearing that his rocket units might be cut off by Allied forces,
relocated them from The Hague to the Gaasterland district of south-
west Friesland. This meant that while London was beyond range of
the V2 the cities of Norwich and Ipswich were now within its reach.
To add further to Duncan Sandys's woes, Colonel Wachtel was
quick to regroup on leaving northern France, and his men were
soon ready to launch the V1 from specially modified Heinkel aircraft
and from ground sites in the Netherlands. The Battle of London was
far from over.

As the V1 and V2 were revolutionary new weapons, the German High
Command were quick to deploy them on sites in Europe in what came
to be known by the Allies as the Continental Crossbow.

No city was to suffer more than Antwerp and, on Saturday,
16 December 1944, more than 1,000 people were packed into the
Rex Cinema in the liberated Belgian city, watching the afternoon
screening. As the cinema audience watched Gary Cooper in an
American western – *The Plainsman* – the SS Werfer Battery 500 fired
a V2 rocket from a forest near Hellendoorn, in an eastern district of
the Netherlands. Landing directly on the cinema roof, the missile
exploded, killing 567 filmgoers and injuring 291. This was to be the
single highest death toll from one rocket attack during the war, one
in which 296 of the dead and 194 of the injured were off-duty British,
American and Canadian servicemen. Antwerp with its strategically
important port – through which the Allies were passing the bulk
of the supplies for their advancing armies – presented the Germans
with a large and vulnerable target. This successful strike for them
coincided with the start of their Ardennes Offensive and the reign of
terror from the skies made the war-weary citizens of Antwerp tense
and fearful, prompting *Time* magazine to christen it 'The City of
Sudden Death'.

In total some 8,564 V1s would be launched against England, of which
just over 1,000 crashed on take-off. The flying bombs killed 6,184 people
in London, seriously injured around 18,000 and damaged or destroyed

750,000 houses. For every V1 launched, roughly one person was killed and three were seriously injured. Added to this some 1,190 V2 rockets reached England. Of these about 500 actually hit the London area, killing 2,724 people and injuring 6,467 – a far higher casualty ratio than that of the flying bomb. To these fatalities were added another 2,000 Allied airmen killed in the counter-offensive against the V-weapon sites, during which some 50,000 tonnes of bombs were dropped. From October 1944 onwards 500 V2 rockets were being built per month, of which 200 were fired against London and 300 against other targets, mostly Antwerp. Despite this death and destruction, the disruption to the war effort from the V weapons would ultimately prove minor. As a forerunner of the future, the V2 might have been a remarkable engineering achievement by Dr Wernher von Braun and his rocket scientists. But since it took the same manufacturing effort to build one V2 as it did to build six or seven fighter aircraft, the rocket proved a very costly method of delivering one tonne of high explosive.

If, however, the Germans had been able to bombard Britain with flying bombs, rockets and long-range shells at the rate they had planned it would have changed the course of the war. London and the major population centres of southern Britain would have been destroyed, which would have meant that preparations for the D-Day landings would have been impossible. To defeat Hitler's V weapons, photographic intelligence was central to an operation that would also use all-source secret information, including one of the greatest leaks of scientific intelligence of the entire war, as well as undercover agents, resistance movements, the interrogation of POWs, clandestine bugging and signals interceptions. By the time of Operation Crossbow, photographic evidence was universally recognised as one of the most important as well as unbiased sources of intelligence. There were squadrons flying throughout the world, dedicated to taking aerial photographs. There were also hundreds of experienced photographic interpreters who were producing detailed information. This had been far from the case at the beginning of the Second World War when the RAF's photo-reconnaissance units had suffered a catastrophic loss of men and aeroplanes. Despite photographic intelligence having proved itself in the First World War on a massive scale, the RAF had failed to develop this crucial function in the interwar years: with the outbreak of war in 1939 they were almost starting from scratch.

*

Sitting prominently in the landscape on an escarpment above the river Thames, with a labyrinth of Nissen huts in its grounds, chalk-stoned Danesfield House near the Buckinghamshire village of Medmenham was perhaps a surprising choice for a top-secret headquarters of photographic intelligence. As one among hundreds of requisitioned country houses it would survive the war unscathed. Danesfield House was destined to become RAF Medmenham and home to the unit responsible for the interpretation of aerial photography, the Central Interpretation Unit (CIU). Administered as an RAF Station, it also had expert personnel from the Navy, the Army, and from many of the Allies, particularly once the Americans joined the war and it became the Allied Central Interpretation Unit (ACIU). Medmenham was to photographic intelligence what Bletchley Park was to signals intelligence. It had similar criteria for the recruitment of personnel and became home to an eclectic assortment of brilliant people. This was its strength: civilians from both the scientific and artistic worlds were put into uniform and brought with them a range of skills, knowledge and experience which were adapted to military-intelligence use. With greater equality than they typically enjoyed in civilian life, women played a critical role in a unit that was more akin to an Oxbridge college than a secret intelligence outfit.

During the war tens of millions of aerial photographs were taken by photographic-reconnaissance squadrons that were operational throughout the world. The brave pilots, who typically flew alone in modified and unarmed fighters camouflaged with seemingly bizarre colour schemes, took aerial photographs that would change the course of the war. Photographic intelligence rapidly became a major and indispensable war-winning asset, with a huge investment in aircraft, other equipment and, crucially, expert interpreters drawn from every walk of life. The quality and frequency of the reconnaissance results allowed the interpreters to accurately and continuously monitor enemy activity. The techniques that were used allowed them to view the war unfolding, with all its triumphs and tragedies, in 3D. Using a Swiss built optical instrument called the Wild – pronounced Vilt – Machine, a vast library of millions of aerial photographs and detailed intelligence reports grew and recorded the Second World War as it happened.

*

At Medmenham, L Section was responsible for the interpretation of photographs of aircraft and aircraft factories, and since its creation, had been run by Flight Officer Constance Babington Smith. Affectionately known as 'Babs' by her friends, she was a well-connected society girl who was fascinated by aviation and who before the war had written articles for *The Aeroplane* magazine, as well as leading a busy social life in London. Slight, pretty, immaculately dressed, she had delicate features and a sparkling deep intelligence. Her favoured perfume was Guerlain's *L'Heure Bleue*. She wore the scent, she said, to counteract the masculinity of her uniform. Her section used aerial photographs to identify prototype aeroplanes, new varieties of aircraft, production facilities and operational activity. Like many other British and American photographic interpreters they were given at times the near-impossible task of analysing the photographs in order to find and identify the unknown.

The thirteenth of November 1943 started as just a normal working day for Constance, but when she was asked by Wing Commander Douglas Kendall – who masterminded all the interpretation work at Medmenham – to look out for aircraft at Peenemunde that might be pilotless, it was the start of a hunt that would prove extraordinary. The search by L section was under way when on 28 November a de Havilland Mosquito piloted by Squadron Leader John Merifield returned with a ground-breaking photograph. When the photograph was analysed by the aircraft specialists, they identified a ramp bearing a tiny cruciform shape on its rails. This discovery, together with the examination of many thousands of photographs of other possible launch sites and storage depots, confirmed that a flying-bomb offensive was being prepared on the other side of the Channel.

For Operation Crossbow – during the hunt for the flying bomb – over 3,000 sorties were flown and 1.2 million prints created for analysis by the Medmenham interpreters. Photographic-reconnaissance pilots flew high-altitude missions in unarmed Spitfires and Mosquitos all across Europe to uncover the vast network of V-weapon experimental, manufacturing, transportation, storage and launch sites. Being little over thirty miles from central London, the Medmenham area was not immune from Hitler's V weapons, and for this reason one night-shift team were not altogether surprised – but were apparently completely transfixed – when on 22 July 1944 the familiar sound of a V1 was heard, followed by the fateful engine cut-out and the inevitable explosion. This came as the flying bomb landed a little over two miles away in Chalk Pit Lane, Marlow, causing casualties and damage to houses.

To celebrate the triumphs of photographic intelligence during the conflict, journalists were invited to visit some of the photographic-reconnaissance squadrons at RAF Benson, Oxfordshire, immediately after the war. Following lunch in the Officers' Mess they travelled the 16 miles to RAF Medmenham – along the same route taken by the despatch riders who had delivered the rapidly printed aerial photographs during the war – to the nerve centre of the camera war. Welcomed by Wing Commander Douglas Kendall – who introduced the top-secret work of the photographic interpreters – their role during Operation Crossbow, when everything was wanted in a hurry, and it was a non-stop race against the clock, was revealed. One of the journalists was clearly excited by his visit and likened the cellars that housed the Wild Machines to the kind of place that an international gang of the fictional thriller variety would use for its secret headquarters. In an official photograph released at the time, Flight Lieutenant Harry Williams – who had worked with Squadron Leader Ramsay Matthews in the top-secret mission to smuggle the machines into the country – is shown operating the photogrammetric equipment, which was already processing photographs taken during Operation Revue. It had been revealed to the pressmen earlier in the day that the pilots who had carried out daring PR missions during the war were now busily surveying Britain for the Ministry of Town and Country Planning to help with post-war reconstruction. In the photograph caption, the vital role of the Wild Machine, operated twenty-four hours a day throughout the war to produce accurate maps from stereo pairs, was revealed. But in a show of classic British understatement, Williams likened its operation to that of a sausage machine, in that two photographs went into the contraption whereupon a handle was turned and a map came out.

During the Second World War the Germans failed to heed the 1938 warning of General Werner Freiherr von Fritsch of the German High Command that 'the military organisation with the best aerial recon-naissance will win the next war'. By 1943 and Operation Crossbow,

Allied photographic reconnaissance and interpretation became universally recognised as one of the most important and completely unbiased sources of intelligence, with photographic evidence being used in the planning stages of practically every Allied military operation. That Britain and the Allies were able to photograph Nazi Europe so thoroughly was down to the determination, application and tenacity of a series of photo-intelligence pioneers who laid the foundations and provided the expertise that Churchill and the Crossbow team were to draw on with such profound consequences in 1943–44. In 1938, with war on the horizon, these operators included a First World War flying veteran recruited by MI6, a maverick Australian and his hand-picked team of reconnaissance pilots flying from a west London aerodrome. After the outbreak of war it included Stephen Spender's older brother, a mapping machine, Constance Babington Smith and a daughter of Winston Churchill. All these – backed by Churchill himself – pioneered approaches to photographic intelligence that allowed the Allies to spy from the sky. Together they created the infrastructure that Churchill harnessed for Operation Crossbow.

CHAPTER 1
Wizard War

In September 1938 the British Prime Minister, Neville Chamberlain, was flown by British Airways to and from Germany via Heston Aerodrome for talks with Adolf Hitler at his Berchtesgaden mountain retreat in southern Bavaria. On his return from Germany to the west London aerodrome on 30 September, Chamberlain addressed the gathered crowds, holding aloft the Munich Agreement signed by Hitler that he would later claim on the steps of Number 10 Downing Street signalled 'peace for our time'. Paradoxically, MI6 were already planning aerial espionage of Germany from the very same aerodrome – with help from British Airways – that would lay the foundations for photographic intelligence throughout the Second World War and beyond. In the years leading up to the outbreak of war, in addition to the Communist threat, MI6 had been assessing the intentions and military capabilities of Nazi Germany. But with intelligence from secret agents drying up and limited capacity to travel unhindered within Germany, the old methods of espionage were increasingly useless.

During the 1930s, the MI6 headquarters was the anonymous-looking Broadway Buildings – 54 Broadway – near St James's Park Underground station, conveniently close to Whitehall. One of the most successful MI6 spymasters, who infiltrated Nazi Germany using an assortment of espionage tactics, was Wing Commander Frederick Winterbotham. Like so many of his generation, he had been inspired by the aeronautical adventures of the Wright Brothers, and during the First World War had joined the Royal Flying Corps. As a pilot, he had been providing air cover for a photographic-reconnaissance aircraft during the Battle of Passchendaele when he was shot down over the

German trenches. Whilst this cost him a broken nose, it gave the Old Carthusian an opportunity to observe the German military mind first-hand during his stay in a prisoner-of-war camp and helped ensure his survival of a war that blighted a generation of Charterhouse old boys. In 1930, and with the Royal Air Force now firmly established as the third fighting force, he joined its senior command structure – the Air Staff – and, whilst officially a liaison officer, unofficially he headed the Air Department at MI6 where he reported to the Chief of the Intelligence Service, Admiral Sir Hugh Sinclair – who maintained the tradition of the Chief signing documents with the initial 'C' in green ink. The immense task facing Winterbotham, in a world increasingly dominated by air power, was the creation of an Air Intelligence service.

Throughout the 1930s Winterbotham travelled regularly to Germany whilst purportedly on leave, and presented himself as a senior member of the Air Staff with Nazi sympathies. He made personal contact with many of the Nazi leaders and in March 1934 this culminated in an audience with the Führer himself in the Chancellery in Berlin. On entering a vast office guarded by two tall black-uniformed guards, he found a 'little man, forelock, moustache and all . . . dressed in the simple brown shirt and black tie, with nothing on his uniform except the swastika armband.' In the course of their meeting Hitler outlined his intentions to develop a first-class modern air force in flagrant breach of the Treaty of Versailles. Winterbotham knew, from his experience as a pilot, the superiority of new aircraft over older ones and realised that, even if Germany was to start from nothing, in a very short time one hundred modern aircraft would be worth three hundred antiquated British ones. The dangers posed by this 'Gathering Storm' and the British policy of appeasement troubled fellow spook Desmond Morton, head of the Industrial Intelligence Centre, who had been tasked with obtaining and collating information on German armaments and factories. A member of the Chartwell circle, Morton was routinely invited for Sunday lunch at the country home of Winston Churchill MP. As a confidant of Winston, Morton took the opportunity to leak information that provided Churchill with the ammunition he needed to ask Prime Minister Stanley Baldwin probing parliamentary questions about the growing strength of the German Luftwaffe.

A fellow member of the Chartwell circle was Frederick Alexander

Lindemann, then Professor of Experimental Philosophy at Oxford University and later Churchill's wartime Chief Scientific Adviser. Whilst on first impression they might have appeared complete opposites, with Churchill counting eating, smoking and drinking among his pleasures and Lindemann a vegetarian, non-smoking teetotaller, they quickly came to appreciate each other's qualities. The Jewish-German émigré feared the growing threat of Nazism, and in a letter to *The Times* in August 1934 – 'Science and Air Bombing' – he questioned the widely held belief that there could be no defence against bombing aeroplanes and that any nation assaulted in this way could only rely on counter-attack and reprisals. If this were the case, he argued, bombers in the hands of gangster governments threatened the whole future of western civilisation. Despite the Air Ministry's earlier advice to Baldwin some of their own scientists agreed with Lindemann and Churchill, and so – fearful that the responsibility for air defence might be removed from them – the Ministry was spurred into action. The result, in late 1934, was the setting-up of the Committee for the Scientific Survey of Air Defence. Typically referred to as the Tizard Committee, it was chaired by the eminent chemist and inventor Professor Sir Henry Tizard, and whilst perhaps best known for its role shepherding the development of radar and the building of the Chain Home Radar Array that proved so effective during the Battle of Britain, it also helped ensure that scientists would play a role in warfare like they never had before. Churchill understood that the next war would be 'between scientists as much as soldiers, sailors and airmen'. In his words, the application of science to warfare, the work of the boffins, was 'The Wizard War'.

As information from earlier sources inside Nazi Germany became harder to obtain, Winterbotham turned to aerial photography to discover what the Germans were doing. Around the time of the Munich Agreement in late 1938, he had been discussing the challenge with Colonel George Ronin, his opposite number at the French intelligence agency, the Deuxième Bureau. The French had been carrying out aerial espionage over Germany using an old bomber, which they had flown over the Rhineland and Siegfried Line. But since hiring the services of a society photographer from Paris – who operated an old wooden camera – was proving to be an 'amateur and hopeless affair' the two men agreed to investigate alternative approaches. For advice they turned to Alfred J. Miranda Jr of the American Armament Corporation who thought the scheme might appeal to a business associate. And so it was that (Frederick) Sidney Cotton, an Australian aviator and entrepreneur who was then based in London, was recruited.

A colourful character with a love of aviation, money and women, Cotton had flown with the Royal Naval Air Service during the First World War, had invented the revolutionary 'Sidcot' (Sid Cot) silk- and fur-lined flight suit, and embarked on a business career that was dominated by aviation and his unique entrepreneurial approach to life. In the 1920s he established an airmail, aerial survey and seal-spotting business in Newfoundland and by the early 1930s was a pioneer in colour photography and an all-round serial inventor. As the owner of the European sales rights of the revolutionary colour photographic film 'Dufaycolor', Cotton frequently flew all over Europe in his own private aeroplane and was well known in aviation circles. This was just the kind of man that Winterbotham and Ronin were looking for, someone who would not arouse any suspicion as he travelled on business in a private aeroplane.

In the meetings that followed, having sworn Cotton to absolute secrecy, Winterbotham introduced the Anglo-French scheme to him, explaining how the French had struggled to take photographs from short-range aeroplanes, how the Germans were increasingly suspicious of French aircraft, why other sources of intelligence had dried up, and why a modern long-range 'civilian' plane owned and flown by a 'British' civilian could be the way to undertake clandestine photography. For Cotton this couldn't have come at a better time, since recent business failures with Dufaycolor had given him financial problems and had led to him sinking into a trough of depression. Agreement was reached with Cotton, who was extremely enthusiastic about the scheme, and with a generous allowance he fronted a new company – the Aeronautical Research and Sales Corporation – from his offices at 3 St James Square, London, SW1.

Cotton was to combine the activities of this cover company, which would take clandestine aerial photographs of military installations and manufacturing sites in German and Italian territory, with his existing Dufaycolor business. Although Winterbotham was frequently frustrated by Cotton and his Machiavellian ways, he recognised completely the value of his discretion and entrepreneurial approach. Concerning Cotton, he remarked that whilst the aviator-entrepreneur's 'strongest instinct' was 'to make money' he nonetheless loved adventure 'for the sake of adventure' and 'had a business approach to everything'. Although Winterbotham acknowledged that Cotton could be 'most annoying to officialdom' he felt that the Australian 'was completely reliable on security' – 'like a clam' – and that he should know as he would regularly 'check up on him'.

After several meetings it was decided that the luxurious Lockheed-12A 'Electra Junior' aeroplane would be the right choice for the top-secret Anglo-French operation. This ultra-modern American aircraft, complete with heated cabin, would fit Cotton's role as the successful businessman. To avoid suspicion, British Airways were approached and agreed to place the order. An isolated hangar was rented from Airwork Limited at Heston Aerodrome, Middlesex, along with rooms in the Aero Club. Situated ten miles west of central London, the aerodrome was part-owned by businessman Nigel Norman, whose cooperation and discretion Winterbotham trusted. But even before the first reconnaissance mission was flown, on the morning of 15 March 1939, the German Army marched into Czechoslovakia; the Czech government gave its official surrender in Prague the following day. With his policy of appeasement in ruins, Chamberlain set in place preparations for war and assured the Poles that Britain would support them if their independence was threatened. Chamberlain's policy of against-the-odds hope shifted to awareness of a reality that Winterbotham and MI6 had long feared.

Whilst MI6 and the Deuxième Bureau carried out a series of missions over Germany in late March and early April 1939, by mutual consent they quickly felt it would be better if they worked on their own aerial-espionage programmes. With the French then taking custody of the Lockheed-12A – which they had already paid for with sackloads of untraceable French banknotes – another Lockheed was bought from the Gillies Aviation Corporation of Garden City, Long Island, New York. Arriving at Southampton aboard the SS *Aquitania* – and passing through Customs courtesy of a note from Admiral Sinclair – it was assembled by Cunliffe-Owen Ltd in Eastleigh, near Southampton, and registered courtesy of British Airways. Aircraft G-AFTL was destined to become one of the most significant aircraft in the history of aerial espionage. Cotton set about modifying the aeroplane with his customary zeal and had two 70-gallon fuel tanks installed to extend the range of the aeroplane from 700 to 1,600 miles. A hole was cut in the entrance door, along with two in the fuselage, to allow cameras to be set up and operated, with small metal panels that made them almost impossible to see when the cameras were not in use. Before turning their attentions on Germany, Winterbotham decided that Cotton and his co-pilot Robert 'Bob' Niven – a tall, zestful and enthusiastic Canadian who was recruited after his short-service commission in the RAF came to an end – should conduct an air tour of the Middle East. With Cotton playing the part of a rich, socially connected English gentleman with an interest in archaeological ruins, they brazenly photographed military targets throughout the region in mid-June 1939. This was followed by a long series of 'business trips' to Germany in the Lockheed that led them to earn the distinction – on 24 August 1939 – of being the last civilian aeroplane to fly out of Berlin before the outbreak of war.

On Tuesday, 19 September 1939, with the Second World War little over two weeks old, Adolf Hitler made a typically vitriolic speech in the city of Danzig – recently appropriated by the Nazis – in which he claimed, the headlines reported, that the Germans had a secret weapon. Chamberlain asked MI6 to investigate, and although they were unable to provide any immediate answers they were 'only too glad to be able to report that they had just appointed a man for the very purpose, and would ask him to'. The man in question was twenty-seven-year-old physicist Dr Reginald Victor Jones, who had just transferred from the Royal Aircraft Establishment (RAE), Farnborough, where he had been working on the problems associated with defending Britain from air attack. Jones had studied for his doctorate at the Clarendon Laboratory, Oxford, under Professor Lindemann. In May that year a member of the Tizard Committee had approached Jones with a job offer. Since MI6 could provide the committee with no information about the extent to which the Germans were applying science to warfare, it had been decided that a scientist should work with MI6 to discover why their agents were producing such little scientific information. Jones happily accepted but, given the importance of the work he was doing at Farnborough, it was agreed to delay his transfer to MI6 until 1 September 1939.

Jones reported for duty at Broadway Buildings that Friday – on the very day the Germans invaded Poland – and joined Winterbotham's team. With guidance from Winterbotham in the art of spotting the difference between good and bad information, and with his scientific knowledge, his first task was to go through the MI6 intelligence files. Since all the pre-war intelligence files had been evacuated for safe keeping to Bletchley Park, Buckinghamshire, Jones found himself at the top-secret Station X, fifty miles north-west of London. For weeks Jones spent his days scouring intelligence documents and his evenings discussing cryptography with Alan Turing and his colleagues. He found the files weak on science and technology and came to the conclusion that the average MI6 agent was a 'scientific analphabet'. After Hitler's speech in Danzig, and as the weeks passed with no sign of a secret weapon, the Führer's claim soon became a national joke, dismissed by the newspapers as a German bluff, the propaganda of Dr Joseph Goebbels.

Jones went back to the start: he requested a colleague to look again at Hitler's Danzig speech and discovered that the alarm might have been sparked by an error in the Foreign Office translation. Jones asked a linguist friend to check the BBC's recording of the speech and found that Hitler was not referring to a specific weapon, *Waffe*, but to the Luftwaffe as a whole. However, what the trawl through the MI6 files revealed was the extent of real and potential threats, which Jones listed in his report:

- Bacterial Warfare
- New gases
- Flame weapons
- Gliding bombs, aerial torpedoes and pilotless aircraft
- Long-range guns and rockets
- New torpedoes, mines and submarines
- Death rays, engine-stopping rays and magnetic guns.

The threat from secret weapons would lurk in the shadows – its importance in official minds depending on other intelligence priorities.

But this was no bluff. For years, in contravention of the Treaty of Versailles, Nazi Germany had been secretly developing new weapon technology. At Peenemunde, 110 miles north of Berlin, a top-secret experimental military facility had been built for the German Army and Luftwaffe, with an uninterrupted 300-mile stretch of German-controlled coastline at its disposal. The first indication that this place existed was to come via Captain Hector Boyes, the British Naval Attaché in Oslo who received an anonymous letter on 4 November 1939 telling him that if Britain wanted a report on German technical developments, all they need do was alter the preamble on the BBC World Service German language broadcast on a certain evening, so as to say, 'Hallo, hier ist London'. The writer would then know that the information was wanted and would send it.

On the basis that there was nothing to lose, the BBC preamble was duly altered and the intelligence services waited. One evening in November 1939 Winterbotham went into Jones's office and dumped a small parcel on his desk, courtesy of the anonymous source. It contained a typed report and a glass fuse. The document would turn out to be one of the most spectacular security leaks in history. Winterbotham asked Jones to unravel the package's contents. The seven-page typed report detailed radar equipment, new fuses for bombs and shells, the progress of dive-bomber research, the development of rocket technology, details of a large experimental facility at Peenemunde, and a description of a radar aid to guide bombers at night called 'Y-Geraet' or Y-Apparatus.

Jones took the parcel's contents seriously: he wrote that

a careful review of the report leaves only two possible conclusions: (1) That it was a 'plant' to persuade us that the Germans were as well advanced as ourselves or (2) That the source was genuinely disaffected with Germany, and wished to tell us all he knew. The general accuracy of the information, the gratuitous presentation of the fuse, and the fact that the source made no effort, as far as is known, to exploit the matter weigh heavily in favour of the second conclusion. It seems, then, that the source was reliable, and he was manifestly competent.

The report in many ways was too good to be true and could easily have been a hoax, or a nicely placed piece of German deception, because 'no man could possibly have known of all the developments that the report described'.

The report had been signed 'a German scientist who wishes you well'. In 'one great flash', Jones later wrote, this individual 'had given us a synoptic glimpse of much of what was foreshadowed'. What became known as the Oslo Report was the first pointer that Peenemunde, on the Baltic Coast, was the seat of rocket development. The author of the leak would remain unknown throughout the war and for many decades after. When he died in 1980, the last will and testament of German physicist and mathematician Hans Ferdinand Mayer revealed that he was the report's author. At the time he had been a forty-four-year-old Director at Siemens Research Laboratory, Berlin, a senior position that gave him access to such top-secret scientific-military information. Mayer was a brilliant scientist, who Gestapo records later showed had committed a scattering of offences against the Reich – including listening to BBC broadcasts and spreading the news he heard. Arranging a business trip to the Norwegian capital at the beginning of November 1939, Mayer booked into the Hotel Bristol, borrowed a typewriter and wrote his first letter to the British. When the preamble was duly altered on the BBC World Service, he placed his second letter and a little glass fuse in the early hours of 5 November 1939, on the window ledge of the Villa Frognaes, the British Embassy, in the affluent Frogner district of the city. Then he returned under cover of his business role to Berlin.

At the time the report was read, digested, and downgraded by British intelligence as a likely plant. Jones, however, recognised it to be an invaluable resource, revealing the increasingly accurate systems of radio navigation that the Germans had developed to ensure more accurate bombing. During the Battle of the Beams, it helped the British Scientific Intelligence fightback to develop a variety of increasingly effective countermeasures, including signal jamming and the distortion of radio waves. Later, reassessing the report, Jones would describe it as 'probably the best single report received from any source during the war'. As the war progressed, and one Nazi terror weapon after another emerged onto the world stage, it became obvious that Mayer's report was both real and largely factually correct. 'In the few dull moments of the war', said Jones, 'I used to look up the Oslo Report to see what should be coming along next'.

In 1943, Mayer was arrested by the Gestapo for 'crimes against the state' but they knew nothing of the full scale of his treachery. He was interned in concentration camps, including Dachau, for the rest of the war. It is likely that he would have been murdered like so many other millions of prisoners had it not been for his former doctoral supervisor, the Nobel-prize-winning scientist and proponent of 'Aryan physics' Philipp Lenard, who, though a keen Nazi and anti-Semite, intervened on his former student's behalf. After being liberated, Mayer went to the United States as part of Operation Paperclip and initially worked in the US Air Force research laboratory at Wright-Patterson Air Force Base, Dayton, Ohio. Later he moved to Cornell University, New York, where he became Professor of Electrical Engineering before returning to Germany, and Siemens, in 1950.

The existence of the Oslo Report would become more widely known after Jones revealed its existence in a 1947 talk to the Royal United Services Institution but it would be a chance meeting in 1953 with Henry Cobden Turner on board the RMS *Queen Mary* that would lead Jones to the discovery of Mayer's identity. The head of Salford Electrical Instruments – who developed proximity fuses during the war – would later reveal to Jones that at the beginning of the war, he had received correspondence from his friend Mayer, from Oslo. On learning of Mayer's technical background and using Turner as a go-between, Jones began writing to Mayer. In 1955 all three men attended a radar conference in Munich. Meeting in private, and at the request of Mayer – who feared reprisals against his family – Jones vowed never to reveal the identity of the Oslo Report's author during the German's lifetime. In his seminal 1978 book on Scientific Intelligence during the Second World War – *Most Secret War* – Jones praised the German scientist and, true to his word, never revealed Hans Mayer's identity.

In an office two floors below Winterbotham's in Broadway Buildings the cryptographers of the Government Code and Cypher School worked under Commander Alistair Denniston. It was from Denniston that Winterbotham learnt about the world of codes and cyphers and the importance of the Enigma machine to the Germans' secret communications. The Germans produced these devices by the thousand before the war and the British worked with the Polish Secret Service under utmost secrecy to obtain a new electrically operated Enigma machine. Denniston collected it from Poland and brought it back to MI6 headquarters where it stayed until August 1939 when the school was relocated to Bletchley Park. Winterbotham would find himself responsible for the security, organisation and distribution of Ultra intelligence to commanders in the field. The intelligence services knew that information derived from Enigma decrypts would be absolutely authentic, since the Germans considered the codes unbreakable and could therefore speak freely with each other.

Having been mentored by Winterbotham, the value of photographic intelligence was appreciated by Jones from the start, and aerial photography would prove a vital source of information for scientific intelligence throughout the war. Throughout 1940, Winterbotham had been receiving requests for information from Robert Watson-Watt, the inventor of the British radar system, about what progress the Germans had made on radar and other scientific devices that might be used by the Luftwaffe. With Enigma decrypts providing them with a geographical reference to the nearest mile of a Knickebein radar installation near Cherbourg in France, Jones requested photographic reconnaissance of the area in order that he could pinpoint its location and study the transmitter and layout of the installation. In the event, however, it is recorded that he was unhappy with the service he received from a certain Wing Commander Hemming, senior officer at the Photographic Interpretation Unit, who he felt was treating Jones's requests for interpretation as something of a sideshow to their monitoring of the Operation Sea Lion invasion ports. But whilst Jones – not for the last time – might have complained about the service he received from photographic intelligence, Hemming had responsibility for prioritising the many and varied requests they received. Once it was actually confirmed that the Germans were using radio navigational beams to direct their bombers over Britain to their targets, the appropriate level of interest was shown.

When Jones was supplied with reconnaissance photographs of the Knickebein site, he vividly records that during a daylight bombing raid on 17 September 1940, and while taking cover in the crowded basement of Broadway Buildings, he made use of the time to interpret the photographs. Whilst accepting that he was in a better position to know what he was looking for in the photographs – either a squat tower or more probably some sort of turntable – he was the one who actually spotted the Cherbourg Knickebein radar installation. The following day he wrote his report on the site, thanking the Photographic

Interpretation Unit for the photography, and later recorded that when Wing Commander Hemming discovered that Jones was in a position to write reports that could be read by the Prime Minister he was swiftly allocated one of the best photographic interpreters. The partnership that developed between Jones and Claude Wavell was to play a major part in unearthing the Germans' offensive and defensive radio systems, and from the outset highlighted the importance of the reciprocal flow of information between Medmenham and Bletchley Park, with each being able to contribute information to the other. The existence of Bletchley Park, however, would be unknown to the interpreters, with the sole exception of Douglas Kendall in his role as Technical Control Officer at Medmenham. Checking Ultra information against what could be seen on the aerial photography was a powerful combination, but since the subject matter in the photography was on occasion considered too dangerous to bring to the attention of the interpreters they were on many occasions deprived of Ultra information that would have been vital to the success of their work.

CHAPTER 2
The Phoney War, Special Flight and Aerofilms

Six days after Neville Chamberlain landed at Heston Aerodrome and confidently declared the settlement of the Czechoslovakian problem, Major Harold 'Lemnos' Hemming – Managing Director of the Aircraft Operating Company – wrote to the Air Ministry with an intriguing proposal. After outlining the work of his company and its various operating arms – Aircraft Operating Company (Africa), Aerofilms Limited, and H. Hemming and Partners – he explained their unique skills, experience and equipment that would be invaluable to the RAF in the event of a national crisis. Hemming was a veteran Royal Flying Corps pilot, who had encountered Sidney Cotton during the First World War and ran his Newfoundland aerial survey unit during the early 1920s. Hemming was known by the nickname 'Lemnos' after the Greek island where the Royal Flying Corps served during the First World War. Well known in aviation circles – and instantly recognisable with his trademark black eyepatch – Hemming was tall, thin and jovial, had a flair for expressing himself and a wonderful memory. An avid pilot, it had been while taking off from the Ensbury Park racecourse during the 1927 Bournemouth Whitsun meeting of the Royal Aero Club that his de Havilland DH37 biplane hit a scoreboard and crashed, costing him the sight of his right eye. Hemming was one of a group of First World War veterans who, inspired by their wartime experiences in aviation, had recognised the business potential of photographing the world from above.

Established in May 1919, Aerofilms Limited was the first British company to specialise in aerial photography. Francis Wills – its founding-father and first managing director – abandoned his pre-war career as an architect after his wartime experiences as an observer in the Royal Naval Air Service. Together with aviation pioneer Claude Graham-White as its chairman, and the cinematographer Claude Friese-Greene, Aerofilms rented a Geoffrey de Havilland DH9 biplane – from AIRCO Limited – based at the London Aerodrome, Hendon, and occupied a suite of rooms in the London Flying Club. To minimise overheads the bathroom functioned as a makeshift darkroom. Their first year of business was dominated by Friese-Greene filming flying sequences for silent films. By the second year – and throughout its entire existence – the main source of the company's income came from oblique aerial photography, as it became increasingly de rigueur for the owners of country houses, factories, royal palaces or municipal buildings to have photographs of the structures from an aerial perspective. Taking the photographs involved the pilot manoeuvring the aircraft – often at seemingly precarious low levels – to ensure the best height and angle for the shot. As in the recent war, taking the photographs involved the camera operator manually loading glass-plate negatives – measuring five by four inches – for each exposure.

Throughout the 1920s Aerofilms gradually expanded the business, but not without a few brushes with authority along the way. On 31 March 1920, during a low-level flight over London, their Avro 504K biplane developed engine trouble. Forced to crash-land, the pilot chose the boating lake in Southwark Park – much to the amazement of the crowds that soon gathered and to the disapproval of the Commissioner of Police. Becoming a subsidiary of the Aircraft Operating Company in 1924 and working from a mansion house at The Hyde on the Edgware Road that they christened 'Aerial House', Hemming and Wills set about expanding their aerial-survey business at home with Aerofilms and abroad with the Aircraft Operating Company. In 1925 the London Metropolitan Police commissioned Aerofilms to carry out a groundbreaking road-traffic survey over the capital. But the lucrative contract that Aerofilms hankered after most was supplying services to the Ordnance Survey. They were understandably delighted when in 1925 a trial survey of Eastbourne from the air was commissioned to determine the potential of using vertical aerial photographs and photogrammetric techniques to produce maps. Aerofilms produced good results and proved that aerial survey could be accurate, cost-effective and quick. But these inconvenient facts weren't about to change Ordnance Survey policy and their dependence on ground survey just yet. This same indifference to aerial photography was echoed in Britain's armed forces, despite photographic intelligence having proven itself in the First World War. In the immediate aftermath of the war – after the Royal Air Force (RAF) was created in 1918 – the Chief of the Air Staff, Air Marshal Hugh Trenchard, presented Winston Churchill with a memorandum for the RAF and its future. As Minister of War and Air in Lloyd George's Cabinet, Churchill approved of the 'Trenchard Memorandum' which identified aerial photography as one of the 'primary necessities' of the RAF. But despite that early post-war recognition, and with photographic interpretation considered the sole responsibility of the Army, aerial photography would come to be regarded as just one of many military functions.

The opposite proved to be the case in the private sector, and no organisation exemplified this more than the Aircraft Operating Company. Following its creation in 1924, over a period of just five years it went from having only three directors, a secretary and a typist to a firm employing 150 people – or 170 if Aerofilms was taken into consideration. Hemming's company had surveyed thousands of square miles in the Copperbelt region of Northern Rhodesia as well as districts of Iraq near Baghdad and the city of Rio de Janeiro, Brazil, for map-making. By 1932 their fortunes had prospered sufficiently for them to take offices in the opulent Bush House at the southern end of Kingsway in central London, later to become the iconic home of the BBC World Service. By the mid-1930s, Aerofilms were benefiting from the Ordnance Survey's inability to supply local authorities with modern mapping – to support town-planning initiatives – and were flying throughout Britain. As this coincided with the Aircraft Operating Company winning major contracts from the Anglo-Iranian Oil Company, all production work was relocated outside central London, to a brand new building at Beresford Avenue off the North Circular Road in Wembley.

In January 1937, the company was floated and within twelve months Sir Percy Hunting – the shipping and aviation entrepreneur – bought a major stake in it. With the entry of the Hunting family into the business, Hemming and Wills had investment capital and made a decision that would have far-reaching implications for photographic intelligence in the Second World War. To process their customer orders they decided to purchase a mechanical plotting machine – with powerful stereoscopic optics that allowed the three-dimensional analysis of photography – to facilitate the mass production of maps. For the previous three years, Wills had been taking summer courses on photogrammetry with the German company Carl Zeiss AG, in Jena, Thüringen. Visiting the Zeiss works he was given a demonstration of the enormous Aerocartograph by Professor Dr Reinhard Hugershoff, who proudly boasted that 'he preferred sleeping with his machine to his wife'. From Germany, Wills travelled on to Switzerland where he visited Wild Heerbrugg for a demonstration of their machine, the 'A5 Autograph'. Whilst the company name might have amused Wills, the Wild Machine was much preferred to the Zeiss device as its ball-bearing movements ensured less wear and tear, and it occupied significantly less space. When the machine – with its hefty price tag of £5,500 – arrived at Wembley towards the end of 1938, it was installed in its own specially air-conditioned room, with a vibration-proof base fifteen feet deep.

To operate the Wild Machine, Aerofilms employed thirty-two-year-old Michael Spender, the elder brother of Stephen, the poet and literary critic. Growing up in the seaside town of Sheringham, in rural Norfolk, the brothers attended Gresham's School in the nearby market town of Holt. Stephen's difficult relationship with his elder brother and his persecution as a 'Hun', along with his younger brother Michael at Charlcote Boarding School in Worthing, Sussex – after school bullies discovered the brothers' maternal name was Schuster – inspired his more or less autobiographical 1940 novel, *The Backward Son*. When later discussing the relationship between the boys and their parents, Stephen recalled his brother's view that 'they [the parents] were emotional and unreliable' and that Michael's reaction was to 'mistrust human contacts, and to worship unemotional efficiency'. At school Michael developed a passion for music and science which he studied at Balliol College, Oxford under Professor Lindemann, graduating with a double first. With a zest for life, he had worked as a surveyor on expeditions to the Great Barrier Reef and East Greenland before joining the 1935 expedition to the Himalayas where he mapped twenty-six peaks. In 1933 he married Erika Haarmann and in 1936 their son John Christopher was born, only for Michael to leave his four-year-old son and German wife after a tumultuous affair with the artist Nancy Sharp, whom he would later marry in 1943. After training from Wild in Switzerland, Spender set about developing the company's expertise in photogrammetric methods. The need for absolute accuracy when operating the machine was said to have appealed to his scientific mind. But with his complete belief in intellectual honesty, coupled with a legendary arrogance and complete absence of tact and diplomacy, he would frequently and unintentionally hurt the feelings of his colleagues when he questioned their results, their usefulness and sometimes their competence.

In his proposal to the Air Ministry in late 1938, Hemming had outlined his company's air-survey experience in Britain and throughout Africa, Australia, Newfoundland, British Guiana, New Guinea, Brazil, Egypt, Iraq and Iran. Highlighting their knowledge gained from operating in far-flung parts of the world – under conditions ranging from arctic cold to equatorial heat – he stressed that not only did they have experienced ex-military men on their staff but also equipment that would be required in an emergency. First and foremost among that equipment was the Wild Machine. Ever the shrewd businessman, Hemming's proposition was that in an emergency, the Aerofilms branch of their business should be taken over and run as an air-survey unit for the RAF. He knew that war would mean Aerofilms would be unable to function in Britain, and that this could spell disaster for the company. But whilst in the aftermath of the Munich Crisis the Air Ministry would show no interest in this early suggestion of a public-private partnership, in a little over twelve months' time Hemming would be working in cahoots with Sidney Cotton and would be receiving the attention of Winston Churchill.

In the days before Britain declared war on Germany, rules of engagement meant that the RAF could not scramble their recon-naissance aeroplanes to take aerial photographs of the German Navy. In desperation the Admiralty turned to Winterbotham for help, and the MI6 unit took photographs of Wilhelmshaven on 28 and 29 August 1939, which informed the Admiralty of the strength of the German Navy closest to Britain, and even included the Führer's yacht, the *Grille*. With the declaration of war on 3 September 1939, Churchill became First Lord of the Admiralty – as he had been during the first part of the First World War – and a member of the War Cabinet. When the Board of the Admiralty first learned the news that Winston Churchill was once again First Lord of the Admiralty, a signal was sent to the Fleet: 'Winston is back.' From the outset, Churchill would depend on the knowledge aerial photography could provide. But since a photograph is just a snapshot in time, the need for a steady supply of aerial photographs so they had comparative knowledge about day-by-day change was required. Churchill's demands for aerial photography at the beginning of the war were to prove decisive in the development of airborne reconnaissance within the RAF.

On 22 September 1939 Cotton's unit moved from MI6 to RAF Fighter Command and became the top-secret 'Special Flight' with authority from the Chief of the Air Staff to operate carte blanche. With Cotton in the acting rank of Wing Commander, and with Britain at war, he swapped his business suit for a uniform and set about developing the unit. Incapable of changing his entrepreneurial approach to life, he ignored the paperwork, sent cables around the world for the people he wanted, and began making enemies amongst the regulars. It was quickly arranged that Niven would be joined by Maurice 'Shorty' Longbottom, Hugh McPhail – who was then working as a pilot in Lima, Peru – and ex-British Airways pilot Denis 'Slogger' Slocum. By October they had taken over the British Airways hangar, the Flying Club and most of the Aerodrome Hotel, and Cotton's little red Hotchkiss sports car and his young cockney chauffeur Cyril Kelson – who doubled as his gentleman's gentleman – became an increasingly common sight at Heston. For his inner circle Cotton had badges made with the inscription CC11 or Cotton's Crooks Eleven – with eleven being his addition to the Ten Commandments: 'Thou shalt not be found out.'

In August 1939, Flying Officer Maurice 'Shorty' Longbottom had submitted a report – 'Photographic Reconnaissance of Enemy Territory in War' – to the Air Ministry, full of ideas that had been crystallising since he first met Cotton and Niven in mid-June 1939, during a stopover on Malta before their Middle East tour. Cotton had taken to the 'young man with a slide-rule mind' from the outset, and when Longbottom returned to Britain on leave in late June, the three men began developing what became known as the 'Longbottom Memorandum'. This proposed that photographic reconnaissance should be carried out so as to avoid enemy fighters and anti-aircraft defences as completely as possible. The best way of doing this would be to use a single small aircraft, which relied on its speed, rate of climb and ceiling to avoid destruction. A machine such as a single-seat fighter – Longbottom suggested the Supermarine Spitfire – could fly high enough to be well above balloon barrages and anti-aircraft fire, and could rely on sheer speed and altitude to get away from any enemy fighters. It would have no use for its armament or radio, and these could be removed to provide an extra payload capacity for more fuel in order to achieve the necessary range, which a fighter did not normally have. As most fighters had a very good take-off speed, due to their great reserve of engine power, they could be considerably overloaded for this purpose with further fuel, giving them an even greater range. In clear weather the aircraft would fly at a great height all the time when it was over enemy territory, and would be too high to be heard or seen with the unaided ear or eye. If detected by sound locators it would still be out of range of any ack-ack guns, and with its great speed and advantage of altitude it could almost certainly elude fighters coming up to intercept it from the ground, particularly as it would be a very small machine camouflaged in a way that would reduce its visibility against the sky as much as possible.

In October 1939, and much to the annoyance of Fighter Command who were desperate for every last Spitfire for the defence of Britain, Special Flight was allocated Supermarine Spitfire N/3071, after the intervention of the Chief of the Air Staff. Working with Harry Springer – Farnborough's expert on aerial photography and 'a natural technician who knew where to drill the holes' – a standard RAF-issue Williamson F24 camera was installed, and by the end of October they were ready to test the idea of high-altitude photography in a photographic-reconnaissance or PR Spitfire. In an effort to preserve secrecy, Special Flight was renamed No. 2 Camouflage Unit on 3 November 1939. After flying to Seclin in northern France on 5 November in the Lockheed and Spitfire – and with the cooperation of the Deuxième Bureau – from 18 November Niven and Longbottom flew sorties over Germany and the Low Countries. Working alongside them on this Special Survey

Flight was Rolls-Royce engineer 'Pinkie' Price, who had been put into uniform, and Dr Robson, who took blood tests after every flight to monitor the effects of high-altitude flying on the pilots. Nobody before had ever operated for sustained periods at such heights so the impacts on the human body were largely unknown.

On 3 January 1940, as the Special Survey Flight was nearing completion of its test missions, Air Vice-Marshal Richard Peck – Assistant Chief of the Air Staff – chaired a conference to reconsider the whole system of photographic reconnaissance. So far in the Phoney War, seven squadrons of Bristol Blenheim bombers – specifically modified for photographic reconnaissance – had flown eighty-nine sorties over enemy territory, of which only half had produced any photographs. This was at a cost of sixteen aircraft and meant that the crews faced a one-in-five chance of survival. The statistics from the field tests in a single-seat Spitfire were sobering, with Niven and Longbottom returning from all fifteen sorties, having photographed successfully on almost two-thirds of them. Alongside Cotton at the conference was Wing Commander Frederick Charles Victor 'Daddy' Laws – who had worked in aerial photography since before the First World War when he had taken photographs from balloons and man-carrying kites. Laws had commanded the RAF School of Photography for much of the interwar period, before working in commercial aerial survey and then as managing director of Williamson's, the maker of aerial cameras. He readily acknowledged the failings of the F24 camera – its tendency to malfunction and, critically, its inadequacy for high-altitude work – and accepted that a camera with a much greater focal length was urgently required, to produce larger-scale photography. But whilst Cotton's theories had come to be accepted, since the Blenheim was the only means of providing most of the photographs – and despite the heavy losses of men and machines – until they could be replaced the Blenheim squadrons would carry on flying during the Phoney War.

However inadequate the approach to photographic reconnaissance would prove to be at the beginning of the war, the resources dedicated to taking the photographs were enormous when compared with that of their interpretation. On the outbreak of war only seven dedicated Photographic Interpreters – habitually referred to as PIs – were employed in the entire RAF. Six of them were based in the

Photographic Interpretation Unit (PIU) at the headquarters of Bomber Command in Iver, Buckinghamshire, where they were commanded by Flight Lieutenant Peter Riddell. The team had been hastily formed in 1938, and with information about key targets in enemy territory 'very sketchy and inaccurate' Riddell's brief was to improve matters using all the information and the few aerial photographs available. The seventh PI – Flight Lieutenant Walter Heath – worked at the Air Ministry in Air Intelligence 1(h) where he maintained the PI training manual, which by his own admission was sorely wanting, and provided basic PI training to Station Intelligence Officers. Heath had first encountered photographic intelligence after being 'yanked out of the trenches' to become an Intelligence Officer in the First World War, and quickly became convinced that it was the 'finest possible training in deduction and reasoning'. On the outbreak of war, he was promoted to Squadron Leader and given command of Air Intelligence 1(h), which moved from the Air Ministry buildings in central London to Hibbert Road School at Wealdstone near Harrow where he would be joined by two Flight Lieutenants – untrained in PI – a draughtsman and an office boy.

On the outbreak of war, Dr Hugh Hamshaw Thomas – known as 'Ham' to his friends – was a Reader in botany at Cambridge University, and since the summer of 1939 had been on field work in Jamaica. A Fellow of the Royal Society, the academic was known for his thoughtful and considerate nature, and instantly recognisable with his bushy grey moustache and eyebrows. On his return to Cambridge in October that year, and despite being fifty-four years old, he immediately got in touch with intelligence personnel at the Air Ministry to offer his services. He had been a pioneer of photographic intelligence in the First World War, and had developed techniques for gun-ranging and map-making in Palestine that were used extensively by General Edmund Allenby, the Commander-in-Chief of the Egyptian Expeditionary Force. While in the Middle East, Hamshaw Thomas came into regular contact with Lieutenant Colonel T. E. Lawrence (of Arabia fame), who had a particular interest in aerial photography that covered his specialist region. Mentioned in despatches, Hamshaw Thomas was made an MBE and awarded Egypt's highest state honour, the Order of the Nile. Instructed by the Air Ministry to report to Peter Riddell at Bomber Command headquarters, he arrived there much to the surprise of Riddell, who knew nothing about his posting. For the time being, he would instead become the Station Intelligence Officer at RAF Hemswell, Lincolnshire, from where Bomber Command flew Hampden bombers.

At the conference in early January, the photographs taken by Special Flight were discussed, and were considered by Bomber Command to be of such small scale that they defied 'interpretation, reduced the effort expended to futility and the writer to tears' and their 'sole achievement is waste of petrol, time, paper, energy and imagination'. Cotton held a different view. In early December he passed photography to his old colleague Hemming, who returned a detailed interpretation report – courtesy of Spender and the Wild Machine – on 11 December 1939. Cotton was sufficiently impressed that agreement was reached for the unofficial supply of film-processing and interpretation services. Although his company was still busily processing work from its surveys in Africa, and for Anglo-Iranian Oil, Hemming was more desperate than ever to work with the government, since their market and much of their skilled labour had disappeared since the outbreak of war.

The impact of war being declared was as immediate as Hemming had feared. In early 1939 Aerofilms had bought two brand new de Havilland Rapide aeroplanes, and when they arrived at Heston – complete with Williamson aerial cameras – the company had top-of-the-range aerial survey equipment, modern aeroplanes, a state-of-the-art laboratory and the latest mapping machine. The aeroplanes were immediately impounded and used to create the No. 1 Camouflage Unit on 9 October 1939 at Baginton airfield – now Coventry Airport – which involved Aerofilms staff taking oblique photographs to assess the effectiveness of camouflage schemes used on factories and military installations throughout Britain. These were supplied to the Camouflage Directorate, then being developed in nearby Royal Leamington Spa. Based in the Regent Hotel and neighbouring Old Museum and Art Gallery, a talented group of British architects including Basil Spence and Hugh Casson, the artist Edward Seago and stage designer Oliver Messel spent their war designing ingenious camouflage schemes.

In an effort to salvage what he could from the situation, Hemming rewrote his 1938 memorandum to the Air Ministry, and this time submitted it to the Deputy Chiefs of Staff. Yet again the proposal was rejected, in the belief that Hemming was losing money and was simply trying to cut his losses. But when Hemming replied on 23 January

1940 he politely pointed out – because of the top-secret work they were doing for Cotton – that their application of photogrammetric methods meant the company could extract intelligence 'even from air photographs on so small a scale as 1/80,000'. Given the problem of extracting intelligence from small-scale aerial photography this claim couldn't be ignored and the Joint Intelligence Committee arranged for the Directors of Naval and Air Intelligence to visit Wembley on 1 February 1940. They immediately recognised that the company could greatly improve the standard of PI, in respect of both the detail that could be observed and the speed with which interpretation could be carried out.

But before any decisions were made an event of the greatest importance was to take place that would perfectly highlight the capability that had been developed. The Admiralty had just received reports that the battleship *Tirpitz* had left dry dock in Wilhelmshaven, but the RAF had been unable to confirm this by either photographic or visual reconnaissance. This coincided with the recent development of a Type B PR-Spitfire, which had been developed by the Royal Aircraft Establishment, Farnborough. An extra fuel tank now brought the ports of north-west Germany within range of airfields in East Anglia for the first time. Flying from RAF Debden, Essex, Shorty Longbottom took off at 12:05 on 10 February and, crossing the German coast at 33,000 feet, he successfully photographed the ports of Emden and Wilhelmshaven in the Ems and Jade estuaries. Landing back at Heston, after a flight of three hours and twenty minutes, Longbottom brought back the first photographs of north-west Germany taken from a Spitfire flown from Britain.

As soon as the photographs were developed, copies were supplied through the normal channels. They prompted a report from Air Intelligence 1(h) that they had no intelligence value because of their small scale. However, copies had also been rushed to Michael Spender in Wembley. Using the Wild Machine he was able to create a plan of Emden port to a scale of 1:10,000 and marked individual vessels on the plan. The Wild Machine, and its expert operator, had established that the *Tirpitz* was still safely in dry dock. Both this confirmation and the speed with which such detailed plans had been prepared impressed the Naval Intelligence Division and Winston Churchill, who was astonished to learn how they had been produced. Cotton was summoned to a meeting that evening in the Admiralty War Room, where their need for aerial photography of the German Fleet was emphasised to Air Vice-Marshal Richard Peirse by Admiral Sir Dudley Pound, Chief of the Naval Staff.

Whilst Peirse readily accepted that the Aircraft Operating Company and its Wild Machine must now be requisitioned, to ensure that it actually happened Churchill wrote to Sir Kingsley Wood, Secretary of State for Air: 'Major Hemming's organisation . . . including the expert personnel, should be taken over by one of the Service Departments without delay . . . if for any reason the Air Ministry do not wish to take it over, we should be quite prepared to do so'. The risk of losing photographic reconnaissance to the Admiralty jolted the Air Ministry into action. Closer working began almost immediately, while discussions continued for weeks that resulted on 21 May 1940 in Secret Directive 155/1940 (418), which formally merged the Air Ministry PI Unit with the Aircraft Operating Company. The very next day Squadron Leader Walter Heath's team relocated from the Hibbert Road School and Wembley became a military-intelligence unit. The Director of Intelligence at the Air Ministry had tried to use the occasion to merge all the PI units but Bomber Command maintained their independence for now, putting themselves on a collision course with the Air Staff.

On 17 January 1940, and whilst the No. 2 Camouflage Unit became the Photographic Development Unit (PDU), the Special Survey Flight in France became 212 Squadron. Since being established in September 1939, Cotton's unit had grown from 8 to 21 officers and from 19 to 171 other ranks. It now also employed 17 civilians. With such rapid growth and with work ongoing in France and Heston, Squadron Leader Geoffrey Tuttle was appointed Cotton's deputy and was posted to Heston on 10 Feburary 1940. With instructions from Air Marshal Peck to 'sort it out' he found himself the only regular RAF officer in a very irregular unit. Tuttle would later recall finding 'an atmosphere of high-priority, top secrecy and glamour' and a place where things were 'pretty free and easy'. Realising it would be a mistake to tighten up everything at once, though, he concentrated on the things that really mattered and turned a blind eye to minor details, such as Slocum's penchant for blue suede shoes and for wearing his field-service hat on the wrong side of his head.

Less than a month after Tuttle arrived at Heston, on 2 March 1940, history was made when Niven successfully photographed the German industrial heartland of the Ruhr. Tuttle would later recall that for eleven days beforehand Blenheims had been sent to get cover and all of them had been lost – eleven Blenheims in eleven days – and, whilst under no illusion about Cotton's Machiavellian streak, Tuttle recognised that his inventive genius saved many lives and gave the British a quite extraordinary intelligence capability. At this time Cotton lived in a rented villa near the headquarters of the British air forces in France at Coulommiers, 37 miles east of Paris – where he reported to its commanding officer Air Marshal Arthur 'Ugly' Barrat – and was near the French Chief of Air Staff at Meaux. Cotton's luxurious Hotchkiss sports car, featuring the registration plate RAF X complete with an illuminated Union Jack flag, could often be found in Meaux. It is no small wonder that Ian Fleming was said to have been so inspired by Cotton – who supplied the Naval Intelligence Division with photography of neutral Ireland in September 1939 – that he based much of the character Q, the fictional head of the MI6 research and development division in the James Bond novels, on him.

In Meaux, Cotton was introduced to Colonel Lespair, commandant of the French PI school, and became convinced that the French had some important lessons to teach the British. The officer chosen to work with them was twenty-five-year-old Douglas Neville Kendall. A British national, Kendall was born in the city of Porto, Portugal, and spent much of his childhood in the country when not boarding at Ampleforth Prep School in Yorkshire. A mathematics graduate of Christ Church, Oxford, he had taken postgraduate courses in surveying before winning his pilot's licence, flying a de Havilland Tiger Moth at the London Aeroplane Club at Stag Lane Aerodrome in Edgware. He had worked for the South African division of the Aircraft Operating Company where he learnt the practicalities of aerial survey on forestry, soil-erosion, geological and road-construction projects. When war was declared, he had been holidaying in Portugal, and after reacquainting himself with Hemming was commissioned as a pilot officer and posted to Wembley.

After attending the French PI course, Kendall joined 212 Squadron, and when the German invasion of the Low Countries came on 10 May 1940 – signalling the end of the Phoney War – a series of military events was set in motion that culminated in the squadron leaving France a little over one month later. As the German blitzkrieg gained ground and momentum, requests for information came more frequently and sorties became of immediate operational significance. Towards the end of the month, when it became clear that a British withdrawal from the Continent was imminent – after which most of their intelligence sources would break down and disappear – the Joint Intelligence Committee reviewed how they could obtain the earliest warning of German preparations to invade Britain. The available sources of information were identified as being: wireless intelligence, the behaviour of enemy agents in Britain, the preparation of an invasion force in enemy ports, military concentrations near enemy ports and the actual sailing of the invasion armada. For the last three, aerial photography was clearly essential but required the regular and systematic photographing of all potential invasion ports. With a shortage of 'Cottonised' reconnaissance aircraft, the summer of 1940 was to prove a period of rapid expansion and change at both Heston and Wembley.

During the evacuation of Dunkirk – beginning on the evening of 26 May and ending on 4 June – the main objectives for 212 Squadron were photographing enemy troop movements and enemy-occupied airfields, as well as naval bases in Holland and the Frisian Islands from where fast enemy naval craft could interfere with the evacuation. After the fall of Paris on 14 June, the ground units retreated south-westwards. The two Spitfires were flown back to England and a Lockheed Superelectra ferried personnel back to England. Cotton caught up with the squadron in the early hours of 16 June 1940 when it was in the Brest peninsula, before taking off and saying he would fly to Bordeaux and would rendezvous at 11:00 hours. When he hadn't returned by 16:00 hours the group decided they had no choice but to destroy their photographic trailers and much of their aerial-photograph archive. Having set fire to several of the vehicles they headed off in the four remaining ones towards the coast where they hoped to catch a boat. Just at that moment Douglas Kendall spotted Cotton landing in Lockheed G-AFTL, which he discovered was full of people. Cotton was disbelieving of the destruction of the aerial photography and flew off in the Lockheed, telling Kendall that another aeroplane would land soon and they should find fuel for it. A troop-carrier aeroplane capable of evacuating the remaining men landed and they made it back to Heston the following day – 18 June 1940 – after refuelling stops in Nantes and the Channel Islands.

At this pivotal time, the relationship between Winterbotham and Cotton broke down irreparably. Despite Cotton's transfer to the RAF, he was still working for MI6 as their 'transport manager' with responsibility for the evacuation of secret personnel and their equipment to Britain when the time came. But when that time did come, Winterbotham would later claim that Cotton was otherwise engaged, helping a collection of Frenchmen with suitcases full of French banknotes out of the country. This included the French entrepreneur and racehorse owner Marcel Boussac, who would later finance the creation of the Christian Dior couture house. In Winterbotham's view Cotton's downfall was a consequence of his loves: 'one was flying, two was money and three of course was women.'

For the Air Ministry, who had been planning how they were going to remove Cotton, this was both the final straw and the opportunity they had been waiting for. When Cotton arrived back at Heston on 17 June he was handed a letter thanking him for his 'great gifts of imagination and inventive thought which you have brought to bear on the development of aerial photography' and informing him of his dismissal from the RAF. Newly promoted Wing Commander Geoffrey Tuttle was now commanding the unit which became known as the Photographic Reconnaissnce Unit or PRU, which was now part of Coastal Command. Cotton and his nonconformity, hatred of red tape and willingness to cut it had made him the despair of the RAF administrative channels, and thanks to the seemingly unending rumours about his activities his dismissal had been made all the easier. With Cotton having 'served his purpose' and with Heston part of Coastal Command, there was never going to be a role for him in the mainstream RAF. At the time Cotton was upset, particularly when notices were posted at RAF stations forbidding his entry. His contributions would be recognised with an OBE in the New Years Honours List – something that Tuttle had energetically lobbied for – and a credit in the official history of the RAF in the Second World War. After the war his buccaneering life continued in oil exploration, with his activities often coming to the attention of MI6. This included the time he spent in the service of Osman Ali Khan, the last Nizam of Hyderabad State, when he used war-surplus Lancaster bombers to airlift arms and supplies ahead of the Indian army taking control of the state in 1948.

After Tuttle took charge, a new intelligence room was established at Heston and it became a hub of activity. It was where the pilots came to be briefed, where they planned their sorties and in turn briefed the intelligence officers after returning from their missions. As soon as the photographs were printed, they were handed over to the team of WAAF 'plotters' whose job it was, by using a library of military maps, to locate and identify exactly the area that each photograph covered. They relied on a sheet of greaseproof paper – called the pilot's trace – on which the pilot had marked his flight path and roughly where he had photographed. By drawing the 'footprint' of each photograph onto a map of the corresponding area, the process of photographic interpretation was greatly speeded up since the plots quickly showed where each print covered.

With the Aircraft Operating Company now part of the RAF, and Hemming now an honorary Wing Commander, he was keen for the interpretation unit to have its own identity. He christened the Wembley building PADUOC House – an anagram of the initials of the PDU and AOC – and the nickname quickly stuck. More formally, on 12 June 1940 PADUOC House became the 'Photographic Development Unit: Interpretation and Intelligence' and its 122 personnel – just over a third of whom were civilians – were responsible for the interpretation of all the photographs taken by the PRU; construction of models, and the maintenance of a library of aerial photographs.

Riddell – like Churchill, Winterbotham and Cotton – believed that photographic interpretation could be made into 'a potent weapon in the armoury of air intelligence'. Growing increasingly frustrated with the narrow approach that was being taken at Bomber Command, he requested a transfer to Wembley which, given the dire shortage of PIs there, resulted in his transfer on 26 June 1940. Like Tuttle at Heston, Riddell found himself in a rather oddball unit, and set about developing approaches to PI with specialist sections on German airfields, aeroplanes, navy and shipping, army subjects, radar and camouflage. After his arrival, Riddell found a wealth of material for army intelligence, enabling him to make a case for acquiring interpreters from that source, and on 25 July 1940 the War Office attached an army captain for the ultimate purpose of creating a section for the land forces.

In Riddell the interpreters had an energetic organiser and a PI experienced in the full range of enemy activities. The Wembley team was a diverse group of military and civilian personnel. Riddell called the Wembley PIs the 'irregulars' and felt it was the 'sense and sensibility of the irregulars that made Photographic Interpretation'. From the Air Ministry they had PIs with experience in specialised interpretation, from the Aircraft Operating Company they had experienced civilian photographic interpreters and photogrammetric specialists, and from 212 Squadron, which had now returned from France, they had PIs with experience under operational conditions. Already containing many RAF and WAAF officers, the involvement of the Army and Admiralty signalled the beginnings of Wembley as an inter-service unit, and many of the Aerofilms staff too would soon be in uniform. There was the danger of each service, and many units within them, trying to set up their own little PI branches. Ultimately the Navy, Army, RAF and Bomber Command would have done so. Riddell worked on the basis that photographic interpretation had to be centralised first: he followed one of the basic copybook principles of war – 'contract to expand'.

As Deputy Director of Intelligence at the Air Ministry from 1940, with responsibility for photography, Daddy Laws considered himself to be part of the 'straightening-out' programme – much in the way that Geoffrey Tuttle and Peter Riddell had been given the task of regularising photographic reconnaissance and photographic interpretation, but on the technical side. A new Photographic Research Committee was established that included technical experts from Eastman Kodak and Ilford Limited. Between 1940 and 1942 the committee would address the myriad of technical challenges posed by photographic intelligence-gathering under the increasing pressure of war needs: to drastically improve the quality of images gathered, and speed up systems of processing and printing; camera mounting on aircraft; optical lens and film design; and how to correctly expose film under combat conditions – it was considered unrealistic for the pilot to control exposure settings himself and so from tests they calculated what exposure would be required, based on the latitude and longitude of the target and the time of the photography. They had to grapple with the technical issues of how to take aerial photographs at night, and considered whether or not to try taking colour or infrared photographs.

This was a time when PI was also being shaped and strengthened by an influx of outstanding people, and one of the most important was Hugh Hamshaw Thomas. He had become increasingly disheartened in the first six months of 1940, with the drudgery of being the Station Intelligence Officer, and found hopelessly inadequate the aerial photography that he used to brief bomber crews, fearing that he was needlessly sending men to their deaths. This was compounded by his knowledge that, whilst many bomber crews claimed to have hit their targets and achieved success, this was not the reality. When one crew reported that their bombs had dropped easily 'within fifty miles', as if they were doing well, it highlighted a problem that the British had experienced since their first attempt at attacking German territory on the night of 19/20 March 1940 when, in retaliation for an attack by the Luftwaffe on the Royal Navy at Scapa Flow days earlier, the British attacked the German seaplane base at Hoernum, on the island of Sylt in the North Sea.

The next day, once all the RAF bombers had returned to base, a statement was read to the House of Commons on behalf of Prime Minister Chamberlain. From information gathered from pilot debriefings, it was reported that hangars, oil-storage tanks and a light railway had all been hit and were on fire. When the press announced this great British triumph, German claims that the bombs had actually been dropped on a Danish island – claims that would subsequently turn out to be true – were dismissed at the time as disinformation. Late that night, when the photographs, taken from a PR Blenheim, arrived with Peter Riddell and his team at the Bomber Command PI unit, they agonised over them because they recorded a perfectly intact seaplane base. This caused bewilderment in the senior ranks of Bomber Command, and fully confirmed the need for much-improved navigational and bombing accuracy. In a British Pathé newsreel item – 'RAF Heroes of the Sylt Raid' – first shown in cinemas on 28 March 1940, the returning airmen are interviewed and confidently explain that not only did they hit the target but they hit it well. When asked how much damage they'd caused, an airman replies that the Germans are the best ones to ask. Words of wisdom from the voice-over man claim that whatever the Germans say has to be taken with a grain of Sylt.

In the summer of 1940, Hamshaw Thomas requested a transfer to photographic-interpretation duties and worked in Air Ministry Intelligence at 14 Ryder Street, St James, London. There he was given the task of compiling a revised version of the manual which had been written earlier by Squadron Leader Heath on the interpretation of aerial photography, and which badly needed updating. After rewriting the manual, Flight Lieutenant Hamshaw Thomas reported to Wembley in October 1940, and was responsible for the interpretation of photographs of enemy industry.

The previous month, the geologist Lieutenant Norman Falcon joined the twenty-two-strong Army Section. Another early arrival was Dr Glyn Daniel, a young Cambridge don who had spent the last few years giving lectures, supervising students and writing a book on the chambered tombs of Western Europe. The Welshman would later become Disney Professor of Archaeology at Cambridge and gained fame as the urbane chairman of the BBC television programme *Animal, Vegetable, Mineral?* Broadcast once a fortnight, the quiz challenged a team of three experts to identify a succession of objects taken from British museums, and introduced many to archaeology for the first time. When Daniel received a letter confirming his commission as an RAF pilot officer, he had instructions to report in three days' time to the personnel headquarters of RAF Intelligence in London. On arrival at the old Charity Commissioners office, on Ryder Street, he found himself in a room alongside twenty other newly uniformed and commissioned officers. They all appeared to be given postings 'to the ends of the Earth – Reykjavik, Singapore, Washington, Colombo, Cairo . . .' When his time came, the young academic discovered that he and Pilot Officer Bill Wager, who would later become Professor of Geology at Oxford, were to report to an address off the North Circular in Wembley.

Taking the Bakerloo Line to Stonebridge Park station in north-west London, Daniel found himself alongside the London, Midland and Scottish Railway main line and the vast Willesden Marshalling Yard. Walking beside the newly built North Circular Road, underneath the arches of seven railway bridges, he passed the American-inspired Ace Cafe and Service Station that served traffic using London's new arterial road. Approaching the PDU building along Beresford Avenue, Daniel walked past on one side 'a row of jerry-built two-storeyed mock-Tudor villas with gables, which looked as though they were built out of cardboard' and on the other a maze of factory buildings. At Paduoc House, he reported to Riddell who was 'sitting in his shirtsleeves and braces working out his bank statement'. The new recruit was told to organise his own accommodation and report for duty at nine the following morning when he would begin training as a photographic interpreter.

When Daniel arrived in September 1940 the war came to Wembley with the first bombing raids. The unit's location in a built-up area, over which enemy bombers passed on their way into and out of the metropolitan zone, and in close proximity to a major marshalling yard, a main railway line, and various bridges and factories certainly wasn't ideal. Nor was the fact many of the Aerofilms workers, and many military personnel, lived in local houses and meant the future of PI was somewhat vulnerable during the Battle of Britain. RAF standing orders soon required Daniel to find a billet outside London for his own safety, and prompted an offer from Aerofilms employee David Brachi to live with his family in Berkhamstead, Hertfordshire. After studying geography at St Catherine's College, Brachi had gone straight to Wembley from Cambridge in 1938 and had worked on the Anglo-Iranian Oil contract. He had always been interested in ships, and whilst Spender used the Wild Machine to measure them Brachi did much of the detailed interpretation and identification.

Paduoc House received its first direct hit on 2 October 1940 just after midnight when the Luftwaffe dropped one of their oil bombs – a large incendiary weapon containing an oil mixture and a high-explosive bursting charge. A thirty-six-year-old RAF policeman, Leading Aircraftman Robert Ammon, was killed and the hut which housed the PI school was completely destroyed. On 17 October two high-explosive bombs were dropped at night and fell between two nearby factories. Since little warning was received about the raid it was fortunate that most personnel were already in the shelter or working in the Wild Room. In the damage report written by Spender, it is recorded that personnel were back at work shortly after and that Hemming 'provided beer (3 cases) for which all ranks would like to thank him tremendously'. With the roof patched together with tarpaulin, a common wisecrack in Paduoc House at the time was that an umbrella was as important to the PI as the stereoscope.

At Heston the strain of the hectic days and sleepless nights during the Blitz was immense, and it was during this time that Tuttle recalled being in his office when his adjutant, Flight Lieutenant Noel Sherwell, came to say that the policeman at the main gate was reporting that the King was there. The importance of photographic intelligence to the war effort had earlier been recognised by a royal visit to the Wembley and Heston teams by George VI and Queen Elizabeth on 22 July 1940. Tuttle was in a state of complete disbelief: the King had been driving to London from Windsor and, realising the main hangar had disappeared, he just looked in 'to see how we were'. Since the October, bombing raids had destroyed the PI school and drawing office, and had damaged the top floor of the factory. The leaky roof meant that the top floor could not be used at night as it was impossible to black out the damaged covering. A safer location was needed for the interpreters.

Kendall was the instructor for the first three PI courses at Wembley, and was responsible for introducing new recruits to the world we live on from an entirely different perspective. A built object or a topographical feature, instantly recognisable when viewed on the ground, very often becomes quite unfamiliar when seen from above. It was this ability to recognise features from a new angle that was essential if someone was to become a PI and it required a peculiar mentality, 'a kind of super-jigsaw mind' that someone, irrespective of their natural-born intelligence, either did or didn't have. The recruits soon discovered that there were five logical steps to the identification of objects in aerial photographs: shape / size / shadow / tone / associated features. Also essential to the PI was the capacity for vision through both of their eyes, as the photographs were taken with the specific intention that the enemy should be watched in 3D. The reconnaissance aeroplanes were fitted with single-lens cameras which could be set to automatically expose the film at calculated intervals – using a device called the intervalometer – in order to have a 60 per cent overlap between each successive frame, with the exposures taken when the camera lens was pointed at an angle as close to the vertical as possible.

Viewing in three dimensions could be achieved not only with the optically powerful Wild Machine but also with a simple pocket stereoscope. This relatively primitive-looking device had two magnifying glasses mounted in a rectangular metal frame and four metal legs that raised the glasses several inches above the prints below. To view a particular location in 3D, the interpreter would take two consecutive prints from a sortie and place them under the stereoscope. These two prints were referred to as 'stereo pairs'. The first print would show one view of, say, a building, while the second print, because of the passage of time, would show the same building from a slightly different viewpoint. By adjusting the prints under the stereoscope in order, each eye would be staring at the same building but from a different viewpoint. The images would fuse together and the building would suddenly appear, as if by magic, like a three-dimensional model. It would often take new recruits a number of attempts before they could see the third dimension, as they strained their eyes and squinted, but when successful it almost always resulted in a childlike excitement.

PI training at Wembley was the model for photographic intelligence for the reminder of the war, and at Paduoc House involved all the recruits being given a lecture by Flight Lieutenant Michael Spender. Interpreters recalled that he likened the job of a PI to that of a motorist driving through a town who suddenly sees a ball bouncing across the road. Whilst the motorist cannot see any children, in an instant the brake pedal is pressed as the indications are there. To illustrate his tuition with a practical example he continued by talking about the approach that had been adopted to the interpretation of shipping since the summer of 1940 and the invasion threat of that period: it was vital to know what was normal and what was abnormal. This explained why, when the Dutch shipyards were photographed crowded with barges having their bows cut off, they knew it was important. They had been monitoring the normal life of the shipyards and knew that this was something different and significant.

Young WAAF officer Eve Holiday attended the PI course at Wembley in December 1940 and later recalled the great panic that recruits had at the prospect of being examined by recently promoted Squadron Leader Riddell. Half sitting on a desk, wearing his flying boots and continually holding a lit cigarette in a long holder, Riddell would pose questions to each in turn; if someone got stuck he would turn to the next person with the same question. Throughout the course, basic questions were fired at the recruits to assess their reasoning, such as: Question: Were these photographs taken over Europe or Britain?

Answer: Europe, because the traffic is moving on the right-hand side of the road. Question: In snowy conditions how can you tell if a factory building is in use? Answer: By the appearance of the roof, as heat would cause the snow to melt. Years later Eve Holiday could still recall the panic after being asked the difference between a naval vessel and a merchant vessel and luckily remembered their lecture from Spender: 'A merchant vessel is like an oblong box with pointed ends, and a naval unit is a very long elongated oval – cigar-shaped,' she answered correctly.

Training to become a PI alongside Eve was twenty-eight-year-old Constance Babington Smith, one of nine children – five daughters and four sons – born to Sir Henry and Lady Elizabeth Babington Smith. After working for the Treasury, in 1894 her father had been appointed private secretary to Lord Elgin (whose grandfather had acquired the Parthenon marbles) and by 1898 had married Elgin's daughter Elizabeth. With her father a director of the Bank of England, Constance spent most of her childhood living at the family home of Chinthurst Hill, an Arts and Crafts-style house designed by the celebrated architect Sir Edwin Lutyens near Wonersh, Surrey. Lutyens's fame had grown through the popularity of *Country Life* magazine, which presented an idealised view of the British upper class, their country houses and gardens, and issues from the period effectively chronicle the privileged life enjoyed by the young Constance. Having been educated by governesses and tutors at home, she enjoyed a busy social life in the 1930s, during which period she worked for the fashionable milliner Aage Thaarup, wrote for *Vogue* and with a fascination for flight wrote articles for *The Aeroplane*.

After the PI school had been destroyed, one of the houses in Beresford Avenue was requisitioned and an upstairs room with a 'sputtering fire' was used as the 'classroom'. Babs passed the examination set by Riddell, so dreaded by aspiring WAAF interpreters, became a PI and was soon specialising in the interpretation of aircraft. The PIs were schooled in three-phase interpretation techniques which had been developed by Riddell the previous year, building on a highly successful approach established by Hemming. Drawing on his PI experiences at Bomber Command and the combined experiences of his civilian and military colleagues at the PDU, Riddell developed and formalised the approach to photographic interpretation that would continue throughout the Second World War and beyond. Riddell devised these phases in the summer of 1940, when everyone was demanding immediate information about the invasion threat. It would have been easy to ignore second- and third-phase interpretation, but Riddell had the foresight to recognise that these additional phases of information would provide an enormous mass of data to create an emerging picture that would have long-term intelligence value, rather than just snapshots in time.

The first phase required immediate reporting of important intelligence – the movement of ships and aircraft, rail and canal traffic, the extent of bomb damage and the position of ammunition dumps. In cases of special urgency this information was to be available immediately, when so-called 'wet' interpretation would be undertaken, which involved a PI interpreting from the negative the moment it had been processed. Normally first-phase reports would be issued within a few hours of the aircraft landing, with the information relayed by teleprinter on a 'Form White' to Coastal and Bomber Commands, the Air Ministry and the Admiralty. Second-phase reports were a more detailed statement of activity, had to be produced within twenty-four hours, and also aimed to provide a view of what was shown in one set of photographs coordinated with what was shown in earlier photographs of the same location. These reports were typed up and distributed to Coastal Command, PRU Heston, the War Office, the Admiralty and various Air Ministry departments. Third-phase reports involved the compilation of even more detailed intelligence by specialist photographic interpreters, for specialist users, on such things as factories, airfields and aeroplanes.

As the value of photographic intelligence came to be recognised, so the workload and pressure on the small unit at Wembley increased. It was operational twenty-four hours a day, with staff working twelve hours on duty and twenty-four hours off, to satisfy the demands. That summer the unit scored some early successes. In July, Michael Spender identified the existence of some German 'modified barges' which during the ensuing months were kept under surveillance in the Channel invasion ports of Le Havre, Dunkirk, Calais, Boulogne and Ostende. On the night of 17 September 1940, in secret session, the new Prime Minister Winston Churchill told Parliament: 'At any moment a major assault may be launched upon this island. I now say to you in secret that upwards of 1,700 self-propelled barges and more than 200 seagoing ships, some very large ships, are already gathered at the many invasion ports in German occupation.' Agonisingly tense days followed as the invasion ports swarmed with activity preparing for Operation Sea Lion, Nazi Germany's planned invasion of Britain. But when the tide turned in the Battle of Britain, and the Luftwaffe failed in its bid for air supremacy – combined with fears about the lack of coordination between the different branches of his armed forces – Hitler abandoned the invasion. The interpreters at Wembley were the first to see concrete evidence of this. Photographs brought to them from Heston, dissected, projected and mapped, showed an evolving situation: the crisis had clearly passed; levels of activity lessened and gradually the barges, and the rest of the German invasion fleet, became dispersed.

The need to move quickly was brought sharply into focus after the fiercest German air raid of the war so far, during the nights of 30 November and 1 December 1940. The Luftwaffe dropped hundreds of tonnes of bombs and incendiaries, hoping to knock Southampton out of the war: the city area was a prime target as the production centre of the Supermarine Spitfire at Woolston and a key military port on the south coast of England. The Ordnance Survey head office on London Road was hit during the Southampton Blitz and, although nobody was killed, entire departments were destroyed, including the negative store, map store, drawing office and – critically for photographic intelligence – the only other Wild A5 Autograph in the country: Instrument Number 54. The Director General of the OS, Major General Malcolm MacLeod, spoke of the 'melancholy task' of disassembling the few remaining charred remnants of the machine. As soon as he heard of the loss of the Wild Machine in the Southampton Blitz, Hemming wrote to Laws saying 'it would be a catastrophe' if their Wild Machine was destroyed and suggested moving the whole operation to a safer location outside a target area immediately, as the machine and his team were running at full tilt, working twenty-four hours a day. Although Hemming was not yet aware of the plan, it had already been decided that the Wembley-based PI unit should be transformed into a normal service unit, to be named the Central Interpretation Unit (CIU), and should be relocated.

Hemming had already attempted to get a second Wild A5 Autograph. He had received a letter, courtesy of the British Military

Attaché in Switzerland, from Jacob Schmidheiny, one of the founders of Wild Heerbrugg. After the fall of France Schmidheiny had tracked down the shipment to Bordeaux, where it was en route to England. Clearly conscious of Swiss neutrality, he offered to recover and store it until after the war. But for the British it was clear that more machines were needed from Switzerland, or at the very least detailed drawings to allow more to be manufactured in Britain or the United States. On 5 December 1940 Hemming wrote again to Laws asking for permission to initiate private negotiations with a Portuguese air-survey company, with whom he had been cooperating in peacetime, as a means to obtain a new Wild Machine from Switzerland.

Using family connections in Portugal, Douglas Kendall approached General José Norton de Matos, the owner of a mapping company in Lisbon who had been the Portuguese Minister of War during the First World War and later ambassador to the United Kingdom. With both the general and Wild Heerbrugg in Switzerland prepared to cooperate, the British hatched a plan in which the general, who conveniently owned a Wild A5, would actually send the Ordnance Survey's A5 with the cover story that it needed servicing and repair, only for a fully overhauled Wild A5 to be returned in its place. The surviving parts of the Ordnance Survey Wild Machine were collected while Flight Lieutenant Harry Williams found engineering companies around the country who could quickly manufacture the parts destroyed during the Southampton Blitz. Working from plans, and with no knowledge of what they were making, the idea was that all the parts could be assembled and, whilst the machine would not work, it would look sufficiently convincing. Once all the parts had arrived and had been prepared, everything was crated and transported to Lisbon where it was repackaged into Portuguese crates and sent on to Switzerland via Germany. The top-secret undercover operation, with transportation courtesy of the Germans, resulted in the refurbished A5 Autograph being delivered back to the general in Lisbon. After being driven to Gibraltar it was shipped to Britain on the battlecruiser HMS *Hood* and was then sent on to RAF Medmenham where it joined the original Wild Machine before being moved in September 1942 to RAF Nuneham Park, where part of W Section was based in case Medmenham should ever be attacked.

CHAPTER 3
The Chalk House

In late 1940, thirty-year-old Ursula Powys-Lybbe was staying
with a friend in her idyllic sixteenth-century cottage near the
Buckinghamshire village of Medmenham, thirty-eight miles to
the west of central London. Born into a family of independent means,
she emerged from her Catholic girls' school education as a self-confident
young lady who expressed herself through her talent as a photographer
and her passion for travel. While many of her contemporaries in
England operated their photography businesses from the fashionable
studios of Bond Street or Berkeley Square, Ursula chose to work in the
avant-garde style, touring the country, visiting clients in their homes.
She became well known for taking stylish photographs of members
of high society in their 'natural habitats' – their country houses and
estates – and whilst Aerofilms photographed them from above Ursula
photographed them surrounded by all the iconography and trappings
of wealth and position. In 1937, the *Tatler* published one of Ursula's
photos of Lady 'Mamie' Lygon, a close friend of Evelyn Waugh's and
the inspiration for Julia Flyte in *Brideshead Revisited* – complete with
pet dog – that led to her being commissioned to contribute a series of
images of high society which ran until the war.

During the winter of the Phoney War, Ursula put her camera to one
side and made the unlikely leap to the Auxiliary Fire Service where
she chauffeured officials around London. By the spring of 1940 she had
grown bored of her driving duties, resigned and, vowing never to be
seen in uniform again, went to stay with her friend in Medmenham.
One day towards the end of October 1940 the Luftwaffe randomly
dropped a couple of bombs on the area, causing slight damage to

houses at Bockmer Farm, and obliged Ursula to dive for cover under a table for the first time.

Around the same time, she happened to read one of several articles that appeared that year in magazines and newspapers, including *The Flight*, that revealed how the information captured in aerial photographs taken on photographic-reconnaissance missions was being used in the fight against Nazi Germany. With her passion for photography, Ursula immediately decided that aerial photography was for her. In November, by extraordinary coincidence, her friend had a telephone call from an old beau who happened to be visiting Medmenham and wanted to meet up for lunch.

They met in a pub. The man strode in, a black patch over his right eye, wearing the uniform of an RAF wing commander. It was Hemming. When introduced to Ursula, he must have been taken aback when the irascible young lady – brandishing a newspaper cutting – explained her knowledge of photography and her desire to work in aerial reconnaissance. Hemming advised her to join the WAAF, serve some time in the ranks, and then they might be able to do something. With a goal in sight, by the beginning of 1941 Ursula was working as a clerk in RAF Records, Ruislip. But as month after dreary month passed, she began to wonder if she had done the right thing. Then the call came: would she like to attend the Photographic Interpretation course? In one of the last groups to be trained at Wembley, Ursula passed, received her officer's commission, the uniform of an Assistant Section Officer, and orders to report for duty on 1 April 1941.

Hemming had visited Medmenham after having been tasked with finding new premises for the Wembley operation that were near enough to RAF Benson in south Oxfordshire – the new home for the Heston operation – and far enough away from London to keep the Wild Machine safe from air attack. After visiting the country houses of Oxfordshire and Buckinghamshire, including Thame Park House, which would become an SOE training centre, Huntercombe Hall in Henley-on-Thames, and Medmenham Abbey, Hemming recommended Danesfield House, located near the quintessentially English village of Medmenham where the Dog and Badger Inn, the parish church and village shop could be found huddled around a crossroads on the route between Marlow and Henley-on-Thames. South of a steep chalk Chiltern Hills escarpment, picturesque cottages and houses lined Ferry Lane on its way to the river Thames where an ancient crossing point and Medmenham Abbey could be found. Surrounded by lush countryside, and with beautiful river views, Medmenham had long been a playground for the nouveau riche.

Sitting on a hill above the village, surrounded by woodland, Danesfield House was in many ways a strange choice for the new home of Britain's top-secret photographic-intelligence unit. Its dominant position on a plateau above the river Thames and its construction from brilliant white stone made it an obvious navigation target for miles around. With its red-tiled roof and distinctive richly patterned chimneys, the neo-Tudor house topped with architectural fancies and with an elaborate Jacobethan interior was considered slightly vulgar by many of the locals who christened it 'The Wedding Cake'. Winston Churchill, however, would develop a great affection for the place, dubbing it the 'Chalk House with the Tudor Chimneys'.

To this setting came a whole range of specialists and academics. The country's universities and Allied service personnel were scoured for experts and almost the entire Archaeology Department of Cambridge University was recruited on the basis that archaeologists know how to piece together information from scraps of evidence. This was a time when photographic intelligence was being shaped and strengthened by an influx of outstanding people from a wide range of backgrounds. Danesfield was to photographic intelligence what Bletchley Park was to signals intelligence. Like the code-breakers at Bletchley, the Central Interpretation Unit placed a high priority on the recruitment of its personnel since many different kinds of experts would be required to work in the specialist teams that would be assembled. This would lead to the enrolment of an ill-assorted collection of geographers, geologists, architects, engineers, artists, actors, cartoonists and even a famous choreographer.

Danesfield House was owned by (Arthur) Stanley Garton and was temporarily part-occupied by Colet Court prep school which had been evacuated there from Hammersmith with around eighty boys, until it was requisitioned by the Air Ministry. Danesfield had the advantage of being large, with central heating and plumbing that worked and a roof that did not leak. It was the third house to have been built on the site and had been built for the Victorian soap magnate Robert William Hudson, of Hudson's Dry Soap fame, who favoured the neo-Tudor style that was popular in the late nineteenth century. One subsequent owner, Mrs Hornby Lewis, was so attached to the place that she left instructions to be buried in the rose garden after her death. In 1938 the house was bought by Stanley Garton, a gold-medal-winning Olympic rower and businessman, who took up residence after Mrs Hornby Lewis's remains were removed and reburied in the nearby cemetery at Hambleden. Garton had been left a fortune by his industrialist father and had a multitude of business interests, including the local Marlow Brewery.

Unhappy about his property being requisitioned, Garton roguishly proposed that he be allowed to retain at least twenty rooms for himself, including two in the basement, and that others would be required to store furniture according to specifications drawn up by Harrods. For the Air Ministry this was completely impractical – given the top-secret work that would be carried out there – and so Garton was forced out. But not before he successfully persuaded the Air Ministry that all the interior walls must be protected with sheets of fibreboard – a move that meant much of the splendour of the house would be hidden from view during its wartime years – and the Ministry would pay for the removal of all his furniture. Harrods duly moved the furniture – at a cost of £1,700 – to the covered tennis courts and swimming baths at Medmenham Abbey. After the English Reformation, the Cistercian abbey had been converted to a private house and was then home to Arthur Bendir – founder of Ladbrokes the bookmakers – and a friend of Garton. In the eighteenth century the abbey and its owner Sir Francis Dashwood gained notoriety when the depravity, debauchery and devil worship of the Hellfire Club – whose members styled themselves the Mad Monks of Medmenham – were exposed.

One of the later Medmenham Station Commanders, Group Captain Peter Stewart, had many memories of bitter clashes with Stanley Garton, who moved to Kingswood House, located just off the Henley Road, opposite Danesfield House's main gate. Garton maintained a close surveillance on his requisitioned house throughout the war. One of many disputes involved some ornamental lead objects that Stewart removed from Danesfield House's gardens during the time of the drive for scrap metal. Although Stewart asked the professional advice of Geoffrey Deely – the distinguished sculptor who ran the model-making section and who felt they didn't have any artistic value – Stewart made the fatal error of failing to consult Garton. After Medmenham received special congratulations from the Air Ministry for having contributed two tons of lead in one month, Garton began a three-year correspondence with the Air Ministry claiming compensation. Ultimately Stewart was ordered to compensate Garton out of his own pocket; when he refused to pay the £50 bill it was docked from his pay.

On 7 January 1941, after a meeting between Laws, Hemming and Riddell, the Central Interpretation Unit (CIU) had been created as a self-contained element, absorbing much of the Wembley operation, the interpretation unit at Heston and a team of model-makers from RAE Farnborough. The key functions of the CIU were to provide detailed interpretation of aerial photographs taken by the RAF from stations across Britain; deploy photographic interpreters to the RAF stations to allow first-phase interpretation to be undertaken quickly; plot aerial photographs using the Wild Machine and prepare accurate plans; maintain a library of aerial photographs from all sources; and build models from aerial photographs of targets as required.

What wasn't revealed to Hemming or Riddell at this stage were the behind-the-scenes discussions – between the Commander-in-Chief of Coastal Command, Laws and Air Marshal Peck – that meant Hemming's days as shaper and facilitator would be short-lived. Riddell was perhaps the obvious choice to lead the development of the new unit, but diverting him to the administrative challenge of moving from Wembley and establishing RAF Station Medmenham was not judged the solution by Air Marshal Peck. Instead he instructed Laws to appoint an outsider as Medmenham Station Commander, in order that Riddell could 'bend his whole ability to the technical aspects' of photographic interpretation. This separation from the outset of specialist PI work from the routine administration of Medmenham as an intelligence unit would prove inspired.

On the bitterly cold morning of 17 January 1941 – a Wednesday – the huge Wild A5 Machine, with its complex optical system that could measure the smallest detail, was carefully dismantled in Wembley and transported by Messrs Baird and Tatlock (London) Limited to Medmenham. The installation of a 4-inch-thick concrete roof above the Swiss-made machine had ensured both its survival and its popularity at Wembley as people huddled around it for safety during bombing raids. At Medmenham a sealed room in the basement at Danesfield was chosen as the safest location, and initially the precious device shared a space with Garton's soda-water machine and great lumps of chalk, stockpiled to replace bits of the building that would fall off when least expected.

Wembley's print library was moved into the hammerbeam-roofed minstrels' gallery above the Great Hall. The Hall itself was used to store thousands of military maps, plans and plots of the major towns and seaways of Nazi Europe. Throughout the war, the stock of the library would continually outgrow the space allocated to it – and would take over more and more of the house, moving staff and their accommodation further into the grounds. By the end of the war it held records of over 80,000 sorties and occupied more than three miles of shelving in a maze of rooms in the main house and surrounding network of Nissen huts. Essential backup copy prints were stored at RAF Nuneham Park and 'Pinetree', the latter being the code name for the HQ of the US Eighth Air Force, based in the requisitioned Wycombe Abbey Girls' School, four miles from the Bomber Command Headquarters at High Wycombe.

On 1 March 1941, a key decision was taken by the Air Staff that put the four PR units which then existed at RAF Stations Benson, Wick, St Eval and Oakington, together with the CIU, under the command of Charles Medhurst, the Assistant Chief of the Air Staff with responsibility for intelligence. With Laws designated technical adviser to the newly created Assistant Directorate of Intelligence (Photographic), Medhurst wanted changes in the existing set-up. Although Cotton had gone, his influence and many of his buddies, such as Hemming, were still around – and needed to be purged. Medhurst approached Group Captain Peter Stewart, who had recently successfully organised the Air Ministry's War Room within the Cabinet War Rooms – the maze of interlocking rooms and tunnels underneath King Charles Street that sheltered Churchill and his cabinet during the Blitz that continued until May that year. When Stewart replied that he knew nothing about photographic intelligence, Medhurst replied that he needn't bother. What he wanted was an organiser: 'someone to beat the bugs out of the system'.

In early March the Aircraft Operating Company's contract with the Air Ministry was cancelled and with Hemming pushed aside, as Cotton had been before, he was replaced by one of the RAF's own, Wing Commander Carter. Hemming was to receive notification by postagram from Coastal Command that his services were no longer required, an action that ensured everyone at Paduoc House that day knew before Hemming did. The cruel way in which Hemming discovered the news came as a bitter blow to him, and the company, however inevitable it might have been. To add insult to injury the treatment he received from his successor, who immediately banished him from Wembley on fourteen days' leave, left Hemming in a complete state of suspense. Laws later criticised Carter's actions as being 'not so sympathetic' as they might have been, particularly given how crucial Hemming and the Aircraft Operating Company had been for photographic intelligence.

Hemming's logistical and administrative brilliance in running aerial-survey programmes around the world before the war, combined with the Aircraft Operating Company's application of new technology, were skills that Bomber Command desperately needed. This ensured that he was engaged on 'flying-control' duties, a drive for efficiency that ensured the ground crews kept the bombers flying and that technical improvements – notably air navigation – made it from theory to application. Hemming's association with Bomber Command motivated him to help finance the creation in 1945 of the Pathfinder Association Limited, an employment and general advice bureau for the Bomber Boys – many of whom had gone straight from school into the air war. Its first secretary, Squadron Leader Edward Brant, explained to the press that from premises in Mount Street, Mayfair, 'we shall keep a central register of members and through contact with prospective employers do all we can to help the boys get jobs'. For eight years after the war Hemming left the aviation world and occupied himself running a fishing hotel in Ireland, but in 1955 was lured back and worked for Burnley Aircraft Products – manufacturers of rockets and guided-missile components – becoming their chief liaison officer.

After working the last night-shift at Wembley, Daniel and Kendall set-off at 8 a.m. on a snowy 1 April 1941 and drove to RAF Station Medmenham where they handed over the unfinished reconnaissance sorties to the day-shift who started work on them immediately. The fluency and continuity of the transfer was the product of months of planning. The bulk of the Wembley staff – draughtsmen, photographers and photographic interpreters, plus any civilians who worked in a clerical capacity and who wished to continue such work by joining the WAAF – moved into Medmenham on April Fool's Day. Also reporting for duty that day was recently qualified PI Ursula Powys-Lybbe, who would go on to head the Airfields Section. But for Hemming and those personnel considered 'rather expensive members of the company' the creation of a supposedly regular RAF unit meant they were superfluous and they were summarily laid off.

When Hemming began looking for a new home for the Wembley set-up in November 1940, he worked on the understanding that the maximum strength of the unit would be about 469 personnel. Little could he have imagined that after the Americans officially joined forces with the British at Medmenham in May 1944, on the eve of the Normandy landings – when it became the Allied Central Interpretation Unit – there would be 566 officers and 1,186 other ranks working at the headquarters of photographic interpretation. By then the unit had been involved in the planning stages of practically every single operation of the Second World War before the Allied invasion of Europe and would be involved in all aspects of the subsequent ground advance. And from the handful of people involved in high-altitude unarmed photographic reconnaissance at the beginning of the war alongside Sidney Cotton, in the European Theatre alone thirteen entire squadrons – five in the Royal Air Force, five in the United States Army Air Force, and three in the Royal Canadian Air Force – would be dedicated to strategic photographic reconnaissance.

To provide the CIU with housing and other facilities the Air Ministry Directorate of Works began to create what would become a small town around Danesfield House. Unfortunately for the Medmenham personnel there was an overall ruling that, wherever possible, buildings should be constructed underneath trees for camouflage purposes. Since the area was surrounded by woodland, it meant they had a four-year battle with damp conditions, and clothes stored in drawers would easily sprout fungus. The conditions at the station came as quite a shock to many of the American and Canadian personnel who would be based there and who were more accustomed to dry, centrally heated atmospheres.

When Flight Lieutenant Glyn Daniel first arrived at Medmenham on April Fool's Day 1941, the construction of accommodation was still under way, which meant he found himself billeted at Spinfield House, owned by the Hood-Barrs family in the neighbouring town of Marlow. Mr Hood-Barrs was not particularly happy with his stockbroker-belt house being used as a billet – particulary after an experience a few nights earlier when Daniel's predecessor had been found drunk in the rockery garden while looking for his bedroom. Allocated a large bedroom with beautiful views across farmland to the river Thames, Daniel was looked after by Jackson the butler. Spinfield House was conveniently close to the Hare and Hounds on the Henley Road, an inn that would become a popular haunt for Daniel and many of the PIs.

One evening in early October 1941, there was great excitement in Spinfield House at the news that 'Uncle Fred' would shortly be visiting from California. The man's true identity would turn out to be Frederick Joseph Rutland. During the Battle of Jutland in 1916, Rutland had flown visual reconnaissance missions and was awarded the Distinguished Service Cross for gallantry (earning him the sobriquet 'Rutland of Jutland'). Resigning his RAF commission in 1923, in the mid-1920s Squadron Leader Rutland became an adviser to the Japanese on naval flying, and in November 1932 was recruited by the Japanese Naval Attaché in London. With the code name Shinkawa – New River – Rutland travelled to the United States where his extravagant lifestyle in Beverly Hills soon brought him to the attention of the FBI. When in June 1941 they arrested a Japanese spy and Rutland was implicated, to avoid British embarrassment the Americans agreed to him being discreetly removed from the country. Arriving in Britain on 5 October 1941, Rutland stayed with his brother-in-law in Spinfield House. Interviewed by MI5, the Naval Intelligence Division and the Air Ministry, Rutland claimed that he had never given nor intended to give any information of value to the Japanese but had accepted their offer because it was easy money and a chance of adventure, before adding that he would be more than happy to act as a double agent, an offer they all politely declined.

On 8 October 1941 Daniel was summoned by the Medmenham Station Commander – Wing Commander Laing – and sent in a staff car to the Air Ministry building in King Charles Street, Whitehall. There he was introduced to officers of MI5 and told to cooperate with them in every way possible. The somewhat mystified Daniel was driven to the Oxford and Cambridge Club in Pall Mall, where in a private room they revealed their interest in the Hood-Barrs' house guest. Asked to keep a discreet eye on Rutland's movements, Daniel was to keep in contact by using a public phone box on the Marlow road. He was given a telephone number to call and a secret code word. With information from Daniel and a 'telephone check on Marlow 543' MI5 kept a close watch on Rutland until 16 December 1941. When Daniel returned to Spinfield House that evening from Medmenham he discovered a shocked, solemn-looking family in the drawing room, and learned that Uncle Fred was now languishing in Brixton prison.

At 10 a.m. that morning Superintendent Frederick Gee of the Buckinghamshire Constabulary had detained Rutland under the wartime Defence Regulation 18B, which suspended the right of an individual to habeas corpus. Since the British knew about Rutland running an operation in Hawaii, and following the Japanese attack on the US Pacific Fleet at Pearl Harbor on 7 December 1941, he was interned and would join the group of British Nazi sympathisers that included British Union of Fascists leader Oswald Mosley on the Isle of Man until he was released from Internment Camp M in December 1943. Rutland failed to confess to being a Japanese agent but there was no doubt that he was a 'paid agent of the Japanese' according to a report signed by A. F. Blunt – the MI5 officer Anthony Blunt who would later be knighted for his work as the surveyor of the queen's pictures before being exposed as a Soviet agent in 1979. Whilst rumours about Rutland dogged him before and after his suicide in 1949, the national embarrassment of a retired RAF officer having such an involvement meant the MI5 papers on the case have only recently been released.

By mid-1941 Danesfield had been transformed into a high-security RAF station with perimeter fences and armed guards on the gates, and had become home to one of the most impressive collections of specialists, technicians, academics and wizard-minds in the Second World War. It was a working environment and community that combined the best of military and civilian thinking. The rules dictated that everyone, including the commanding officer, had to stop at the gatehouse, identify themselves and present their papers to the armed guards before they could enter. Camouflaged soldiers hid in the woods, defending this vital intelligence unit. The high-security status of Medmenham was brought home to WAAF Officer Pamela Bulmer one day while walking in the woods at the station, when she came across a camouflaged soldier. When she asked what on earth he was doing, he simply replied that he was guarding her.

In late 1940 Michael Spender was tasked with establishing a first-phase capability at the 1PRU satellite stations of RAF Wick in Caithness and RAF St Eval in Cornwall, where the reconnaissance effort was focused on shipping in the occupied French ports. The standard procedure at St Eval when a PR aircraft landed was for the pilot to be debriefed by an intelligence officer while the photographs were being developed – a process that would take about an hour. The Form White would be written by the PIs within two hours of the aeroplane landing and sent by teleprinter to the various intelligence commands before the film was flown to Benson, from where prints were sent on to Medmenham.

Just after Christmas 1940, WAAF Eve Holiday transferred from Wembley to RAF Wick where she was joined by David Linton, a geography lecturer at the University of Edinburgh – later to become Professor of Geography at Sheffield and Birmingham Universities – who had completed his PI training at Wembley in October 1940. With Wick a coastal airfield on a high plateau in bleak, bare countryside with no trees, close to dark rocky jagged cliffs, this proved a dramatic contrast to suburban Wembley. From Wick, German shipping and Schnellboote – motor torpedo boats – that were operated from German-occupied Norway were monitored for the Admiralty. The airfield was frequently bombed, and Eve vividly recalled years later the time when a Heinkel He 111 was shot down and the crew landed safely by parachute. When they were brought into the Ops Room, where she happened to be, the *Hauptmann* – the officer in charge – boldly gave a Heil Hitler salute, only for the mild-mannered Commanding Officer to counter with a polite 'Good afternoon'.

It was from Wick, on 21 May 1941, that one of the most famous early photographic-reconnaissance flights – sortie N/183 – would be flown. The events that followed would prove a great triumph for photographic intelligence, and following so shortly after the creation of Medmenham helped confirm its status as a war-winning intelligence unit. Days earlier, on 18 May 1941, the battleship *Bismarck*, accompanied by the *Prinz Eugen*, had sailed with over 2,000 men on board from Gdynia on the Baltic coast. British intelligence had been monitoring the progress of the *Bismarck* having noticed increased amounts of photographic reconnaissance by the Luftwaffe in the North Atlantic, which suggested that the German fleet would soon break out into the Atlantic Ocean. The first indication came two days later from Captain Henry Denham, the British Naval Attaché in Stockholm, who had received a report from Major Carl Petersén of C Bureau – the ostensibly neutral Swedish Secret Intelligence Service – that the *Bismarck* and *Prinz Eugen* were sailing through the Kattegat (the strait connecting the Baltic and North Seas) accompanied by a number of destroyers and aircraft. The following day members of the Norwegian resistance sighted the ships off the south coast of Norway.

With this news the only two serviceable Spitfires at RAF Wick were made ready. Two experienced pilots were chosen to fly them, Pilot Officer Michael Suckling and Flying Officer Greenhill DFC. Linton and Greenhill, who were senior to Suckling, agreed on the plan that Greenhill should cover the coastline towards Oslo because it seemed most likely that the ships would be there, whilst Suckling should photograph the Bergen area. The son of a market gardener and then only twenty years of age after recently celebrating his birthday in Wick, with fair hair, blue eyes and a boyish face, Suckling had been given the nickname 'Babe'; the PR pilot Noel Sherwell, who shared a room with him, said he looked about sixteen and a half. Suckling took off at 11:05 and after refuelling at Sumburgh, Shetland, to give himself an extra thirty minutes' range, headed towards Norway. After Greenhill set off thirty minutes later, Linton and Eve Holiday lunched together in the mess. Then Eve returned to the PI office and Linton stayed in the mess to write letters.

Around 14:30 Eve spotted Suckling's Spitfire landing, far earlier than expected, and sped across to the dispersal point in a van they used to get around the airfield. Airmen quickly removed the camera magazines before setting off to the photographic section to immediately process the film. Years later Eve would recollect Suckling's childish face beaming with excitement: 'It's there! I've seen them, two of them – I think they might be battleships or cruisers!' Desperate to let Linton know about his discovery, Suckling raced over to the mess where Linton was startled to hear the door of the anteroom fly open. Still wearing his flying kit (a breach of mess etiquette), the young pilot stood there with a beaming smile and called out – 'I've found it!'

The film was soon developed and although it hadn't even dried yet, the 'wet interpretation' created much excitement. Telephoning the news straight through, Linton followed up with his Form White in which he said the ships were probably about to move. This was met with an immediate response from Coastal Command headquarters asking for copies of the prints. The transportation of films from Wick to Medmenham was usually undertaken by train because the weather could be too uncertain for flying, which meant that films could sometimes take two days to arrive. Given the extreme urgency in this case this presented a slight difficulty because the only aircraft available to take the prints was Suckling's Spitfire and he had just completed a three-hour sortie. Suckling nevertheless took off as darkness fell. He flew south until he found himself short of fuel – at night – on the outskirts of Nottingham, near his home town of Southwell. Determined to complete his journey, he landed his aircraft and made his way to a friend who owned a garage and had a car. From there Suckling and his friend continued the journey together – averaging a speed of over fifty miles an hour – and by driving through the night eventually managed to deliver the prints to Coastal Command headquarters in the north-west London suburb of Northwood in the early hours of the morning. The prints were examined, and the Admiralty was able to confirm that Suckling had indeed photographed the *Bismarck* and *Prinz Eugen*.

By coincidence, the very day when Suckling had been taking his photographs of the *Bismarck*, Captain 'Jock' Clayton, the Deputy Director of Naval Intelligence, had been at Medmenham to meet David Brachi who had transferred to Medmenham from Wembley. Brachi worked in A Section, a specialist third-phase section responsible for the interpretation of naval and merchant shipping. Among the more important subjects dealt with was enemy shipbuilding, particularly U-boat construction, ship repair, ship armament, port facilities, concrete constructions such as U-boat and E-boat shelters and minesweeping. From aerial photography and pre-war *Jane's* merchant shipping manuals, drawings and silhouettes were produced in the section, from which models were produced for identification and training purposes.

A Section was based in various parts of Danesfield House including the Long Gallery, a room that the PIs nicknamed 'Hell's Kitchen'

– in reference to the activities of the Hellfire Club – and a bathroom that served as an office. From his bathroom Brachi was responsible for providing a strategic overview of enemy shipbuilding down to individual ships in shipyards. He recorded the date when each keel was laid down in a shipyard and allocated it a reference – for example, Rostock R1 – and through subsequent aerial photography would monitor the ship's construction. When he monitored a shipyard and its activity for a period he was able to calculate accurately the launch date from the appearance of the hull. With every vessel given a unique identifier, it was then possible to track their movements. This meant that when a ship was attacked but not destroyed, its repair in a shipyard would also be detected by the work of A Section.

The day after Suckling had flown his sortie over Norway a copy of the aerial photographs that he had taken were being analysed by second-phase interpreters at Medmenham. At eleven in the morning Brachi received a telephone call from Clayton at Naval Intelligence and was asked whether the *Bismarck* was going raiding in the Atlantic, whether it was being prepared for a seaborne invasion, or whether it was just on a cruise in Norwegian waters. Using his third-phase interpreter subject specialism, Brachi was asked to provide a verbal report as soon as possible. By 12:30 he telephoned back to say the evidence showed that the *Bismarck* was going raiding and his reasons were as follows. The merchant vessels with the *Bismarck* were not troop carriers. The ships with the *Bismarck* and *Prinz Eugen* had made rendezvous with them off the coast of Norway, not in harbour, so the transfer of soldiers would not have been likely; the presence of a tanker suggested refuelling for a long trip; the presence of merchant vessels suggested they were taking on board supplies. Brachi had observed other naval units preparing for Norwegian cruises and they had not been refuelled by special tanker. Some of the merchant vessels had come out of Bergen and that was not routine.

The way Brachi was able to quickly and accurately analyse the photographs that Suckling had taken at 1:15 p.m. the previous day from 25,000 feet over the Norwegian coast while reconnoitring the approaches to Bergen allowed the orchestration of a pursuit that culminated within days in the sinking of the German battleship *Bismarck*, on 27 May 1941. For Britain and photographic intelligence this was a triumph, and the exploits of Michael Suckling would feature in a Ministry of Information publication the following year that explained the objective of Coastal Command as being: 'Find the enemy; strike the enemy; protect our ships.' At Medmenham the PIs enjoyed a varied social life, snatched between shifts: there were frequent parties held to celebrate Allied successes in the war. Specially notable was the one held after the sinking of the *Bismarck* when morale at Medmenham was particularly good.

A dedicated entertainments officer ensured a varied social calendar which included concerts by ENSA (the Entertainments National Services Association – though sometimes a popular translation of its acronym was 'Every Night Something Awful'). There were also galas and fancy-dress dances; RAF Gang Shows; lectures, Brains Trusts, general knowledge contests and visits from the RAF Griller String Quartet (one of which was conducted by Sir Malcolm Sergeant) and once from Glen Miller and his Dance Band. According to Shirley Komrower, a WAAF Section Officer who worked in the Press Section, the impact of the Americans was felt immediately. She was livid when, on the first night of their arrival at nearby Wycombe Abbey, US airmen visited the WAAF mess at Medmenham bearing bananas and nylon stockings. This had the effect of dividing the women straight down the middle: between those who would go out with the Americans to get their nylons and those who thought the others were prostituting themselves. Medmenham was unique: when later commenting on her time there Komrower observed that 'practically every officer was a character. They were professors, they were scientists, they weren't military people at all and they didn't comply in a military way, they were really a bunch of nuts – you might say'. With such characters it's not altogether surprising the locals called the inhabitants of RAF Station Medmenham, the 'Mad Men of Ham'.

In such an environment it was not surprising that close relationships developed. Pamela Bulmer later recalled working on the same shift as Frederick Ashton, the dancer who would become the founder choreographer of the Royal Ballet, Sarah Oliver (née Churchill) and archaeologists Glyn Daniel and Stuart Piggott and how social life would have to be snatched between work. The RAF Medmenham Players performed eight full-length plays, two revues and a pantomine during the course of the war. Sarah Oliver appeared in several of their productions while her husband Vic once headed a cast of artistes at the Odeon Cinema in Marlow. Racier options were available, too: they included the Windmill Girls from Mrs Henderson's Windmill Theatre: nude *tableaux vivants*, completely naked women standing motionless like living statues. Inspired by the Folies Bergères and Moulin Rouge in Paris – and on the basis that nude statues could not credibly be considered morally objectionable – their ruse sidestepped the Lord Chamberlain's obscenity laws with the maxim: 'If you move it's rude'.

At the tamer end of the scale entertainments included art and handicraft exhibitions, a competition to design 'the best security poster' and a photography exhibition in 1943 for which almost 200 photographs were submitted. With many established artists at Medmenham, including William Mann and Clarence Woodburn in V Section, it was not long before giant murals could be found decorating the different station messes. Leslie Durbin also worked in the model-making section until the summer of 1943 when he was given leave of absence for six months. As one of the most admired silversmiths of the twentieth century – who would go on to design parliamentary maces for newly independent former colonies, and coins for the Royal Mint – he designed and fashioned the gold- and silverwork on the hilt and scabbard of the Sword of Honour to commemorate the defence of Stalingrad that was presented to Stalin by Churchill, on behalf of King George VI, at the Tehran Conference in late 1943.

Central to the social scene at Medmenham and frequently cited as one the unit's most colourful and beloved characters, whose antics helped everyone's morale no end, was the middle-aged Lady Charlotte Bonham Carter. After enjoying life as a debutante before the First World War, she served in the Foreign Office and MI5 during it, before being assigned in 1919 to the British delegation attending the Paris Peace Conference. Her love of the arts prompted her in 1926 to become a founding director of the Rambert Dance Company, now Britain's oldest dance company, and this was followed in 1927 with her marriage to the fifty-seven-year-old Sir Edgar Bonham Carter. At the age of sixty-two she chose to forgo her life of privilege, and instead billeted in one of the Nissen huts in Medmenham Woods, finding the chill dampness tiresome, especially when it came to keeping clothes dry. Whilst seemingly living in a world of her own, Section Officer Bonham Carter had an acute social, political and historical awareness. Instantly recognisable, with strands of grey hair flowing from under her WAAF cap, one of her habits was to take a small wicker basket to breakfast in the mess, into which remnants of her breakfast would be added, wrapped in toilet paper, so that she could eat them for elevenses. Walking the half-mile to Danesfield House, she would frequently take a short cut across the playing fields – much to the annoyance of the sports officer – with her regulation shoulder bag and carrying her basket in one hand and umbrella in the other. On one occasion, Sarah Oliver arranged through her mother Clementine – who headed the Red Cross Aid to Russia Fund – for the Ukraine-born concert pianist Benno Moiseiwitsch to give a concert. Concerned that the thumped-out canteen piano would not be suitable, Sarah was relieved when Charlotte happily agreed for her grand piano to be disgorged from wartime storage. When the concert day came, Moiseiwitsch started playing, only for a string to snap and begin waving wildly in the air. When the pianist finished, he stood up and bowed, only for Charlotte to sweep onto the scene to thank 'Mr Moiseiwitsch, for ruining my piano'.

One of Sarah's favourite people at Medmenham was Flight Lieutenant Villiers David, who served as a go-between with MI6 and managed copies of their records at Medmenham. When the barrister first arrived for wartime duty with the RAF he was accompanied by his valet, who dutifully laid out his employer's pyjamas in a twenty-bed dormitory. With a first-rate mind and entertaining personality, Villiers David was part of the Sassoon family and was a talented amateur artist. Throughout his time at Medmenham he lived with his sister at Friar Park, in Henley-on-Thames. Subsequent owner George Harrison bought the luxurious architectural fantasy and its thirty-five acres of grounds after the Beatles broke up. He began rejuvenating the house and rediscovered a garden with subterranean caverns, and even found a stone crocodile in the lake. Villiers David would frequently entertain his colleagues and friends, including Glyn Daniel who enjoyed delicious dinner parties and overnight stays at Friar Park. The worldly-wise Villiers David was admired by many of the younger interpreters, for whom he wrote a little book, *Advice To My God Children*, with which he would regale his 'godchildren' in their breaks and keep them amused and alert.

The propaganda value of aerial photography during wartime was exemplified by Suckling's photograph of the *Bismarck*, which was issued to the press by the Ministry of Information with the caption 'the photograph that sank a battleship'. In the same month when Suckling took the photograph and the *Bismarck* was sunk, J Section was created at Medmenham to identify photographs of a spectacular or topical nature that could be released for use in newspapers and weekly periodicals and for public exhibitions. In March 1943 a touring exhibition called *Bomb Damage on Germany* was devised and during its showing on the first floor of the Woolworths store in Hastings, Sussex, the Ministry of Information secretly observed the public reaction. Visitors were recorded to have shown a 'grim satisfaction' and, whilst amazed at the extent of damage to the German cities, welcomed being able to see and study the damage in detail, rather than merely reading about it in a newspaper or hearing about it on the radio. J Section was also responsible for compiling albums of aerial photography covering important and spectacular incidents, such as the Dambusters Raid, for presentation throughout the war to the likes of the King, Prime Minister, Chiefs of Staff and senior politicans. But after a while the major part of their time was spent on the production, layout and editing of *Evidence In Camera*, a magazine that was the brainchild of Group Captain Peter Stewart. During a visit to Bomber Command headquarters one day, he went into the anteroom after lunch in the mess and saw that all the picture papers such as the *Illustrated London News* were being read whilst on the table the text-heavy broadsheet newspapers such as *The Times* and the *Daily Telegraph* were ignored.

The first number of *Evidence In Camera* was issued in October 1942. The title was chosen by the Section and was derived from the historical and legal professions' use of the term to indicate evidence that is heard in private for the common good. Readers were warned that their copy should be kept under lock and key, and should not be treated as reading matter for the train or as a topic of conversation down at the local. The magazine included strike-attack photographs, night photographs, bomb damage, camouflage, identification of shipping and aircraft, communications, high- and low-altitude obliques, and port identification. It effectively chronicled the war as it happened. For every edition a different cover was designed and cartoons were drawn by the *Daily Mail* cartoonist Julian Phipps and *Punch* magazine's John Langdon who both served at Medmenham. On the last page of every issue was a problem picture that showed something weird or wonderful to challenge and educate the reader. Produced weekly and later fortnightly, 103 numbers were published, including special issues about D-Day, the Flying Bomb and Mulberry Harbours, with a final issue on how photographic intelligence had developed throughout the war.

Although pilots from nearby RAF Benson were a familiar sight at Medmenham and were often invited to social events, the daily interaction at the airfields between the first-phase PIs and PR pilots meant that the loss of fliers' lives was all the more immediate and personal. A bitter-sweet reminder of the dangers faced by the PR pilots was felt in the Nottinghamshire home of Michael Suckling when, two months to the day after his delivery of the vital reconnaissance images of the German battleship, a telegram was received by his family. After being awarded the DFC for his bravery, Suckling had been transferred to RAF St Eval – from where daily observations were made of the French ports – and it was during a mission over La Rochelle while flying at low level that he was shot down and killed.

CHAPTER 4
Target for Tonight

One evening in the Hare and Hounds, Glyn Daniel spotted his old schoolfriend Bryan Hopkin at the other end of the bar. The two men had studied at Barry Grammar School and St John's College, Cambridge, where Hopkin had been an economics student under John Maynard Keynes. In 1941 he had been plucked from the Ministry of Health to become one of Churchill's statistical team under his Chief Scientific Adviser Professor Lindemann – raised to the peerage as the first Baron Cherwell of Oxford in July 1940. Cherwell lived nearby at 'The Heights' on Henley Road, Marlow, and for his own safety was forbidden by Churchill to spend the night in London. When Daniel explained that he worked at Medmenham, Hopkin knew immediately that his old friend worked in photographic intelligence. On discovering that Hopkin had a personal Medmenham connection, Glyn Daniel and subsequently Douglas Kendall were soon invited for dinner. Daniel recalled Cherwell waving his hands at the red despatch boxes that were piled up in his drawing room and saying 'these and my word will protect you' before asking for a personal briefing on what photographic intelligence was actually revealing. Churchill wasn't convinced that the Cabinet were getting all the facts, and he suspected Bomber Command of overestimating the effectiveness of their bombing campaign.

In 1941 the Air Ministry were anxious for the Ministry of Information to make a propaganda film at the earliest opportunity, one that recorded Bomber Command 'cutting the arteries of the enemy'. Giving it top priority, the Air Ministry offered John Betjeman – the future Poet Laureate who then worked for the Crown Film Unit – every cooperation and access to all their facilities. The result was the documentary film *Target for Tonight*, which featured a bombing attack on Germany and Medmenham characters Constance Babington Smith, Glyn Daniel and Peter Riddell. It was directed by Harry Watt who had co-directed the 1936 documentary film *Night Mail*, which used the rhythmic words of W. H. Auden and the music of Benjamin Britten to tell the story of the overnight mail train from London to Scotland. Since the outbreak of war he had been busily directing propaganda films, including *London Can Take It* and *Christmas Under Fire*, and now in *Target for Tonight* he featured unnamed RAF personnel, playing themselves, who revealed how a raid was planned and executed, from the photographing and identification of the target to the return of the last bomber – 'F for Freddie' – to its base. With music courtesy of the RAF Central Band, the film was shot at Medmenham (which was never identified in any way) and the fictional Millerton Aerodrome – aka RAF Mildenhall, Suffolk – and begins with a Blenheim reconnaissance aircraft flying low over countryside. A package drops attached to a small parachute, which is then collected on the ground by an aircraftman. Heading into woodland, he disappears into a bunker and through a door marked 'Photographic Section – Bomber Command – Secret'.

Cutting to a photographic darkroom, a technician is shown working with a developing tray as the outlines of an aerial photograph begin to appear through the solution. After small talk with another aircraftman about the first-rate quality of the photography, and how the Squadron Leader will want all the photographs sharpish, Peter Riddell appears on screen inspecting a wall map of the German Ruhr region while confidently scribbling notes onto a clipboard. With a cigarette nonchalantly held in his hand, he walks across to a young WAAF officer – Constance Babington Smith – and asks whether there are many aircraft in the photographs that she is interpreting. She replies in her cut-glass English voice, 'Oh, crowds, sir. Five JU52s and quite a lot of 109s'. Asking to see the photographs when she is done, he walks back to his desk, and the office with its rows of desks, and officers busily examining aerial photographs, surrounded by paperwork and maps, is revealed.

When he sits down the newly developed reconnaissance photographs are handed by the aircraftman to the Squadron Leader who is soon examining them through a magnifying glass rather than a stereoscope. With the words 'I think we have something here' he begins a conversation with the Flight Lieutenant at the next table (Glyn Daniel). Asking for the Freihausen file, Riddell intently inspects the photographs and says to Daniel: 'They have done a tremendous amount of work here. You remember we saw it about three months ago. Quite a small place – just a siding in a wood. No activity of any kind, and now look at it. Greatly increased sidings, pipe lines running along the side of the wood down to the barges in the river, oil tanks at the side of the sidings. It's a colossal installation'. To which Daniel replies – much to his later embarrassment – 'It certainly is a peach of a target, isn't it, sir'. While Daniel completes the photographic analysis, Riddell immediately begins writing his report for the Commander-in-Chief, who on receipt of the report decides that whilst the city of Kiel will be the main bombing target for that night, one squadron will be diverted to this new target.

With the order passed to Millerton Aerodrome, news spreads among the crews about that night's target. With the men briefed on their mission, and told that the target is hidden in a wood but can be easily found by following a canal system, Dickson from the crew of F for Freddie is named captain. After scenes of the men preparing for their flight, mixing banter with information about the mission, the Wellington bombers are filmed taking off. Nearing their target, the crew of F for Freddie descend under the clouds and identify their target with the aid of the pre-dropped incendiaries. After unloading their bombs successfully, they are suddenly hit by ground fire from the German defences, damaging the port engine and wounding the wireless operator. Having lost contact, the officers at Millerton wait anxiously for news of the plane as the weather worsens. F for Freddie finally makes it back to England but is hampered by fog. Given the option of bailing out or attempting a difficult landing, the crew unanimously vote for trying to bring the plane down safely, which they succeed in doing. On landing the crew give a report of their mission to the station intelligence officer, on how their first four bombs dropped short of the target but the last one was a direct hit which 'caused a hell of a great big fire, buckets of smoke, visible fifty miles away'. *Target for Tonight* would provide inspiration for the US Eighth Army Air Force, who in 1944 commissioned the First Motion Picture Unit of the Army Air Force to produce *Target for Today*, which featured the men of the

Mighty Eighth 'doing their day-by-day jobs, on the ground and in the air' and showed the execution of a daylight bombing raid on several German cities including Brandenburg and Danzig.

Agnes Agatha Robinson of Pinner watched *Target for Tonight* in Leicester Square, and wrote to the Ministry of Information that it was

> doubly satisfying: we saw how things are run and the story was good . . . If all the British men who were 'stars' for that famous night could have witnessed the second film shown with it and could have realised the contrast in their faces – clean cut, strong, sincere, earnest – with the face of the principal start of the second film – with its flabby emotional weakness, they would go on their way thanking heaven for their own faces. Their action, I do not want to say acting, it was too good for that, was super. Not a flaw in the film . . . Bless 'em!

But the use of real personnel from the RAF during wartime had an inevitable consequence that was duly reported in *Variety* magazine in January 1942: 'A severe shock came to Mrs Gwyneth Hindle of Windsor, Ontario. While sitting in her local picture house, she saw her husband who had died while serving with the Royal Canadian Air Force the previous year'. When brought to the attention of the Ministry of Information by the Associated British Picture Corporation, the Crown Film Unit in reply lamented the inevitability of the situation and observed that they were

> frequently asked by parents or relatives of those who have been killed to send them stills of the scenes in which the men appear. I fancy that the severity of the shock experienced by the lady in Canada may have been present only in the imagination of the reporter. The usual reaction in such circumstances is one of consolation rather than pain.

In *Target for Tonight* the cinema audiences were told that 'the one word that can never be applied to British bombing is haphazard. Every bomb has its mark worked out for it perhaps weeks in advance'. By the time the film was distributed in October 1941, Lord Cherwell and Churchill knew this to be categorically untrue, and mere propaganda.

In 1941 Lord Cherwell assigned one of his statistical investigators, the economist David Bensusan-Butt, to study the effectiveness of the bomber offensive. Circulated on 18 August 1941, the Butt Report revealed that on 100 separate raids on Germany, in June and July 1941, on average less than one-third of the bombers navigated to within five miles of their target. But this shocking average figure was positively impressive compared with the analysis from aerial photographs which showed that of the bombs that were dropped on Germany by night only five per cent had hit the target. This made sobering reading and prompted Churchill to prioritise the development of radio navigation systems including Oboe and Gee, the H2S ground-scanning radar system, and to implement a series of changes at Bomber Command.

On 3 September 1941 the Photographic Intelligence Section' at Bomber Command finally transferred to Medmenham from Bomber Command HQ – then located at Naphill near High Wycombe – and became K Section, a team of specialist third-phase intepreters dedicated to damage assessment. The failures in the bomber offensive also resulted in Arthur Travers Harris being appointed Commander-in-Chief of Bomber Command in February 1942. In time he was nick-named 'Bomber Harris' by the press and 'Butcher Harris' by many in the RAF. He set about rectifying the deficiencies with energy and zeal. When those changes were combined with a strategy of area bombing rather than precision bombing, the destructive power of Bomber Command would become immense. The Area Bombing Directive of 14 February 1942 was strengthened by Cherwell in his 'de-housing paper' – circulated on 30 March 1942 – that aimed to address the deficiencies identified in the Butt Report. Not for the first time Tizard disagreed with Cherwell, arguing that the only benefit of strategic bombing lay in inconveniencing the resources used to defend Germany, and that the same results could be achieved with a far smaller bomber force.

From the information contained in aerial photography it was possible to analyse and identify the physical make-up of German towns and cities. This meant that places such as Lübeck, with its medieval centre of wooden buildings that would burn well, could be mapped block by block and building by building. Aerial photography allowed Bomber Command to create the most destructive cocktail of high-explosive and incendiary bombs, to calculate how many and what type of bombers they should deploy, and in what sequence and where the bombs should be dropped to achieve maximum damage. With night photographs during bombing raids taken by Bomber Command using shutterless cameras, the effectiveness of this bombing and the resultant spread of fires could be accurately calculated once the unique 'language' of shutterless photography was understood. In addition to the actual damage to buildings and installations, interpreters also monitored the time taken for factories to become operational after attack and the methods of repair and reconstruction. For accurate bomb-damage assessment large-scale photographs were required, taken as soon after a raid as possible in order that the full disorganisation created by it could be assessed. Within hours of a bombing raid, it was common for work to begin on clearing debris from roads, or for repair work to begin on railways. With strategic aerial-reconnaissance photographs taken before and after an attack Medmendham had the information to assess accurately the damage caused by a particular raid and systematically calculate the impact of any destruction.

K Section was by a long way the largest Section, and grew from ten interpreters on transfer to Medmenham to a peak of fifty-four in 1944, reflecting the dramatic growth in scale of the mass destruction of mainland Europe. Studying the unfolding war in 3D from the relative serenity of an English country house came to distress Michael Spender, whose brother Stephen would later recall Michael's horrified comments on the carpet bombing that he witnessed. An indicator of the levels of detail captured in the aerial photography used for bomb-damage assessment can be seen in its use for propaganda purposes. In a similar way to that in which Lord Haw-Haw would broadcast from Germany in an attempt to discourage and demoralise the British population, the Allies would broadcast accurate information to the Germans in minute detail, courtesy of information from K Section. Using pre-war street plans, telephone directories and gazetteers they were able to establish who or what had occupied a destroyed building, sometimes achieving such fine-tuned detail as the location of a corner shop. The potential for this to have a demoralising effect on the Germans was obvious, but was also a salutary lesson to the Allies of the psychological power of reconnaissance.

When Hemming allocated R. V. Jones one of his best interpreters, he established a partnership that was to play a major part in unearthing the Germans' offensive and defensive radio systems. Claude Wavell had worked for the Aircraft Operating Company since 1928, when he worked in Brazil as their Chief Surveyor on an aerial survey of the municipality of Rio de Janeiro. With the city concentrated in the valleys between the mountains and the sea, creating the cartography of Rio was to prove technically challenging but did lead to the creation of impressive aerial photographs, particularly when viewed in 3D through a stereoscope. Working through the Brazilian Revolution of 1930 was to prove daunting and risky, but with Rio the original beach-party town the company's visitor book soon featured many luminaries including Noël Coward, who visited them during his 1931 stay at the Copacabana Palace.

Returning to his home on the Isle of Wight in 1932, Wavell was approached by Hemming at the beginning of the war for help, but since he ran the local Home Guard he decided not to oblige. When the Dunkirk evacuation came and the 'little ships of Dunkirk' included boat crews from the island, Wavell wired Hemming 'If you still want me I'll come', to which he received the immediate reply 'Come at once'. Arriving at Wembley on 3 June 1940, while the evacuation was ongoing, Wavell worked with Spender, and whilst later acknowledging his brilliance found that he couldn't stand him. Dedicated to the interpretation of radar and wireless transmitters from January 1941 – when it came to be appreciated that one WAAF interpreter working part-time on the detection of enemy wireless and radar installations was insufficient – that same month Wavell telephoned Jones after discovering an installation at Auderville on the Cherbourg peninsula. From two successive aerial photographs, taken nine seconds apart, Wavell noticed a subtle but definite difference in the shadow pattern from the installation. As an experienced interpreter, he knew this indicated movement on the ground in the time between the first and second photographs being taken and suggested that Auderville could be home to a rotating transmitter and receiver. With much trouble, at the beginning of 1941 Jones requested a low-level sortie, which was successfully flown by Flying Officer William Manifould on 22 February 1941. When the superb low-level oblique photographs landed on Wavell's desk the following day, the interpreter now affectionately nicknamed 'Wavey' by his colleagues had the confirmation that they had discovered the first Freya radar station, used by the Germans to detect hostile aircraft. By the middle of 1941 his team had developed a good knowledge of the Freya stations on the Channel coast.

From November 1941, Squadron Leader Claude Wavell ran G Section as a specialist third-phase team with responsibility for compiling information about all enemy radar, navigational beams and wireless stations. This work over the remainder of the war was to result in huge amounts of information, all of which had to be managed. With information gathered from POW interrogations, secret agents, resistance agents and, unbeknown to the interpreters, Ultra, this mass of data was collated and systematically recorded on maps, target dossiers and card-index systems and arranged in such a way that anyone in the team could access it quickly. This meant close cooperation between Wavell and his interpreters in G Section and the newly promoted Assistant Director of Intelligence (Science) – or ADI (Science) for short – R. V. Jones and his colleagues in the scientific intelligence unit.

When the need to determine the height of masts supporting aerial arrays became essential, this created something of a problem for G Section as calculating heights from shadows would occupy a skilled mathematician for at least half an hour. But with the heights of wireless masts urgently required by Fighter and Bomber Commands for low-flying information and ADI (Science) also interested in heights and the distances between installations in order to calculate radio frequencies, Wavell returned to an idea he had first had at Wembley. Based on the principles of spherical trigonometry he built a machine of his own invention – the Altazimeter – that got the results in a few minutes. Building the prototype himself from a number of narrow circles of plywood which fitted and slid over one another, Wavell succeeded in making an instrument that was accurate to within 1.5 per cent of those results obtained by computation. In 1954, long after he had retired, Wavell contacted the Ministry of Supply to enquire about the later success of his invention, and was rewarded with a £25 cheque enclosed with the reply.

But the interpreters had failed to find evidence of a smaller installation whose wavelengths could be heard, and that Jones knew had been described in the Oslo Report. With information sanitised for security purposes, Wavell was given data about the appearance and possible layout of the Würzburg radar installation. The break came when the Army Section – who had been investigating a suspected gun position at

Cap d'Antifer, on a clifftop near the village of Bruneval, decided it wasn't and questioned whether there could be a radar connection. When PR pilots from nearby RAF Benson visited Medmenham, Wavell took the opportunity to discuss his ideas about the Bruneval photographs with Squadron Leader Anthony 'Tony' Hill, and jokingly quipped: 'You pilots annoy me! You go over this place time and time again and never turn on your cameras in time.' Despite Jones submitting an official request for aerial photography of the installation, it was Tony Hill who opportunistically managed to take a low-level oblique photograph of the installation at Bruneval, on 5 December 1941, which showed perfectly the radar installation beside the large house. As Wavell had seen Hill's name on the board of operational sorties for that day, with the target of nearby Fécamp against his name, he was jittery until news of Hill's return came through. Wavell telephoned Benson and was told by Hill that it looked remarkably 'like an electric bowl fire', a commonplace 1930s electric heater that looks – to modern eyes at least – remarkably similar to a radar dish. From that moment on the Würzburg installations were given the nickname the 'bowl fire'.

From Hill's oblique photographs, a scale model of the site, exact in every detail, was made by the model-makers, while the Army Section used vertical aerial photographs to study every square inch of the local terrain to locate defences and understand the local topography. With this photographic intelligence and with ground intelligence from the French Resistance, Operation Biting was planned and successfully executed on the night of 27/28 February 1942 by British Combined Operations. In *Target for Tonight*, the Wellington bomber 'F for Freddie' was piloted by Percy Charles 'Pick' Pickard. He was then commander of 51 Squadron that successfully parachute-dropped commandos from specially modified Armstrong Whitworth Whitley bombers. After his starring role, Pickard became a well-known figure and this, combined with his imposing physical stature, is recorded as having in no small part reassured the commandos involved in the daring raid. After being dropped a few miles from the installation and under the command of Major John Frost, the commandos advanced to the target where, after a brief firefight, the Würzburg installation was captured. Dismantling key parts of the radar system, the raiding force was then collected from a nearby beach by landing craft and after transferring to several motor gunboats (MGBs) returned to Britain. This triumph for Combined

Operations, which benefited so much from the detailed photographic intelligence and a model provided by Medmenham, helped with the development of radar countermeasures and was a welcome morale boost for the British public.

Percy Charles 'Pick' Pickard went on to fly with 138 (Special Duties) Squadron, which had been created to support the work of the Special Operations Executive whose function was to promote sabotage against the enemy. SOE required agents, ammunition and equipment to be dropped inside enemy territory. Whilst modified bombers were used for supply drops, the Westland Lysander was used on 'cloak and dagger' missions to deliver or collect agents or equipment. Flying without any navigation equipment other than a map and compass, Lysanders would land on farm fields or clearings in woods, which had been carefully studied by Medmenham interpreters but were only marked for the pilots – if they were lucky – by members of the local resistance on the ground holding a couple of torches. Identifying where the Lysanders could safely land required detailed interpretation by Medmenham X Section, which was responsible for topographical information.

X Section had its origins in late 1942 when Kendall tasked WAAF Section Officer Irene Marsingall-Thomas with identifying key landmarks on a variety of air routes. This proved so successful in informing RAF Ferry Command – which transported new aircraft from factories to operational units – that a series of briefings folders covering North Western Europe were compiled. Marsingall-Thomas was joined by Section Officer Ruth Langhorne and in the first half of 1943 they created a series of thirty-two mosaics with which bomber crews were trained about the landmarks they would see on their missions over Europe. Langhorne had read geography at St Anne's College, Oxford before the war and later volunteered for service as an interpreter in India, on the grounds that travel at government expense seemed too good an opportunity to miss. She was not dissuaded by the warnings in the RAF India handbook about the need to inspect shoes carefully before wearing them, given that they were a favourite resting place for scorpions. Leaving Medmenham for India with dozens of other WAAF officers, she would, in due course, meet Glyn Daniel, her future husband.

When on Boxing Day 1940 the No. 1 Photographic Reconnaissance Unit (PRU) relocated from Heston to the South Oxfordshire airfield of RAF Benson – 15 miles from Medmenham – it became the hub of photographic reconnaissance throughout the remainder of the war and directed PR activities at RAF Stations including St Eval in Cornwall, Leuchars in Fife and Wick in Caithness. Whilst the RAF interpreters based alongside the reconnaissance squadrons at the airfields undertook first-phase interpretation and completed a Form White, at Medmenham the Army Section repeated this process for the sorties that were flown from Benson and covered areas that were of immediate overnight concern to the Army. This involved working through the night, going on duty at 8 p.m. when copies of the first sorties came in from Benson and staying at their posts until all of them had come in. If the weather had been good over Europe, there might be upwards of a hundred sorties, with thousands of individual photographs to be scanned rapidly for items of immediate Army interest, such as the formation of a military train carrying armour, indicating transfers from eastern to western fronts or reinforcements moving up. Once such intelligence was found the interpreters had to draft a report and pinpoint the location on maps created by the Geographical Section of the General Staff (GSGS) – referred to as GSGS maps. Since the sortie had not yet been plotted, all the Army interpreters had to work with was the pilot's trace. As this was not always an easy task, very often they would still be hard at work at eight the following morning.

As a soldier, Geoffrey Stone was trained in photographic interpretation at the School of Military Intelligence, which occupied Smedley's Hydro, a commandeered hydrotherapy hotel in the Derbyshire town of Matlock. With the bowling greens and tennis courts maintained and kept available for use whenever they could be, many friendships were made there before soldiers were deployed around the world. It was on the tennis court that Stone was to meet Frederick 'Fred' Mason, who would become his best friend and who encouraged him to take the course in photographic interpretation. As Stone had been working as an instructor, teaching the same course every three weeks, his life had become somewhat monotonous. Since the likelihood of being posted increased progressively and could come out of the blue, he was painfully aware of people around him disappearing to far-flung parts of the world and never returning. So he considered there to be wisdom in taking the initiative when a chance offered itself. Since the whole idea of working with aerial photographs and maps appealed to him he applied for the course, was accepted, and passed in October 1943.

From June 1943 the Army course at Matlock had been expanded, once it had become appreciated how important PI would be for D-Day and the land advance through Europe. Although dominated by RAF personnel, the Army had a particular interest and were present at both Wembley and Medmenham given the inter-service nature of both places. The Army was particularly concerned with military defences that soldiers would face on the ground, enemy weaponry, and the 'going' of the terrain. With topographical knowledge crucial to an army, since it dictates which areas can be traversed, in which military vehicles and how quickly it can happen, every Army formation from a division upwards would have an Army Photographic Interpretation Section (APIS). But the headquarters for photographic interpretation would remain the CIU at Medmenham.

After the routine of being an instructor, Geoffrey Stone was looking forward to his new posting at Medmenham in the Army Section (B Section) – which was allocated the primitive working conditions of the Danesfield House basement by the RAF and was run by Major Norman Falcon. The first task Stone was given was to use aerial photography to determine the accuracy of intelligence produced by agents in the field. He soon discovered that many of them produced duff information or embroidered facts, all of which helped determine how well an agent could be trusted. With so many German airmen shot down over Britain, there was also a regular stream of information from POW interrogations which also needed to be ratified by the use of aerial photography. At the start of the war, the Wilton Park estate near Old Beaconsfield in Buckinghamshire was leased to the War Office for use as a top-secret interrogation centre. In addition to supplying Medmenham with interrogation reports, on occasion interpreters would provide their expertise to the interrogators.

When Stone carried out first-phase interpretation on the night-shift, one of the compensations for the hectic nights were the lovely walks he took through the woods or by the river Thames in the early morning before he had to give the commanding officer a briefing at 8 a.m., updating the situation map before going for breakfast. If the weather had been bad, with 10/10 cloud cover almost everywhere, Benson might ring through at midnight and say there were no more sorties coming. Stone and his colleagues would go off shift and have the whole of the following day free, when he would like to go for a cycle run to Marlow or Henley. Being in such a beautiful setting, cycling and walking in the surrounding countryside were popular, while the sports facilities meant that cricket, tennis and football games were common pastimes. With Medmenham being alongside the Thames, boating and rowing were also popular, particularly among those interpreters who had rowed for their university colleges. This environment contributed in no small measure to the good morale that was reported at Medmenham and, given what was recorded on the photography they interpreted on a daily basis, the interpreters knew this was a worthwhile posting.

The academics at Medmenham had a reputation for being punctilious in their PI work, but for many of the Station Commanders they contributed to an atmosphere more akin to that of an Oxbridge college than a military intelligence unit. In early October 1941 Daniel managed to upset the Medmenham Station Commander – Wing Commander A. T. Laing, a regular RAF officer who found himself in charge of an intelligence unit staffed by some very irregular people. After a particularly gruelling night-shift, Daniel had walked down to the village for breakfast at the Dog and Badger and afterwards sat on the churchyard wall with his uniform unbuttoned and without his hat. While driving past, Laing spotted Daniel and duly summoned him and reprimanded him on his discipline and dress code. Still, the Dog and Badger was a welcome alternative to the mess and if Mr Nye the landlord was in a generous mood customers from Medmenham would get some bread and cheese, or that wartime standby the spam sandwich.

Such un-officer-like behaviour prompted one later commanding officer to arrange a parade, which many of the academic personnel attended with much consternation as they congregated in a car park that was going to be used as a makeshift square-bashing ground. Stone witnessed the attempt, which went down in Medmenham folklore. With the Navy personnel insisting on being first in line, followed by the Army, eventually everyone was organised in threes, present and correct. The intention was to march around a corner where the commanding officer was waiting with his acolytes to take the salute. On instruction from the parade adjutant they started but soon chaos ensued and it was decided that marching would not work at Medmenham.

*

When the aerial photography arrived at Medmenham from the network of airfields, it became the responsibility of Z Section to undertake second-phase interpretation. Z Section comprised a number of sub-sections which had teams specialising in particular subjects or geographical areas and who watched the enemy on a day-to-day basis. Working around the clock, there were three shifts which started and finished at 8:30 and 16:30 and 00:30 when personnel would overlap for fifteen minutes. Continuity of work was achieved by means of job cards and log sheets which provided a record of what had been done during each shift. The Coverage Sub-section was responsible for examining all incoming sorties to decide which targets had been successfully photographed and which had to be overflown again. This involved compiling Locality Sheets which identified the targets covered and the corresponding frame references and allowed the second-phase interpreters to identify locations quickly. The Shipping Movements Sub-section kept an index of all the larger enemy vessels on index cards that contained photographs to aid identification and a detailed log of their movements. Some interpreters had a responsibility for a particular geographical area, including those working in the 'Norwegian Corner' – men and women recruited from the Norwegian armed forces.

The work of the Airfield Section was watched over by their mascot Percy, a statue of an eagle, complete with a pearl necklace, which sat on the mantelpiece as they produced the Daily Airfield Report. This section provided intelligence airfield by airfield of the type and number of aeroplanes present and meant that with comparative cover it allowed unusual activities to be noticed. Z Section reported on the relocation or appearance of new enemy balloons and smokescreens that could prove a menace to Allied bombing raids. Unlike many sections that were based in Nissen huts in the grounds of Danesfield, the Railway Sub-section was based in rooms in the main house and was known to have the best views at Medmenham, from a fine bay window of the grounds and the Thames Valley beyond. This group of men and women produced daily reports comprising intelligence about the movement of railway traffic.

The Ground Intelligence Section at Medmenham had been formed in June 1941, three months after the move from Wembley. It typically consisted of seven officers – with a similar number of clerks – five of whom worked as area officers with knowledge and responsibility for particular geographical areas. This explained the presence of a Norwegian officer in the section confirming that personnel from across the Allied spectrum had long been a feature of Medmenham, even before its evolution into the Allied Central Interpretation Unit in May 1944. In addition to personnel from across the British Commonwealth, there were also Free French and Polish officers. The different social attitudes of the various nationalities, and differences in their sense of humour, helped to create a unique atmosphere that was amply reflected in a story that was often recounted by the interpreters about a visit from the Secretary of State for Air – Sir Archibald Sinclair – when he spotted a Polish uniform. Having been instructed that their work was top secret and should not be discussed with anyone, and despite the presence of the Medmenham Commanding Officer, when Sinclair asked, 'And what is this representative of our noble Polish allies doing here?' Bieńkowski stood to attention and, with a twinkle in his eye, replied in a loud whisper: 'I'm making the secret waste!'

The Progress Section had to decide the relative importance of demands for photography that were received from all the services. At peak periods there could be 350 demands for photography per day and over a million photographs could be copied each month. When a film was delivered to Medmenham, first-phase interpretation had already been undertaken at the airfield but more detailed second-phase reporting had to be undertaken as soon as possible. This required the photographers to quickly produce two copies of the photographs from each sortie: they were equipped with the most modern automatic multi-printers that were capable of producing a thousand photographs per hour. And whilst much of the machinery was operated automatically, keeping the machines running around the clock and operating the labour-intensive photostat and lithograph copying facilities required on average more than 250 photographers. The section was also responsible for the highly skilled creation of rectified mosaics: this was where individual aerial photographs were stitched together to create a seamless image. This involved the Mosaic Section in the Print Library identifying shots from the most suitable sortie or sorties, which the photographers would then print and paste onto boards, a process that typically involved enlarging or reducing the scale of prints to fill gaps when results from multiple sorties were used. The photographers had to carefully develop prints in order that the print tones matched as far as possible. Once completed, a mosaic would be annotated by Library staff before being photographed and printed.

During the war, 4,500 mosaics were recorded as having been made to provide an overview of target areas for mission planning and briefings. They often covered entire industrial plants, ports and cities. On some occasions it was necessary to produce rectified mosaics, which involved creating map-accurate aerial photographs. When the reconnaissance aircraft was flying, it was normal for there to be camera tilt and a slight fluctuation in flying height between frames, which was of little consequence given how mosaics were generally used. When it was necessary to take ground measurements from mosaics a much higher level of accuracy was required, which necessitated the creation of rectified mosaics.

For all strategic reconnaissance sorties flown from Britain, Gibraltar or Western Europe before VE Day, it was common practice to run off two sets of prints: one went straight from the Progress Section to Z Section for second-phase interpretation, the other went to the Library – which had Sub-section for Plotting, Tracing, Maps and Mosaics. Creating a sortie plot was the first stage in cataloguing a sortie and involved drawing the area covered by each aerial photograph onto a map of the area. The WAAF plotters had access to GSGS mapping and took copies of the appropriate maps that covered the photographed targets. The Map Section was responsible for keeping a stock of maps and town plans that might be required by the unit, which required them to work closely with the GSGS who also published, on behalf of the unit, place-name gazetteers that were essential to the standardised identification of places in reports throughout the intelligence community. The plotter's relied on the pilot's trace, a piece of tracing paper on which the pilot had marked in blue chinagraph his flying route and in red chinagraph where he had photographed. These helped the plotters to understand the sequence of the photographs, and in 1944 matters were improved again when small-scale printed maps were provided to the pilots to mark up instead of them just sketching a rough outline of where they had flown. Notwithstanding cloud cover, which frequently obscured targets, an expert plotter could plot on average a hundred prints per hour. Once all the photographs had been plotted, and the finished plots had been cut out and mounted onto card with the appropriate latitude and longitude reference information copied from the map, along with a photostat copy – created by the photographers and inserted into the print box – which allowed third-phase interpreters to easily and quickly identify exactly where each photograph covered.

With later photo-reconnaissance sorties capable of producing more than 1,500 aerial photographs each, it was essential for every sortie to be fully recorded, in order that Library staff should have clear knowledge of what geographical areas were covered, when, and at what scale and quality. When the sortie photos were delivered to the Print Library by the Progress Section, they would normally be contained in a transit case and if the prints were curled – which they frequently were because they had been dried in a hurry – they were manually uncurled. A photostat copy of the sortie plot would be obtained from the Plot Library, and every photograph would then be checked off against the information on the plot. Ensuring that every print was arranged in order, the photographs – which typically measured five inches square, eight by ten inches, ten inches square, and ten by twenty inches – were placed in appropriately sized boxes. The photostat plot was then gummed inside one or more of the box lids, while the pilot's trace and booking-out card were inserted loosely. Using Indian ink the Medmenham Print Library reference and the sortie reference were marked onto each box and the boxes were then arranged in sequence, using a system which had evolved from letters and numbers at the beginning of the war to a simple number.

The types of aerial photography held in the Print Library included photographs from the strategic reconnaissance sorties taken by the high-altitude photo-reconnaissance Spitfires and Mosquitos; tactical reconnaissance sorties of specific targets often near the front line; map-revision sorties, which were typically small in scale; strike-attack sorties either at high level by the likes of the Eighth Air Force during bombing raids or at low level by the Second Tactical Air Force during Mosquito attacks; night-reconnaissance sorties taken by photographic flash from medium or low level; Coastal Command sorties of shipping reconnaissance or strike attack; night sorties taken during Bomber Command attacks; photographs taken during meteorological sorties; photographs taken for the creation of mosaics; survey photographs; and experimental sorties flown using infrared film. Since the value of aerial photography is considerably enhanced if it is possible to study a particular location over a period of time, one of the challenges which faced the Library as it began to grow dramatically in size was how best to catalogue the photography.

Comparative cover was vital in order, for example, that damage caused by a bombing raid or changes to a military or industrial target could be accurately assessed. For this reason it was decided early in the war that a system of cataloguing the comparative cover was essential, that this would maximise the use of the photography and ensure that photo-reconnaissance pilots did not fly unnecessarily.

A number of different approaches were taken, mainly involving complex card-reference systems, before a new system – using GSGS maps, tracing paper and a pantograph – was adopted in the autumn of 1941. Once a sortie had been plotted by the WAAF plotters in the Sortie Section, it was forwarded to the Tracing Section where WAAF tracers copied the geographical area covered by each sortie onto tracing paper which could be overlaid on the relevant map to indicate quickly what coverage there was. However, since the maps used for plotting ranged from 1/25,000 to 1/250,000 it was necessary to use a pantograph to transfer accurately the information contained on the sortie plot. The tracer had to set the pantograph, consisting of four brass arms connected by sliding joints in the form of a quadrilateral, to the appropriate scale reduction, with a pencil marking it on the other end. This system worked well for rural areas and meant that the operator could quickly discover what photography existed by selecting the relevant 1/250,000 scale map and overlaying the relevant traces. But for the more intensively photographed urban areas, more detailed mapping and traces were adopted.

In early 1944 Geoffrey Stone was moved upstairs out of the basement to F Section, which had responsibility for the third-phase interpretation of enemy communications. It was home to specialists on railways, waterways and roads who reported on the movement of traffic and the location, layout and vulnerability of railway marshalling yards, depots, bridges, aqueducts and locks. After Allied attacks they assessed how transport infrastructures had been affected and, when they were repaired, reported that they were ready to be attacked again so as to maximise disruption. As well as being able to tell anyone the time of the next train from Marlow to London Paddington, F Section staff undertook detailed studies of the different types of railway wagons and, with models created by the model-making section to aid identification, maintained intelligence about the location of rail transport, which included railway guns and flak wagons, throughout enemy-controlled territory. Many F Section staff (including Stone) were avid trainspotters; others had been recruited after an approach to the War Office to secure the selection and release of suitable people from British railway companies. This resulted in Captain R. J. Moody being transferred from the London & North Eastern Railway Company. His responsibility was to build a library of information about Continental railways comprising handbooks, maps, diagrams and timetables.

In F Section, Stone found himself part of the most remarkable group of interpreters, including his friend Fred, Dorothy Garrod, Robin Orr and Sarah Oliver. The first female professor at Cambridge, Dorothy Garrod held the prestigious Disney Chair of Archaeology. But when her tenure was interrupted by the war, it meant there was nobody to teach and, with little enthusiasm for research under the circumstances, she was delighted when her Cambridge colleague Hamshaw Thomas recruited her. In 1942 she joined the extraordinary collection of people at Medmenham which at one stage included herself as the reigning Disney Chair and her successors, Grahame Clark and Glyn Daniel. Working alongside Sarah Oliver, daughter of Winston Churchill, was Flight Lieutenant Robert 'Robin' Orr, composer and organist of St John's College, who would later become Professor of Music at Cambridge. In addition to Dorothy Garrod, Glyn Daniel was able to recommend other friends and colleagues when asked whether there were any of his cronies hidden away doing worthless jobs. Stuart Piggot and Terence Powell were known to Glyn – and would both go on to become eminent professors of archaeology – but at the time were serving as private soldiers in the Army. Within three months they would find themselves commissioned officers in the Intelligence Corps based in Medmenham's B Section.

In 1941, Sarah Oliver joined the WAAF and started in the ranks as an Aircraftwoman Second Class. Determined to conform, she removed all the obvious make-up and had her hair cut to the regulation length. An actress and dancer and the third of Winston and Clementine Churchill's five children, Sarah was married to the comedian and musician Victor Oliver von Samek, who starred in the BBC radio show *Hi, Gang*. But it was an unhappy marriage. Vic would visit Medmenham most weekends and sometimes provided entertainment in the officers' mess. Winston's thoughts on the matter were summed up in his comments at a dinner party at which Oliver was present, when in response to a question about who he most admired he replied Mussolini, 'because he had the good sense to shoot his son-in-law!'

Stone and Sarah Oliver worked at F Section at a large table in the middle of the room with a counter along the wall. At times the Prime Minister would call the section in order to speak with Sarah. Through official channels, visits and conversations with his daughter, Churchill was acutely aware of the work being undertaken at Medmenham. During the preparations for Operation Torch, the Allied invasion of North Africa in November 1942, Medmenham interpreters were requested at short notice to work with the military planners to select suitable landing beaches: they studied the road network to identify any logistical issues and identified all strongpoints. Sarah often visited her father at Chequers – the official country residence of the Prime Minister in Buckinghamshire – and she was there when at the end of dinner her father rose to announce that in a few hours Allied forces would be landing in North Africa, and offered a toast to the success of the Second Front. With many of the guests greeting the announcement with suitable surprise, the Prime Minister was a little disappointed that Sarah had taken the news so casually. When he asked her why, she told him that she had been working on the planning for the last three months and knew all about it. Impressed at how his daughter, whose career in acting had always been of some concern to him, was playing a serious part in the war effort Churchill reputedly asked her why she hadn't told him. When she said she hadn't really thought about it, he smiled and muttered, 'Suppose you thought I didn't know'.

In the summer of 1941, Lieutenant Commander Robert Quackenbush – known as Q-Bush by his Navy friends – spent three months at Medmenham learning the British approaches to interpretation. The American Navy observer in London, Vice Admiral Robert Ghormley, had been alerted to the importance of photographic interpretation and secured this introduction for the Americans. The Americans had had the same indifference as the British to photographic intelligence during their interwar years, and despite the strong endorsement from Quackenbush on his return from Medmenham, it wouldn't be until after the attack on Pearl Harbor that meaningful amounts of American PI training would begin. In starting to develop interpretation – which would ultimately result in his reputation as the Godfather of Navy PI – he faced very similar challenges as the British. This was borne out during planning for the Guadalcanal campaign, the first major offensive by the Allies against Japan, when Quackenbush travelled to New Zealand and reported to Admiral Kelly Turner on 16 July 1942. Turner reputedly said, 'Young man, all I want from you is complete cover of all Guadalcanal'. To which Quackenbush replied that he was entirely on his own, without photographic equipment, photo-reconnaissance aircraft and pilots, without interpreters. Turner responded, 'Don't bother me with the minor details'.

When the United States Ninth Army Air Force reconnaissance aircraft arrived in Britain, en route to North Africa, they initially considered setting up their own version of Medmenham. By 1943 Douglas Kendall had been promoted to Wing Commander and as the Technical Control Officer was responsible for the operational direction of Medmenham. With the Group Captain above him as Commanding Officer responsible for the administration of a large RAF station, Kendall was left to run and direct the interpretation effort that produced enormous amounts of intelligence in the Second World War. He was unique at Medmenham in being the only person Ultra-cleared, and was responsible for the reciprocal flow of Signals and Photographic Intelligence. On many occasions known information could not be shared with the interpreters on the basis that it was too dangerous to share, even though this did deprive them of data that would have been useful to their work. Much as Peter Riddell had fought in 1940 to avoid the dispersal of photographic interpretation across the Army, Navy and Air Force, Kendall now fought to keep the British and American interpreters as one integrated unit. He knew how effective the centralisation of PI functions at Medmenham was, with the involvement of personnel from all three services and a wide range of countries. His views were not, however, shared by John Winant, the American Ambassador to Britain, who had been to Medmenham on a number of occasions and agreed with the plans of the Commanding Officer of the American Photographic Reconnaissance Wing, Colonel Elliot Roosevelt, the son of President Roosevelt. Sensing the potential problems, the Ambassador telephoned Sir Archibald Sinclair, the Air Minister, asking him to lunch to discuss the matter. He also asked him to bring along Kendall.

Ordered to report to the Minister the following morning, Kendall gave a briefing before going with Sir Arthur Street, the Permanent Undersecretary, to a private room at Claridge's to meet the Ambassador and General Carl Spatz, who had command of the USAAF in Europe. Commenting on the plans by Elliot Roosevelt to create another unit, Kendall pointed out that the integrated unit at Medmenham was already working well and that section heads were a mixture of British and Americans, with people chosen on the basis of their competence rather than their nationality. With the scale of bombing operations then being undertaken and the established success of photographic intelligence, there was no technical need to alter the arrangements and there would be a high risk from doing so. This argument was accepted by General Spatz and the plan was dropped, but it did mean that Elliot Roosevelt barely spoke to Kendall for about six months, until he cornered Kendall one night after a mess dinner and accepted that he had been wrong. As the President's son, Elliot Roosevelt had enormous power to make things happen but was almost universally loathed and detested by the people at Medmenham. The arrangements from this point were that more Americans would steadily arrive at Medmenham and Kendall would be joined by an American Lieutenant Colonel from the USAAF, which ensured that there would be no conflicting priorities between the different countries, and meant joint operational instructions could be issued to the inter-service sections of the Unit. The scale of interpretation both at Medmenham and in the field grew. As the work progressed new sub-sections were created as the demand required, something which would lead, for example, to the creation of Sub-section B4 which concentrated solely on Southern Germany and Austria.

From its creation the Aircraft Section (L Section) was run by the inimitable Constance Babington Smith. Responsible for the identification of new types of hostile aeroplane, L Section had to provide as much information as possible when any such machines were first spotted by aerial photography: their design, dimensions, where they were being developed, where and how many were being manufactured, and what they were designed for. This involved L Section looking at photo-reconnaissance shots of enemy aircraft factories and airfields and meant that the interpreters built up a comprehensive understanding of the German aircraft industry, its varying phases and trends, throughout the war. Initially based in the Long Gallery of Danesfield House, WAAF Hazel Scott recalled that while she was responsible for aircraft interpretation, RAF officer Charles Sims concentrated on aircraft factories. Whilst Babs was 'very much the boss' the team worked very harmoniously together.

Sims had served as an aerial photographer during the early 1920s with the RAF in the Middle East, and had experienced first-hand how little the techniques used by the RAF developed after the First World War. By 1927 he was working for *The Aeroplane* magazine and took photographs that required him to identify the best angle from which to shoot. In those pioneering days air-to-air photography of aircraft – taking a photo of one airborne aeroplane from another – was not recognised as de rigueur. Instead, Sims had to stand in the middle of an aerodrome while a test pilot flew around him. This required absolute trust in the skill of the pilot, particularly when the grand finale was usually a head-on photograph. Having served with the British Expeditionary Force in France – in one of the Lysander squadrons the British planned to use for tactical aerial reconnaissance – Sims was under no illusion as to the inadequacies of this aircraft for the task. He had also attended the Brussels International Air Show in August 1939, a few weeks before war was declared, where he had seen and heard much about the power of the new German Luftwaffe. After Dunkirk, Sims was posted to Wembley, where his detailed knowledge of aircraft resulted in him being posted to the Aircraft Section.

V Section and its model-making at Medmenham was to prove vital to many of the conclusions reached and decisions made during the war. The highly accurate three-dimensional models were the ultimate means of briefing everyone from the Prime Minister to the infantry soldier with information gathered from a variety of intelligence sources: agents, maps, charts, vertical and oblique aerial and ground photographs and ground information. V Section evolved from an experimental unit formed under the Director of Inspection of Camouflage at the Royal Aircraft Establishment, Farnborough, in August 1940. Much as the Camouflage Directorate at Royal Leamington Spa had appreciated the importance of models to the design of camouflage schemes, photographic intelligence came to understand their importance as a means to uncover and understand enemy activity. Models could not only display the three-dimensional intelligence the interpreters could see through their stereoscopes, but could also include additional information about a location or object that could be gathered from other intelligence sources. This distillation and accurate presentation of data resulted by December 1940 in eight civilians being recruited to work under an RAF flight lieutenant. Since this was considered such a departure for aerial intelligence, it was decided that they should wherever possible be from an artistic background. One of the eight was Geoffrey Deeley, a sculptor and teacher at the Polytechnic School of Art, Regent Street, London. Arriving at Farnborough in September 1940, he discovered a unique group of talented artists were gathering, a group that included a number of professional and commercial artists, sculptors, architects, and architectural model-makers.

In May 1941 the team were transferred to Medmenham, where their workshops were based in the vast cellars of Danesfield House – separate from the Wild Machine – and since it was typically a quiet place, as their work required great concentration on a few square inches of a model at a time, visitors could often be mistaken into thinking that the work looked easy and was more like a hobby. Initially the Air Ministry categorised the model-makers as draughtsmen and later as the comparatively lowly Pattern Makers Architectural (Group II). Even though they were trusted with top-secret information, the perception that model-making was a merely manual activity was reflected in their grading and their somewhat derisory pay of two shillings and sixpence per day. The model-makers themselves considered their craft as 'no more manual than writing with a pen'.

Their pay and status remained a bone of contention until September 1944 when the model-makers were finally given Group I status. But this only followed their work on a series of enormous models that allowed for the detailed planning of Operation Overlord and that prompted the Supreme Commander of the Allied forces in Europe, General Dwight Eisenhower, within days of the successful D-Day landings to write and thank them on behalf of the troops who had benefited and continued to benefit from their work. Since 1942, V Section had mounted a major model-construction programme that involved working long hours and adjusting to frequently shifting priorities. Although many of the creative people in the model-making section did not mesh too well with the regular military, Eisenhower thoroughly appreciated the importance of them, their work and its 'contribution to our ultimate victory'. He came to increasingly value the worth of photographic intelligence during the war, and as President of the United States would authorise the deployment of PR aircraft and satellites for intelligence-gathering on an unprecedented scale during the Cold War.

At Medmenham, the increasing demand for models throughout 1941 led to the expansion of V Section and by 1942 the creation of a training school for model-makers at nearby RAF Nuneham Park. Training involved spending a month at Nuneham on practical instruction before undertaking directed theoretical and practical work at Medmenham under workshop conditions. The types of models they produced fell into two main categories: topographical representations of a geographical area, and scale replicas of particular objects such as individual ships, aircraft, military vehicles and railway rolling stock. The models of objects, and photographs of them, were used by the Medmenham PIs to help with their interpretation, while the topographical models were used by military planners during the planning of operations and for briefing purposes. After a suitable base had been made in the carpenters' shop, the first stage involved cutting contours by positioning map enlargements showing the contour lines on a whole series of pieces of hardboard, in order that each contour line could be traced onto a separate piece of board. Then a machine was used to cut along each contour line and the process repeated until the result was a stack of contours similar to terraced land, each shelf representing a rise of a certain number of feet; these were then glued and nailed together. Using a spatula, the model-maker then filled in the space between the contours with 'jollop', a mixture of plaster, paper pulp and glue. Throughout this process they referred to maps and aerial photographs to ensure that river flows and natural features were accurately modelled. The next stage was to skin the model, and involved using an aerial photograph which had been rectified in order that it was map-accurate. The photograph would often comprise multiple vertical aerial shots which had been mosaiced together. While the print was still wet it had to be stuck onto the contour model, with the model-makers ensuring that the photograph matched exactly the same point on the mock-up below.

Now, with the skin fixed, a closer comparison was made by

comparing the view of the model with the view seen when looking at stereo pairs under the stereoscope. This allowed finer points of land form not defined by the contours to be included and involved sections of the photograph being removed. Following subtle additions or subtractions of contour, a new section of photograph was glued on top. Now shadows that were shown on the aerial photographs were manually painted out, while roads and railways were carefully marked to aid their identification. With detailed information about their dimensions, buildings, trees and power lines were cut from plaster, wood, linoleum or perspex and glued in place on the model. Models of large trees were constructed individually and a mechanical device called the hedging machine – somewhat like a mechanised cake-decorating device – was developed by the section and used controlled air pressure to spray materials onto the replicas to create realistic-looking textures. To ensure that everything was accurately modelled, there was a particularly close cooperation between the specialist third-phase interpreters and the model-makers, which included ensuring that replicas were painted in colours that matched the seasons of the year. Once completed, the mock-up was then ready to be photographed in a studio where various lighting conditions – be it moonlight or early-morning sun – could be recreated to help with the planning of operations or briefings. During a visit to Medmenham, Churchill was recorded to have been transfixed by some models that had been arranged for him to view during his visit to the damage-assessment interpreters in K Section. In his excitement he pointed out to his wife Clemmie little pylons that had been constructed from perspex. Model-maker Edward 'Ted' Wood – or 'Woody' as his fellow model-makers called him – was on hand to answer any questions, and later recalled: 'Churchill was delighted with them, and stooped right down to look into the detail, spilling gold pencils from his pockets'.

When talking to Constance Babington Smith in December 1956, when she was researching material for her book *Evidence In Camera* about wartime photographic intelligence, Ted Wood – who was then Publicity Manager for the Boy Scouts' Association – explained that he 'enjoyed his Medmenham work more than anything he'd ever done'. Wood worked alongside Reynolds Stone, the noted English wood engraver, designer, typographer and painter who would go on to design the Royal Arms for Queen Elizabeth II – still reproduced on

UK passports, banknotes and stamps. Whilst V Section was known to have Communists working in it, Ted Wood was pleased to report that there was never a security leak from the model-makers. As V Section was very much out of bounds, Wood set up a little outer room with display models that 'satisfied most people' who were understandably fascinated with their creations. In the hope that the models created at Medmenham could be viewed, not only by military planners and commanders who would be giving the orders but by the personnel who would be carrying them out, in the summer of 1943 Group Captain Peter Stewart investigated, with the Air Ministry Film Unit, whether Pinewood Studios in Iver Heath, Buckinghamshire, could film them. To ensure that security was maintained it was considered necessary for 'CIU Models to be constructed under existing arrangements, packed in screwed-down cases as at present and despatched by M.T. with escort to Pinewood'. At Pinewood it was proposed that a large studio should be kept entirely for secret work, that whenever secret models were on site the studio was to be kept locked and under guard, and measures were to be taken to ensure that technicians remained ignorant of the true location depicted by any particular model. Although the top-secret nature of the models meant that Pinewood was considered too risky a place from a security perspective for filming to take place there, tests at the very least appear to have been carried out. A memorandum from the headquarters of Bomber Command to Wing Commander Hamshaw Thomas, dated 13 June 1943, thanked Medmenham for a briefing film of a model of Pilsen, Czechoslovakia.

During its existence more than 1,400 different models were produced by the section and these were used by all three services. From the original creation of the unit at Farnborough, the models were increasingly used by the Combined Operations Command who used them whilst planning operations and briefing those, including commandos, who took part in them. In early February 1942, V Section was tasked with making a model of the Würzburg radar installation on the French coast at Bruneval. They were provided with maps, as well as vertical and oblique aerial photographs including the low-level oblique dicing shot taken by Squadron Leader Anthony Hill in December 1941, which clearly showed the radar dish alongside a house that was occupied by the Germans. The chief purpose of the Bruneval model was to show the house and radar installation, the possible parachute-dropping zone, and the escape route which had to be taken down a gulley to the beach. One of the problems that the model-makers faced was that the coastline from which the escape was planned faced north, so the foot of the cliff was always in shadow on their photographs and they couldn't fully interpret a kiosk-like building on the beach. If that had proven to have been a defence point it could have spelt disaster for the raid, but fortunately one of the pre-war snapshots held by the Inter-Services Topographical Department showed the building and allowed the kiosk to be constructed on the model, revealing the position of windows from which there might be gunfire. In addition to the model that was used during the planning of Operation Biting, some of the architects in the section were asked to supply building elevation drawings and then – based on the position of the windows and doors – to surmise what the internal layout of the house at Bruneval was likely to be. This was so successfully executed that the returning raiders reported that only one interior door proved to be out of position.

By this period in the war Nigel Norman – friend of Frederick Winterbotham and part-owner of Heston Aerodrome – was the Group Captain appointed to coordinate the training and preparation of the RAF units for the Bruneval Raid. He had visited V Section in the cellars of Danesfield House to inspect the model that was to prove so critical to the success of the raid. Models followed for Operation Chariot, the audacious Combined Operation raid on the port of St Nazaire in German-occupied France on 28 March 1942. The commandos involved in that raid had been able to carefully study one of the section's models before the raid, which meant they were able to move around the docks with far more confidence than they otherwise would have. The importance of V Section to the success of the operation was noted in the wartime Ministry of Information publication *Combined Operations* in no uncertain terms: 'For obvious reasons no publicity can be given, but your unheralded effort is no less valuable to the war than the blazoned exploits of determination and courage in active operations'.

Throughout the war there was increasing demand for models. Later highlights included the Combined Operations raid on Dieppe and for the planning and execution of the North African, Sicilian, Italian and Normandy landings. For Operation Chastise – the attack on the German Möhne, Sorpe and Eder dams – the aircrew involved in the

Dambusters raid were instructed by Wing Commander Guy Gibson to study the models until they could walk away and visualise the dams perfectly in their heads, until their recollection of what they would be facing was perfect. The first models for planning the D-Day landings were begun as early as July 1942 and from September 1943 almost the whole output of the section was dedicated to this project. In all 109 original models – each approximately fifteen square feet in area – with numerous copy models of each, were constructed and delivered to the Supreme Head of the Allied Expeditionary Force.

An important development during the war was the reproduction of models in synthetic rubber. This meant that for big operations such as Operation Overlord, each headquarters could have a model in front of them for planning and briefing purposes, rather than just photographs of the models. In order to create them, boards were mounted on the sides of a completed model and plaster of Paris was then poured over it, with care taken that the plaster covered all the crevices. With this mould, synthetic rubber copies could then be produced in quantity and, after curing in an infrared oven, were lightweight, flexible and durable. By September 1942 the growth of the section necessitated a move from Medmenham to the commandeered Phyllis Court Country Club, Henley-on-Thames. When they returned the following year, the section was housed in huts in the grounds at Danesfield.

The model-makers were supplied with all measurement data for the production of their models by W Section, who were responsible for photogrammetry and the drawing office. Named after the Wild A5 Autograph machine (Instrument No. 50), that came from Wembley and around which the section had originally been formed, the photogrammetric intelligence from aerial photography that was carried out under Squadron Leader Ramsay Matthews could easily have been called 'mathematical interpretation'. For the first few months of the war, the photography obtained by high-flying Spitfires was the only intelligence available of the German ports, embarkation areas and airfields. Without the Wild Machines the photography would have been of little use, and so in assessing their importance at that stage it is no exaggeration to equate the Wild Machines to the 'Bombes' and 'Colossus' used for decoding at Bletchley Park later in the war. Whilst improvements to the focal length of cameras and to reconnaissance aircraft meant the Wild Machines became less vital as tools for identification, they remained indispensable over the course of the war for the production of detailed town plans that were used by the air forces and advancing armies. For Operaton Chastise the section assisted Barnes Wallis, and used the Wild Machines to calculate the volume of water held within each of the dams.

The Wild A5 was originally designed as an instrument for purely practical map-production using survey-standard aerial photographs. Spender had modified various settings and processes in order that the information contained within the aerial photography taken during combat could be extracted. Operating the machine involved taking two consecutive aerial photographs – a stereo pair – which were mounted inside the machine. Looking through a pair of binoculars the operator saw a three-dimensional image and a 'floating dot' which could be manually controlled. This made it possible to trace the outline and contours of an object captured in the photographs, and by means of complex mechanical linkages and a plotting table, the creation of highly accurate plans could be achieved. In addition to the Wild A5, the Wild Company also built the Wild A6 Stereoplotter, and although not as versatile a machine, the volume of work required of the two A5 machines was so great that two new Wild A6 machines were bought by a Swedish businessman living in Stockholm, ostensibly for his own use but actually for the British.

A central figure in the top-secret operation to secure the two Wild A6 machines was Sir (Frederick) George Binney, the Assistant Commercial Attaché to the British legation in Stockholm who had masterminded Operation Rubble, a blockade-busting manoeuvre that resulted in vital *materiel* – including ball bearings and machine tools – being delivered to Britain in January 1941 and that earned him a knighthood. A noted Arctic explorer, while an undergraduate at Merton College, Oxford, Binney had been inspired to become a pioneer in the use of seaplanes for survey work, chronicled in his 1925 book *With Seaplane and Sledge in the Arctic*. Working for the Hudson's Bay Company and United Steel before the war, where he was responsible for establishing an extensive export network, it was these buccaneering skills that resulted in his being recruited by the Ministry of Supply in 1939.

In a plot worthy of a James Bond spy novel, the two machines were transported from the factory in neutral Switzerland and smuggled via a tortuous route from Sweden to Britain. Squadron Leader Ramsay

Matthews, who had been in charge of the small air-survey department at the Ordnance Survey, personally flew in a PR Mosquito to Sweden to collect the Wild Machines. When Matthews arrived in Stockholm he found the two machines stored in enormous wooden crates. Too large to be accommodated in the relatively small Mosquito, he was forced to dismantle the machines and put the innumerable parts into a series of 'diplomatic bags' that were loaded into the aircraft. With Matthews strapped into the bomb bay the pilot took off from Sweden but ran into difficulty over Skaggerak, the strait running between Norway and the Jutland peninsula of Denmark, where they were intercepted by enemy fighters. During the ensuing evading action, and unbeknown to the crew, the bomb-bay doors opened and remained open until they landed back safely in Britain. Matthews survived, having experienced a terrifying and freezing journey over the North Sea.

From 1943, Mosquitos in civilian markings were regularly flown to Sweden by the British Overseas Airways Corporation – the British state airline – between Leuchars and Stockholm. Ostensibly to carry mail and legitimate freight, it became nicknamed the 'ball-bearing run' because of their frequent transportation of ball bearings that were desperately required by British industry. As with Matthews, on occasion they are also recorded to have carried passengers who were strapped into the bomb bays, given oxygen masks and wished good luck. One of them was the Danish physicist Niels Bohr who escaped from Denmark via Sweden and Leuchars to Los Alamos, New Mexico, where he worked on the Manhattan Project. From there being only one serviceable Wild A5 machine in December 1940, through clandestine activities photographic intelligence acquired the refurbished Ordnance Survey Wild A5 in 1941 and in 1943 would aquire a further Wild A5 (Instrument No. 81) that followed a similarly 'interesting journey through occupied and neutral territories until it arrived in Gibraltar' before being installed at RAF Nuneham Park. These Wild Machines, alongside the two Wild A6 Stereoplotters, worked around the clock throughout the war and, whilst some would return to peacetime use after the cessation of hostilities, during the Cold War they would prove their worth again.

CHAPTER 5
Peenemunde and Bodyline

Throughout 1943 Medmenham gradually expanded and became concerned with the planning stages of practically every operation of the war and every aspect of intelligence. It was led by Group Captain Peter Stewart and organised into three squadrons, commanded by Wing Commanders Douglas Kendall, Hugh Hamshaw Thomas and Tom Muir Warden, who had transferred to Medmenham from the Bomber Command PI Unit. The processing of aerial photography, making of the models, photogrammetry, the drawing office, Print Library, Map Library and Press Section were the responsibility of A Squadron. B Squadron was responsible for the Duty Officers, the first-phase interpreters posted to the PR units, the plotting of sorties, second-phase interpretation and the pool of interpreters destined for overseas duty. C Squadron was responsible for third-phase interpretation, Combined Operations and the Intelligence Section.

Tales about the exploits of Peter Stewart, who was seen as a rebel, wickedly mischievous and often great fun, were numerous at Medmenham. Once, when he was reportedly considerably the worse for drink, and disconcerted that Glyn Daniel was being posted to India, he appeared in the middle of the night while Daniel was working his night-shift and invited him for a drink in the mess. With Daniel pointing out that in view of the hour they would all be closed, Stewart proudly produced a set of keys that provided access to every mess bar at Medmenham, whereupon whisky was quickly secured and consumed. As President of the Mess Committee, Hamshaw Thomas had to frequently manoeuvre the commanding officer's attempts to secure most of the mess's whisky allowance for himself. On another occasion, during a party attended by prominent members of the Buckinghamshire establishment Lord and Lady Hambleden – owners of the stationery and bookshop empire W. H. Smith & Son – Stewart appeared on the dance floor pirouetting around on a ladies' bicycle, surprised to be greeted by silence rather than the applause he was expecting. Hamshaw Thomas regarded Stewart as a menace, as 'a round peg in a square hole', who had no idea how a commanding officer should behave.

The often amusing behaviour of Peter Stewart is all the more peculiar given that his own father – Squadron Leader Walter Stewart – was serving as a Medmenham interpreter. ADI (Science) found in Stewart senior a very useful contact: according to Jones, 'Pop Stewart', who had been a Brooklands racing driver before the war, supplied him with copies of aerial photography covering Peenemunde when he was later marginalised from the hunt for the secret weapons. During wartime, and in his later publications, Jones had a rather low opinion of many of the interpreters, with the noted exception of Claude Wavell – affectionately known as 'Uncle Claude' by Jones and his fellow scientists – who had played a critical part in the Battle of the Beams early in the war, and led the ongoing search for, and study of, German radar.

By April 1943 Medmenham had proven its value as a key part of the intelligence machine and was home to hundreds of photographic interpreters. Photographic intelligence had become firmly established as an indispensable source of intelligence, but it was about to face a new challenge, the outcome of which would not only prove how effective it was but would play a major part in determining the outcome of the war. The role of photographic intelligence in the battle for information about Hitler's *Vergeltungswaffe*, or revenge weapons, was to prove how vital it had become. With opinion divided among Britain's top scientists over whether a liquid-fuelled rocket could be developed, the interpreters were given the near-impossible task of finding the unknown. The challenge amply reflected Winston Churchill's belief that the Second World War was a battle as much between scientists as between sailors, soldiers and airmen. With intelligence from resistance groups indicating that the Germans were putting enormous resources into the development of secret weapons, the rockets represented a credible threat. It led to the appointment of Duncan Sandys MP, Joint Parliamentary Secretary to the Ministry of Supply on 20 April 1943, who began the investigation into the rocket-weapons threat. The hunt for the V weapons was on.

In December 1942 Jones had received a telegram from the MI6 station in Stockholm that reignited concerns about the German rocket threat. His position as head of scientific intelligence at the Air Ministry and scientific adviser to MI6 meant, in his own words, keeping a 'constant vigil for new applications of science to warfare by the enemy'. In an echo of Mayer's Oslo Report, a Danish chemical engineer had reportedly overheard a Professor Fauner of the Berlin Technische Hochschule (a Professor Forner was known to exist) discussing a new German weapon. It could fly, contained 5 tons of explosive, had a maximum range of 124 miles and could be accurate to within 4 square miles. Jones was then given sight of a report from a Swedish source, dated 12 January 1943, claiming that the Germans had built a new factory at Peenemunde, where one witness claimed to have seen a rocket being fired on a testing ground. Both accounts prompted Jones to look again at the threat posed by rocket technology.

The Danish report was given particular attention and was to prove a lucky break for the British, since the Germans had only successfully launched a test model of their *'Aggragat 4'* – fourth in a group, or *Aggragat*, of rockets – on 3 October 1942. Developed under the direction of Major-General Dr Walter Dornberger, the A4 would come to life, and into mass production, in 1943 having been rechristened by propaganda chief Goebbels the *Vergeltungswaffe 2*. After years of research and development, in 1942 the first of the German A4 rockets came out of the Peenemunde model shops, and after a period of ground testing were ready for test firing. The A4 was 14 metres long, 165 centimetres in diameter, and weighed 4 tons when empty. When fuelled and armed it carried 5 tons of liquid oxygen, 4 tons of alcohol and three-quarters of a ton of high-explosive, a total weight of nearly 14 tons.

The first A4 test model was ready on *Prüfstand VII* (Test Stand 7) at Peenemunde on 25 February 1942. During a test firing on 18 March, the engine exploded and the test model was destroyed. This was followed by another attempt on 13 June, which also resulted in failure and the frustration of Albert Speer, the Reich's Armaments Minister, who was visiting Peenemunde with senior commanders from the Wehrmacht, Luftwaffe and Navy. Weapons scientist Dornberger had hoped to impress the decision makers, who instead witnessed the rocket going out of control, crashing and exploding two seconds after launch. On the third attempt the following month, the rocket reached a peak height of 7.3 miles, and having reached Mach 1.9 – a supersonic speed – it exploded. People living in the Baltic region around Peenemunde witnessed for the first time 'frozen lightning', the name for the white trail left behind as the rocket climbed. On the fourth attempt, on 3 October, after a flawless vertical take-off from Test Stand 7, the revolutionary weapon reached a peak height of 52 miles, flew 118 miles in 296 seconds and landed only 2.5 miles from its intended landing point. For many of the Peenemunde rocket pioneers this might have been a 'first leap into space', but for the German High Command the military possibilities of this technology, against which there could be no defence, were clear.

The Danish source, Aage Andreasen – code-named 'Elgar' by the British – had overheard the conversations about the rocket in a Berlin restaurant. But at first his reliability could not be judged. Stockholm was no different from any of the other capital cities of neutral European countries in that it was full of dubious characters that included friendly sources such as Elgar, enemy agents, double agents and people anxious to get rich from selling information to anyone. The British did not have the resources to investigate every report of secret weapons that they received but, whilst some could be easily dismissed as absurd, others such as this most recent one from Elgar could not. When Elgar sent detailed photographs of the Lichtenstein aerial system used on German night-fighters – something the Germans would never willingly divulge – British Intelligence were able to rule him out as being a possible German agent. In December 1942, MI6 recruited the chemical engineer and Agent Elgar became a frequent visitor to Stockholm, where his company had an office, while other business connections took him to Germany, Finland and Romania. Creating his own network, Elgar sourced his information from fellow countrymen who worked for German companies manufacturing defence materiel and from technical workers employed at Peenemunde.

In the autumn of 1943, Elgar ingeniously smuggled three hundred reports on microfilm concealed in glass bottles, hidden in one of three barrels of acid that he imported to Sweden for business purposes. Aage Andreasen was among the first secret agents to report on V weapons, and by the time he was caught by the Germans in January 1944 he had even supplied a photograph of a projectile that had landed on the Danish island of Bornholm. Transcripts of his interrogation at the hands of the Germans subsequently revealed that Elgar gave descriptions to them of some of the personnel then working in Stockholm – action that his spymasters later considered reasonable under the circumstances, in order that he could 'save his skin', particularly since he successfully duped his German interrogators with false information that ensured he would survive the war. The evidence from Elgar pointed towards a very real – and imminent – long-range rocket threat. Reports to British intelligence about secret weapons had suddenly increased in frequency: in 1939 there had been two, one in 1940, none in 1941 and one in December 1942. But by March 1943 there had been another five. Questions were growing about German activities at Peenemunde.

It was well known that Germany, like many other countries, had an interest in rockets. After the 1923 publication of *Die Rakete zu den Planetenräumen – By Rocket into Planetary Space* – the subject of space travel and rocketry gained popularity in the country. As early as 1927, the futuristic idea of using rocket projectiles to travel through space to other planets resulted in the creation of *Verein für Raumschiffahrt (VfR)*, or the Association for Space Travel. But with no interest from industry or government, rocket experiments were at this time confined to the drawing board, private workshops, individual enthusiasts and the rocket-club members who fired increasingly powerful rockets. During the late 1920s Fritz von Opel, of the Opel car company, worked with the club and commissioned rocket cars, trains, and even an aircraft – the Opel RAK.1 – primarily as publicity stunts for his company. But as the 1930s progressed, and German scientists considered developing liquid-fuelled engines, the German military began to take much more of an interest in the *Verein für Raumschiffahrt*.

With Germany's armaments production strictly controlled by the Treaty of Versailles, the Army Weapons Agency – the *Heereswaffenamt (HwaA)* – was constantly looking for new arms developments that could circumvent the conditions of the treaty. This prompted one of the most scientifically advanced nations in the world to go all out for new secret weapons. The rocket scientist Rudolf Nebel was given a grant of 5,000 marks in 1930 and access to military training grounds at Tegel, to the north of Berlin, to build and test rockets. While still a university student of engineering, a 'technology enthusiast, Wernher von Braun, nineteen at the time, had joined Nebel's team of inventors'. By 1934, von Braun had completed a doctorate on 'the liquid-fuelled rocket' at Berlin's Friedrich Wilhelm University and began working for the Army Weapons Agency. Under the command of Captain Walter Dornberger, the Agency put theory into practice at their new test facility at Kummersdorf, to the south of Berlin.

In 1933, Hitler had decided to make the Army Weapons Agency responsible for all rocket development in Germany. The political and military decision-makers hoped the technology would give them a head start in the emerging arms race and, as Dornberger would later remark, in the creation of 'weapons of unique significance'. From Tegel and Kummersdorf the trail was to lead by the mid-1930s to Peenemunde. Located on the Baltic island of Usedom in northern Germany, this small fishing village with 96 houses, one school and 447 inhabitants was destined to become a top-secret – and the world's first purpose-built – centre for the development of ballistic weapons. The decision to relocate there had followed a visit in 1936 to Kummersdorf by General Werner von Fritsch of the German High Command. Dornberger and von Braun demonstrated their new rocket engines and secured finance from von Fritsch providing they could 'make a useable weapon out of the rocket engine'.

A meeting soon after between the Wehrmacht and the recently established Luftwaffe agreed on Peenemunde as an ideal location for the creation of a joint test centre, with separate sections for the army and air force. From a military perspective Peenemunde had a number of advantages. There was ample space to launch rockets along the surrounding coastline and up to 200 miles of open water to fire them across, while its isolated location made it much easier to keep secret. The Peene Estuary and Achterwasser Bay separated Usedom from the mainland to the south and, with only three bridges, access could be easily controlled. The Greifswalder Oie, a small island off Usedom, provided an ideal target for projectiles and the whole area could be connected to the surrounding road and rail network. Municipal and private land was quickly and compulsorily purchased, and over 10,000 builders worked on the north-western tip of Usedom from the summer of 1936 until the outbreak of war, under contract to civilian companies, the *Organisation Todt* and the *Reichsarbeitsdienst*. Newspaper adverts were used to recruit construction workers for an unspecified building site in beautiful surroundings on the Baltic coast, somewhere that promised work, bread and plentiful leisure activities.

R. V. Jones had been deliberately keeping the team who were in the know as small as possible for security reasons and in the belief that a few clever people, even a single individual, can be more effective and efficient than a large organisation. But since all the MI6 intelligence had also been going to the War Office's MI14 who specialised in intelligence about Germany, it would be the Army who first turned to Medmenham for photographic evidence of what was happening. The first information suggesting that the Germans were planning to bombard Britain was sent in a memorandum to Major Norman Falcon of the Army Section on 13 February 1943. A graduate of Trinity College, Cambridge, Falcon had worked as a geologist for Anglo-Iranian Oil before the war. An expert in identifying oil and mineral deposits using aerial photography, this had brought him into contact with the Aircraft Operating Company during their work mapping Iran for the oil company. Posted to Wembley when called to wartime service, his technical experience, patience, persistence and personal modesty earned him the respect of his contemporaries and helped ensure the Army Section at Medmenham worked harmoniously across the large inter-service RAF unit.

When Falcon received the War Office memorandum, it was sent for the eyes of B Section only, and was written in the most cryptic of terms. But despite it being unavoidably circumspect Falcon was now at least aware of this new possibility:

- There have recently been indications that the Germans may be developing some form of long-range projectors capable of firing on this country from the French coast;

- There is unfortunately little concrete evidence on the subject available, except that the projector may be similar in form to a section of railway track;

- It is obviously of great importance to obtain the earliest possible warning of the existence of any such device, and we should be grateful if you would keep a close watch for any suspicious erections of rails or scaffolding, and consult MI10 [who had responsibility for weapons and technical analysis] through us, on any doubtful cases;

- We will of course pass you any further information which may come into our or MI10's hands.

The following month more evidence about the existence of a German rocket programme was to emerge from an unusual source. Two high-ranking Wehrmacht generals, Wilhelm Ritter von Thoma and Ludwig Crüwell, had been captured at the second battle of Alamein in November 1942. They had been kept apart for four months but were brought together specifically in the hope of gathering intelligence at the prison for captured German staff officers – Trent Park House, Enfield – on 22 March 1943. The rooms of this grand late-nineteenth-century mansion house were fitted with hidden microphones and listening devices. The two generals greeted each other as old friends: von Thoma told his compatriot that he knew their prison was near London, but since there had been no large explosions there must have been a hold-up with the rocket programme. A full transcript of the recording was sent to the offices of ADI (Science) at the Air Ministry. It was there on a Saturday afternoon, 27 March 1943, that physicist Dr Charles Frank, one of the talented team of scientists now working for R.V. Jones, read the translation of their conversation. A conversation in which von Thoma described to Crüwell his visit to a special site housing rockets that could travel through the stratosphere, and said that the hopeful officer in charge there – presumably Dornberger – had said then that within a year there would be no limit to their range. With Jones regarding von Thoma as an 'intelligent pessimist' it was enough for him to conclude that the rocket threat must be treated very seriously, and this heralded a new phase of urgent War Office and Air Ministry investigations.

Now that multiple intelligence agencies were involved in the hunt for secret weapons, a difference of opinion was emerging about how far the inquiry should be spread, who should know about the threat, and who should be in charge. Jones strongly believed that although he and his team had become aware of something, since they were not in a position to provide the operational commands with information on which they could take countermeasures it was his duty to begin the intelligence chase as urgently as possible, while letting key figures, such as Lord Cherwell, know what he was doing. Whilst the interpreters were having to put together the pieces of the jigsaw, just to discover what they were supposed to be looking for, this lack of information ensured that the relationship between the interpreters and Jones was destined to become increasingly fraught, especially when Jones's involvement in the hunt for the secret weapons came to be marginalised.

Whilst Jones might have wanted to keep matters close to his chest, to collect and collate information and only release it in a meaningful and complete form so that decisions were not based on half-truths, the War Office didn't agree and the whole issue was put before Lieutenant General Archibald Nye. The Vice Chief of the Imperial General Staff immediately summoned two British scientists to advise him: Charles Ellis, Professor of Physics at King's College, London and scientific adviser to the Army Council, and Dr Alwyn Douglas Crow, who as Controller of Projectile Development at the Ministry of Supply was responsible for British rocket research. The scientists concluded that unless an extremely accurate method of directional control had been developed, the rocket would require a launch ramp about a hundred yards long. No attempt was made to investigate the means of propulsion for such a rocket, which would have determined its effective range and payload, and since British military rockets up to this point had been based on solid-fuel propulsion the War Office appears to have been advised that the standard fuel of cordite would have been used. The result was a memorandum entitled 'German Long-Range Rocket Development', circulated by Nye on 11 April 1943, which outlined all the various pieces of intelligence that had been accrued up to that point, including the reports from Agent Elgar. On the basis that such a large projector ramp would be hard to conceal in the Channel coast region of France, Nye recommended that aerial reconnaissance should be used to uncover them.

When the Vice Chiefs of Staff met on 12 April 1943 they discussed Nye's memorandum and agreed that Churchill and Herbert Morrison, the Minister of Home Security, should be warned of the possibility of southern England being attacked by German rockets. Before this could be done, however, Brigadier Hollis of the War Cabinet Secretariat observed that since so many intelligence committees, agencies and their scientific advisers were now involved, and given the importance of the subject, the Vice Chiefs of Staff might want to recommend to the Prime Minister that one individual, who could devote the time that would ensure the investigation could proceed quickly and ensure that no aspect was overlooked, should handle the case. It was left to General Hastings Ismay, Chief of Staff to the Minister of Defence, to write the following minute to Churchill on 15 April 1943:

> Prime Minister,
> The Chiefs of Staff feel that you should be made aware of reports of German experiments with long-range rockets. The fact that five reports have been received since the end of 1942 indicates a foundation of fact even if details are inaccurate. The Chiefs of Staff are of the opinion that no time should be lost in establishing the facts, and, if the evidence proves reliable, in devising countermeasures. They feel this is a case where investigation by one man who could call on such scientific and intelligence advisers as might be appropriate would give the best and quickest results . . . They suggest for your consideration the name of Mr Duncan Sandys, who, they think, would be very suitable if he could be made available.

Sandys was the thirty-five-year-old Joint Parliamentary Secretary to the Ministry of Supply. He had been elected as Member of Parliament for South Norwood in 1935, and was Churchill's son-in-law, having married Winston's eldest daughter Diana that same year. While an MP, he had served with an anti-aircraft regiment during the failed Norwegian Campaign of 1940, and after his return to Britain and three months in the War Cabinet Secretariat was anxious for military deployment. As the commanding officer of the first British 'Rocket Regiment' based at the Projectile Development Establishment at Aberporth in Ceredigion, Wales, his unit would later be credited with hitting the first enemy aircraft to be shot down with a rocket, designed by Dr Alwyn Crow and propelled by slow-burning cordite. As a result of the experiments at Aberporth, the first operational rocket anti-aircraft Z Battery was established in Cardiff, and Sandys grew accustomed to spending his days with the unit in Aberporth and nights waiting for enemy attacks on Cardiff. It was on one of the journeys between the two places that his driver fell asleep at the wheel and, crashing the car, inflicted severe injuries on Lieutenant Colonel Sandys.

After three months in hospital Sandys returned to the House of Commons and after a government reshuffle found himself as Financial Secretary to the War Office before being appointed Joint Parliamentary Secretary at the Ministry of Supply where he was responsible for all weapons research, development and production, which meant he had all the scientists employed by the Ministry at his disposal. With his position in government and his military experience he was considered uniquely qualified, and Churchill appointed him on 20 April 1943. But the appointment of a politician rather than a scientist would cause dismay to two people in particular – Lord Cherwell, the Paymaster General, and Brendan Bracken, the Minister of Information – who believed that Churchill was giving his son-in-law increasingly preferential treatment. Cherwell felt slighted and this, coupled with a long-standing lack of interest in rocket technology – dating back to his time at Farnborough during the First World War – created a toxic situation where his personal feelings would override his scientific reasoning to the point where he would occupy an increasingly indefensible position.

The effects of Sandys's appointment to chair a committee that involved so many intelligence committees, agencies and their scientific advisers was felt immediately. Unaware of the resentment among the scientific community that his appointment had caused, but working on the maxim that 'experts should be on tap, not on top', Sandys began work immediately. One of his first instructions was for all areas within 130 miles of London and Southampton to be photographed by British and American photographic-reconnaissance squadrons (based at RAF Benson and the neighbouring RAF Mount Farm respectively) to ensure that every square mile of the French coastal area from Cherbourg to the Belgian frontier would have been photographed since the beginning of the year.

On 19 April 1943 the Air Ministry had directed the CIU at

Medmenham to investigate the German secret-weapons programme and had instructed them to send reports to the Air Ministry and War Office and to the office of Mr Duncan Sandys. Peter Stewart appointed Wing Commander Hugh Hamshaw Thomas, who was then in charge of all third-phase interpretation, to lead the investigation. Medmenham was informed that the German rocket had a maximum range of 130 miles; that British experts had accepted that such a weapon was technically possible; that experiments with the weapon had been reported in the Peenemunde area; that the Germans were known to be intensifying production of smaller rocket projectors already in service; and that development and use of the long-range rocket must, therefore, be seriously considered. Since the obvious target for such a long-range weapon was London, it was noted that areas in France within a radius of 130 miles from the centre of the capital should be watched most carefully for suspicious activity. All photographs taken since 1 January 1943 within range were to be analysed and any gaps in coverage were to be filled in by the PRU. The interpreters were told that the weapons could include: a long-range gun; rocket aircraft controlled on the Queen Bee principle [a radio-controlled pilotless aircraft]; a tube located in an unused mine out of which a 'rocket could be squirted'.

The highest possible priority was to be given to the investigation, and Hamshaw Thomas and Falcon were under strict instructions that any evidence of such a weapon obtained through photographic interpretation should be treated as 'Most Secret', that Duncan Sandys's office should be supplied with copies of the relevant photography with an interpretation report immediately, and that no mention of this subject should be made in any routine interpretation report. But from the outset the investigation was seriously complicated by the fact that they did not know exactly what they were looking for. The initial focus of the Sandys Committee was Peenemunde. This prompted a visit to Medmenham on Easter Sunday 1943 by two of Sandys's scientists, Dr William Cook, the Assistant Controller of Projectile Development at the Ministry of Supply, and Dr H. J. Phelps from the Ministry of Economic Warfare, where they met Flight Lieutenant André Kenny.

André Kenny worked in D Section, a third-phase team responsible for supplying information on all types of industrial plants in enemy and enemy-occupied territory. Like Hamshaw Thomas, he was an alumnus of Cambridge University, having studied at Trinity College where he had written a doctoral thesis on classical hydraulic engineering that he claimed was not accepted by the university on the basis that they had neither an engineering examiner with enough knowledge of Greek and Latin nor a classical scholar with sufficient knowledge of engineering. To develop their understanding of a particular industrial process, D Section gathered information from wherever they could, including talking to war refugees who had worked in particular factories, specialists from industry and academia, visits to libraries, and visits to industrial sites in Britain. The sheer quantity of photography covering industrial sites meant teams were created within D Section specialising in iron and steel, light metals, electric power, fuel, chemicals and explosives, synthetic rubber, engineering and textiles.

When the scientists visited Medmenham, Kenny was in the midst of analysing all the aerial photography then held that covered Peenemunde. This included photographs taken speculatively during sortie A/762 on 15 May 1942 by Flight Lieutenant Donald Steventon, who turned on his cameras for a short run over the site after spotting the airfield while on his way to photograph the nearby port at Swinemunde. Constance Babington Smith later recalled looking at the photography in the summer of 1942, when it did the rounds among the specialist sections responsible for third-phase interpretation. But whilst three large circular embankments were identified as unusual and puzzling, the other pressures on the interpreters' time meant they could not dwell on these photographs and they were stored in the Print Library for future reference. Whilst the site would have been easily identified by a Cold War interpreter as a ballistic missile site, the Medmenham interpreters could not possibly have known such a thing could exist in 1942. By the time of the scientists' visit to Medmenham on 24 April 1943, this remote part of the Baltic coast in northern Germany had been photographed during four sorties:

A/762 – 15 May 1942
N/709 – 19 January 1943
N/756 – 1 March 1943
N/807 – 22 April 1943

Kenny and the scientists examined the Peenemunde site from end to end and decided that certain installations at the airfield that would subsequently be identified as launching ramps for flying bombs were instead pumping machinery related to land reclamation. Kenny's subsequent report provided detailed interpretation of the installations at Peenemunde under the following headings: The Factory Area; Possible Nitration Houses; Power House; Steam Pipes; Structures Near the Tip of the Peninsula; and The Circular Emplacements. During the most recent sortie N/807 – flown two days previously – it was observed that on frames 2007–2009, 5010 and 5011 a large cloud of white smoke or steam could be seen drifting in a north-westerly direction from the area, while on frame 5010 an object about 25 feet long could be seen projecting in a north-westerly direction from the seaward end of the building which ran seaward from the platform. When photograph 5011 was taken four seconds later this object had disappeared, and a small puff of white smoke or steam was issuing from the seaward end of the building. What André Kenny described but could never have been expected to realise, especially given how little information he was working with, was that they had in their hands an image of the twenty-first A4 rocket (the V2) being run up on Peenemunde's Test Stand 7. By the time the PR Mosquito landed back at RAF Benson, the rocket had been test fired into the Baltic.

On 29 April the first D Section report on secret weapons – Report DS1 – was completed and with copies in his briefcase Kenny travelled with his colleague WAAF Section Officer Winifred Bartindale to a meeting at the Ministry of Supply with Duncan Sandys. Twenty-three-year-old Bartindale had studied at Somerville College, Oxford, and would become a medical doctor after the war, specialising in paediatrics. Arriving at the imposing Shell-Mex House on the Strand – the Ministry's wartime home – Kenny met contemporaries from his time at Cambridge: aerodynamics expert Harry Garner, who was then Chief Scientist at the Ministry, and the geophysicist Dr Edward Bullard. Kenny found he was able to forge a solid working relationship with Duncan Sandys, who was fascinated at the extent and detail of the photographic intelligence. By that evening the Sandys investigation decided, based on the photographic evidence of Peenemunde, that:

- The whole site is probably an experimental station (not yet in full use);
- It is probably an explosives works;

- The circular and elliptical constructions are probably for the testing of explosives and projectiles (if in fact rocket projectiles are being tested here);
- Use of the site has not gone beyond the experimental stage so far;
- The whole area should be frequently monitored by photographic reconnaissance;
- In view of the above it is considered that a heavy long-range rocket is not yet an immediate menace.

Detailed plans of the Peenemunde installation were prepared at Medmenham – by W Section using Wild Machines – and were forwarded on to the Ministry of Supply. On 7 May 1943 the investigation became formalised under the banner 'German Long-Range Rocket Detection' and the initials GLRRD began to appear on official documents.

As the only person at Medmenham to be Ultra-cleared, Wing Commander Douglas Kendall had to undertake many top-secret interpretation tasks and directly provided assistance to Bletchley Park. It seems astonishing now that the great responsibility of being the Technical Control Officer fell on a man then still only in his twenties, but merely highlights 'the great responsibility that was put onto many young shoulders in those days' at Medmenham. As a foretaste of the kind of analysis that would come to dominate photographic interpretation during the Cold War, in 1943 Kendall attended a highly secret meeting with R. V. Jones at which he learnt about 'tube alloys' – the code name for the British nuclear-weapon project – which became subsumed within the American-led Manhattan Project. The American project had its roots in the Einstein–Szilárd letter, which had been sent to President Franklin Roosevelt in 1939, signed by Albert Einstein but largely written by Leo Szilárd and fellow Hungarian physicists, warning that Nazi Germany might be developing nuclear weapons. It was this double fear – that Hitler might have both a rocket *and* an atomic weapons programme – that prompted Churchill to order photographic intelligence to check for *any* slight indication that the Germans had the capability.

Intelligence indicated that German atomic scientists from the

Berlin-based Kaiser Wilhelm Institute had been relocated to the small towns of Hechingen and Bissingen in Wuerttemberg, in the Black Forest region of south-west Germany, when the heavy bombing of Berlin began in 1943. Kendall was informed that if the Germans had an active atomic-bomb programme then facilities for vast new water and power supplies would have been built. A complete check of that area was requested: reconnaissance squadrons were ordered to photograph everything within a hundred-mile radius. Medmenham's Industry Section interpreters were then tasked with studying all the power facilities and industrial plants, but were never told why they were doing the work. The success of the plan depended on the skill of the interpreter to recognise the unusual: interpreters were asked to plot, date and analyse all power lines, industrial plants and water supplies captured on thousands of images. Fourteen new factories, all identical in shape, were identified in an area to the south of Stuttgart, and each site was located next to mining operations and featured a grid pattern, pipes lying on the ground, a number of chimneys and storage tanks. The discovery that all the factories were located on the same contour, and adjacent to mining operations, prompted the idea that there might be a geological rationale behind their location. Kendall sent one of the D Section geologists to the Geological Museum in South Kensington – now part of the Natural History Museum – to consult their records: their data showed that a German scientist had once recorded a thin bed of low-grade oil shales in the locality.

With this clue, and the knowledge that by 1943 the Germans were acutely short of oil supplies, it became apparent that they were faced with exploiting this low-grade and expensive source of fuel. It transpired that oil shale was being burnt at the sites, with the gases drawn into the pipes and brought into a refinery for treatment. With this information Kendall was able to give a qualified but reasonably optimistic report to a member of the Prime Minister's staff that these factories did not represent an 'atomic menace' and, given the reasonably complete photographic coverage of the whole of Germany that existed by this stage in the war, he could advise that there were no suspicious works elsewhere either.

While Flight Lieutenant André Kenny and interpreters from D Section were assigned the task of searching for information about experimental rocket and manufacturing work at Peenemunde, Major Norman Falcon concentrated the efforts of the Army Section on the German deployment of the weapons against Britain. This involved the staff of B Section, notably Captains Neil Simon and Robert Rowell, who systematically interpreted tens of thousands of aerial photographs from the Print Library that covered the entire French coastal region within 130 miles of central London. The challenge they faced was considerable, particularly bearing in mind the vague, almost science-fiction-like briefings from MI10 and the War Office that described tubes in disused mines from which rockets could be squirted and long-range guns.

Neil Simon had been studying to become an interpreter at the Army School of Military Intelligence in Matlock when France fell, and found himself posted to Medmenham when the interpreters moved there from Wembley in April 1941. Being an army officer, he was among the first to know about the German preparations to bombard Britain with revolutionary weapons when, in February 1943, MI14 briefed the Army Section. One of the first tasks for the peacetime stockbroker was to update the Admiralty plans to record changes by the Germans to coastal defences throughout Europe. From Medmenham Simon witnessed the growth of Fortress Europe, and together with his colleagues he plotted all the coastal defences from aerial photography onto the defence maps that would prove vital to the Normandy landings in June 1944.

On 9 May 1943 Sandys visited Medmenham and met Kenny in D Section and Captain Neil Simon of B Section who was then heavily burdened with interpreting aerial photography of northern France. Simon had been searching for disused mines (from which rockets could be squirted) and long-range guns. B Section were particularly interested in gradients as they had been advised by the scientists that a long-range rocket would need to be fired from an inclined gradient, rather than vertically. This prompted Simon to show the Minister an example of a gradient under the stereoscope. On viewing the 3D image Sandys observed: 'It's awfully steep, isn't it?' When told that a stereoscope exaggerated depth, Sandys replied, acknowledging that he had cottoned on: 'Yes, of course, because your eyes aren't 200 feet apart'.

On 14 May a photographic-reconnaissance mission was flown over

Peenemunde – sortie N/825 – from which a more detailed study of the elliptical earthwork and associated structures was possible. In his report Kenny records having spotted a column of five vehicles and that 'the middle vehicle appears to carry a cylindrical object thirty-eight feet by eight, which projects over the next truck'. He states that on re-examination of an earlier sortie a similar object could be seen and that activity over the whole site was high. On 17 May, Sandys circulated a report to the War Cabinet. In it he stated: 'It would appear that the Germans have for some time past been trying to develop a heavy rocket capable of bombarding targets from very long range. This work has probably been proceeding side by side with development of a jet-propelled aircraft and airborne rocket torpedoes.' Sandys recommended that further information on the subject should be obtained from secret agents, prisoners of war and further aerial-reconnaissance photography, and that: 'The experimental establishments and factories which appear most likely to be connected with the development and production of this weapon in Germany and German-occupied territory, together with any suspicious works in the coastal region of north-west France, should be subjected to bombing attack'.

With hindsight, that day, 17 May 1943, was a huge turning point and the first report was issued by Medmenham on activity at Watten in the Pas-de-Calais region of northern France. Although no definite identification was then possible, since it could not be attributed to any conventional military activity, it was highly suspect as being associated with German rocket firing. It was now clear that the Germans were putting an enormous effort into installations at Peenemunde and Watten, and they were highly unlikely to have done this without good reason. As a consequence of these developments Sandys felt compelled to report to the War Cabinet that in his opinion a long-range rocket probably existed and might already be in limited production. Re-examination of all territory within 130 miles of London was again ordered.

The reaction of Lord Cherwell, however, to Sandys's report was the beginning of a long struggle that was to dominate the scientific enquiry from then on. Lord Cherwell argued that a cordite- or solid-fuel rocket would be too large to be a practical proposition and that something such as a liquid-fuelled rocket was beyond possibility at the time. He claimed that the cylindrical objects shown in the aerial photographs were in fact a giant hoax, intended to conceal another secret weapon such as a flying bomb. But, working on the assumption that since the British had not managed to produce a liquid-fuelled rocket then how could the Germans, Cherwell was to make one of the greatest intelligence mistakes anyone could ever make: the classic '*we* can't do it, therefore *they* can't do it'. Whilst the distinguished scientist Cherwell was to be proved correct about the existence of flying bombs, he was to become increasingly frustrated by Sandys leading what he considered to be essentially a scientific investigation.

By now Peenemunde was being photographed as frequently as twice a week – weather permitting – and much activity could be seen by the interpreters on every occasion. By comparing the photography it was noted that the crane-like structures in the centre of the circular embankments moved around and on 2 June – sortie N/853 – it was recorded that an area on the foreshore had been levelled, 470 feet seaward from the elliptical earthworks, and that a 'thick vertical column about 40ft high and 4 feet thick' could be seen. From the shadow cast by the object, and with knowledge of the aircraft's flying height and the camera's focal length, Kenny was able to calculate the thing's dimensions. When R. V. Jones was provided with copies of the aerial photographs from sortie N/853 he claimed to see 'the outline of a rocket, somewhere around 38 feet long, with a tail fin perhaps 10 feet long and perhaps 10 feet across at the near end'. A re-examination of earlier sorties revealed the existence of similar objects at Peenemunde, but the unwillingness of the interpreters, and André Kenny in particular, to identify them as rockets was condemned by R. V. Jones and many later writers on the subject. This reflected the often vexed relationship between the scientists and the interpreters and arguably the principal reason for it was later spelled out by the interpreter Ursula Powys-Lybbe. She stressed that as interpreters they were specifically trained not to make definitive statements about objects they might see in aerial photographs until it had been established what they were, no matter what they might have thought personally. Convention dictated that Kenny had no choice but to describe the mysterious thing he saw at Peenemunde as either an object or vertical column, but in consequence was thereafter to be pilloried by most writers on the subject for not spelling out the word 'rocket'.

Churchill was interested enough in the work of Medmenham to visit the place several times during the war, including on 16 June 1943, when he was shown this most recent photography. Such was his interest in the rocket investigation that he invited Field Marshal Jan Smuts, the South African Prime Minister, who would later visit Medmenham on Churchill's behalf to check up directly on progress. After meeting Hamshaw Thomas and when out of earshot, Smuts apparently turned to Douglas Kendall and said: 'Do you know, that fellow is the world's leading palaeobotanist?'

Disappointed not to be leading the investigation, R. V. Jones nevertheless continued to play a key role and received copies of relevant aerial photographs from Medmenham. His dual function as head of scientific intelligence at the Air Ministry and scientific adviser to MI6 meant that he was uniquely positioned to remain informed and advise on developments. Since reports from secret agents went through his office before anything relevant was copied to the Sandys Committee, he was in a unique position to draw his own conclusions from the raw intelligence.

To prepare for future discussion on the rocket, Jones wrote a report on 26 June that followed another photographic-reconnaissance mission flown on 23 June by Flight Sergeant Ernest Peek, which showed a rocket so clearly that nobody could dispute its existence. This meant that R. V. Jones had both his report to fall back on and the most recent aerial photographs that were made available for a meeting of the War Cabinet Defence Committee on 29 June 1943. In his 1943 report on the long-range German rockets Jones suggested that the Germans must be developing them and that if this was not true the activity detected could be an elaborate German hoax. Alternatively, the information gleaned from both secret sources and photographic intelligence had been misinterpreted and what appeared to be a long-range rocket must in fact be something else. To judge between these different possibilities he returned to the anonymous Oslo Report of November 1939 and, considering it the first serious evidence about German rockets, he observed that nearly all the secret technology outlined in the report had been subsequently used by the Germans, and that in addition it was implausible that the Germans could have been contemplating a hoax four years ahead. He examined the information supplied by the Danish agent Elgar from a conversation overheard in a Berlin restaurant about rocket testing at Peenemunde and reckoned that the Dane could be trusted because he had supplied drawings and photographs about an unrelated matter that no German agent would have supplied. Whilst accepting that four other reports in the first quarter of 1943 could have been 'plants' he considered that the bugged conversation between General Wilhelm Ritter von Thoma and General Ludwig Crüwell at Trent Park on 22 March 1943 had resulted in information of 'good faith'.

The aerial photography certainly confirmed the existence of a very large experimental station at Peenemunde and, given his access to a far greater range of intelligence sources than the Medmenham interpreters had, Jones observed that the Luftwaffe station there ranked second only to the testing centre at Rechlin near Berlin. He considered the most interesting things to be seen at Peenemunde to be the very large torpedo-shaped objects about 40 feet long, and reckoned it was implausible that the Germans would carry a hoax so far as to incite the Allies to attack the site. In addition, there was evidence from foreign labourers who had been conscripted to work there and who began providing reports from early June 1943. One of them was twenty-year-old Luxembourger Leon Henri Roth, who was exiled there after being caught starting a resistance cell. He successfully got letters through to his father, a member of a Belgian network, describing the rocket testing as sounding like 'a squadron at low altitude'. Unfortunately for the resistance hero, Roth would be killed by American army fire in 1945 while escaping with two Frenchmen in a German military car.

From another Luxembourger at Peenemunde, the bacteriologist Dr Fernand Schwachtgen – later Director of the the Laboratoire Bacteriologique de L'Etat – came a report and sketch map of Peenemunde that was supplied to MI6 in Switzerland, courtesy of the Famille Martin resistance network. The information contained on the very dirty sketch plan was microfilmed and passed to MI6 in London while the station head was away. On his return Count Frederick 'Fanny' Vanden Heuvel – the cosmopolitan aristocratic diplomat of Italian descent who had been brought up in England – chided his colleagues for wasting time and resources on something that was clearly nonsense, only for London to send their congratulations for such useful information and to request that anything further should be sent 'Most Immediate'. In the associated telegram dated 1 July 1943 sent by the Geneva station, the sketch's value is questioned with the words: 'Do not think you can count too much on plan as it is rough pencil sketch not to scale, probably drawn from memory by P of W. Have photographed in two negatives two inches by one inch'. It continues: 'Plan shows railway from Zinnowitz running to north-east tip of island. At railway terminus there is submarine and aerial experimental station. Factory making rocket projectiles, part above and part below ground, located to south. Comparing sketch with atlas one concludes projectile factory is on extremity west of Peenemunde bay . . .' Whilst it might have been only a filthy sketch plan, its reference to a 'pilotless aeroplane' that had a range of 150 miles meant it was not surprising that this information, which correlated so neatly with other human intelligence and aerial photography, caused such interest.

On 29 June 1943 the War Cabinet met in the safety of the Cabinet War Rooms underneath Whitehall to discuss the German rocket threat. There was a strong political presence: it was attended by Prime Minister Winston Churchill who sat at the head of the table flanked by his loyal lieutenant Anthony Eden, the Deputy Prime Minister Clement Attlee, the Home Secretary Herbert Morrison, the Minister of Aircraft Production Stafford Cripps and Duncan Sandys. The military attendees included the three Chiefs of Staff and Churchill's chief military assistant during the war, General Hastings Ismay. The scientists, in addition to Lord Cherwell and Jones, included radar pioneer Sir Robert Watson-Watt and Dr William Cook, the Assistant Controller of Projectile Development at the Ministry of Supply.

In laying out his case, Cherwell stated that he doubted whether the Germans had a sufficiently powerful propellant to power a rocket and argued that the aerial-reconnaissance photography of Peenemunde actually showed torpedoes or wooden dummy rockets that had been deliberately painted white so that they could be easily identified. In his view the rockets were a German hoax, an attempt to divert attention away from what he considered to be the key secret weapon – the pilotless aircraft – that he reckoned would be vulnerable to Allied countermeasures. Churchill was now ready to hear from Jones, who referred to his recent report and emphasised that the amount of evidence pointing towards the rocket as a credible threat was significantly greater than any he had ever had about German radio-navigation systems earlier in the war. At this statement Churchill is reported by Jones to have called out 'Stop!' and, turning to Cherwell, to have said, 'Hear that? That's a weighty point against you! Remember it was you who introduced him to me!'

Despite Lord Cherwell playing devil's advocate, the War Cabinet agreed that the rocket represented a credible threat and that consequently Peenemunde should be attacked. But the great distance to Peenemunde, and the short summer nights, meant that an immediate attack was not realistically possible. While Bomber Command planned its assault and waited for longer nights, the Medmenham Aircraft Section, who were responsible for every aspect of enemy aircraft and related industry and had been asked to watch for 'anything queer' at Peenemunde, would soon add an extra dimension to the hunt for Hitler's V weapons. In April 1942 Constance Babington Smith had been puzzled by the circular embankments at Peenemunde. Now, one year on, she was again puzzled, but on this occasion by the small tailless aeroplanes that were visible in the photos from the sortie flown by Sergeant Ernest Peek on 23 June 1943, in which two 'torpedo-like objects thirty-eight feet long' had also been clearly identified. In the meantime, L Section had been pursuing another vital brief – watching out for new types of aircraft, especially jets. 'Keeping an eye on Peenemunde,' Constance later said, 'was a minor task compared to the everlasting watch for new German aircraft'. When Group Captain Frank Whittle, inventor of the jet engine, paid a visit to Medmenham, he was much impressed by what they had found and by Constance.

What L Section had now discovered was the Messerschmitt Me 163 – the revolutionary German rocket-powered fighter aircraft – which was provisionally named 'Peenemunde 30' since its wingspan had been measured as 30 feet wide. The photographs showed the aircraft and the first 'jet marks' that Constance had ever seen. It meant that with this information L Section could begin the process of going back over previous sorties to investigate whether there was any more evidence to be found. This highlighted one of the most common challenges for photographic interpreters: the difference between when something was 'first seen' and when it had actually been 'first photographed'. With the knowledge that now existed about the Peenemunde site, the sorties of 1942 provided many 'signatures' that the site was being used for rocket development. This was equally the case with the tailless aircraft and meant they could now look through all the photography of Peenemunde and establish, based on the telltale signatures in the photography, when the Peenemunde 30 had been first photographed. After Peenemunde, they extended their search to the manufacturing sites and the experimental and operational airfields – which involved working with their colleagues in C Section.

While L Section was responsible for the interpretation of aircraft and aircraft factories, C Section was home to third-phase interpreters, led by Ursula Powys-Lybbe, who provided detailed interpretation of the construction and development of airfields, landing grounds and seaplane bases in enemy territory. As the Luftwaffe expanded, it became apparent that considerable intelligence could be derived from the observation and analysis of airfield construction and operation by the Germans, and this led to the creation of the section. The serviceability and capacity of the enemy's network of airfields and landing grounds was routinely tabulated, and indications of future enemy plans were often revealed by modifications to or the expansion of existing airfields. To conceal an entire airfield was extremely difficult since constant use of the runway – particularly if it was long – wore off the camouflage. The Luftwaffe nevertheless went to extreme lengths to hide their aircraft, often creating lengthy taxi-tracks that led to suburban-looking housing estates where their planes were hidden. This certainly made attacking the aircraft harder for any Allied aircrew but they almost always remained visible to the interpreters. One of the great advantages of photographic intelligence was comparative cover, and with the systematic build-up of coverage in the Medmenham Print Library over the years the interpreters could quickly study scenes that frequently revealed where nefarious activity had taken place.

To better understand what was happening at Peenemunde and the construction site at Watten in the Pas-de-Calais, Hamshaw Thomas arranged for the construction of particularly detailed scale models. With precise measurements from the aerial photography, courtesy of W Section and the Wild Machines, the model-makers set about bringing the world's first ever ballistic-missile test site to life. By this stage in the war, key models built had included: Bruneval for Operation Biting; St Nazaire for Operation Chariot; and Dieppe for the unsuccessful commando raid on the German-occupied French port that nevertheless helped influence Operation Torch, which required models of the North Africa landing zones.

Since August 1942, the model-makers had occupied the requisitioned Phyllis Court Country Club in Henley-on-Thames, an elegant Georgian mansion beside the river that looked over the finish line of the world-famous Henley Royal Regatta. Without knowing anything about the nature of Peenemunde, Flight Lieutenant Geoffrey Deeley and American model-maker Lieutenant Howard Bahr analysed the aerial photography and data, and spent day after day assembling a model of the Peenemunde site (Model M399) and a large-scale detailed model of the elliptical earthwork and gantries, complete with a large torpedo-shaped object on its cradle (Model M400).

On Sunday, 11 July the model-makers were ready, and the models were transported to Sandys's office in Shell-Mex House. The following Thursday two model-makers were summoned and asked how on earth had they reached the conclusions they had when constructing them. Explaining that the models were built based on known facts and not conjecture, following detailed analysis of aerial photography, accurate scaling of features and objects, and lengthy discussions between Hamshaw Thomas and André Kenny, the model-makers' answers were accepted. The models quickly became the focus of attention – doubtless to the annoyance of Cherwell and the scientific doubters – and helped to clarify the scale, purpose and seriousness of the activities being carried out by Hitler's rocket scientists. In recognition of the important part that these models played in both the planning of the attack on Peenemunde and the political and military decision-making that surrounded it, the Imperial War Museum ensured their permanent preservation.

The first meeting of the 'Bodyline Scientific Co-Ordinating Committee' held on Monday, 26 July 1943 in Shell-Mex House was attended by Hamshaw Thomas and André Kenny, along with R. V. Jones, other senior scientists and representatives of the Army and RAF. Under the chairmanship of Duncan Sandys, all members of the committee were asked to examine the aerial photography and models of Peenemunde as soon as possible, and to put forward any suggestions they might have about what was happening at the experimental station. While Dr Alwyn Crow was tasked with producing a 'scientific explanation' of how the Peenemunde site worked, Major General Clarke was asked to consider the possibility that the Germans might be developing a long-range gun, and had to consider what form its emplacement might take. To access the aerial photography and models, the committee members were advised that they should make an appointment with Dr H. J. Phelps from the Ministry of Economic Warfare, who would provide an explanation of the evidence held under lock and key in Room 185, Shell-Mex House.

Days after the Peenemunde models were completed, and while work continued on making models of the Normandy landing zones, a scandalous act by one of the model-makers ensured that V Section would be hastily relocated back to Medmenham, into the labyrinth of Nissen huts in the grounds, in September 1943. By this stage in the war, a Model-Makers Detachment of the US Army Corps of Engineers – who trained at RAF Nuneham Park – worked alongside the British and were a familiar sight in the affluent south Oxfordshire town. Given the top-secret nature of their work, the American model-makers took it in turns to stand guard in the entrance hall of Phyllis Court, armed with a .38 Smith and Wesson revolver and twelve rounds of ammunition. Private John Waters created great alarm on Wednesday, 14 July 1943 when he deserted his guard post, particularly when Phyllis Court received a phone call from Henley police station reporting that 'one of your Yanks just walked into a shop in town and shot and killed his girlfriend' – and that he was now refusing to surrender.

A thirty-eight-year-old married man from Perth Amboy, New Jersey, Waters had made the acquaintance of thirty-five-year-old local woman Miss Doris Staples, who worked in a draper's shop at 11a Greys Road, half a mile from Phyllis Court. Since February 1943 they had been having a tempestuous relationship but when her interest waned, and her attentions strayed, Waters became intensely jealous of rumours that she was seeing other Americans. This culminated in his shooting Doris five times. In the siege that followed – during which the local fire brigade attempted to flush him out with their hoses – Waters fired two shots, one of which shattered the window of a shop opposite. With the US Military Police also in attendance, Superintendent Hudson authorised the use of tear gas, and as the police entered the shop wearing respirators they found Doris's corpse on the shop floor and

Waters in the outside toilet leaning against the wall, having made a botched suicide attempt. Waters was taken to hospital for treatment and despite the bullet having shattered his jaw, mouth and palate before coming to rest just in front of his brain he recovered sufficiently, despite the partial loss of sight in one eye, to stand trial for murder.

An Act of Parliament decreed that, subject to certain exceptions, no member of the American military or naval forces would be prosecuted in front of a British court. It was for this reason that a general court-martial was convened in Watford on 29 November 1943, when Waters was charged with murder, with leaving his guard post and with making himself unavailable for military duty by wilfully maiming himself. Sentenced to death, Waters was returned to Shepton Mallet prison in Somerset, which served as the American military prison in Britain during the war. When a petition for clemency – which included model-makers' signatures – was rejected he was hanged on 10 February 1944 by Thomas Pierrepoint. Thomas's nephew, Albert Pierrepoint, would go on to execute other American soliders at Shepton Mallet and over 200 Nazi war criminals, following post-war trials in the British Zones of Occupation in Germany and Austria. These would include Josef Kramer, the Commandant of Bergen-Belsen concentration camp, and William Joyce, who used his Lord Haw-Haw alias and *Germany Calling* radio-propaganda broadcasts to unsettling effect during Germany's deployment of V weapons against Britain.

Inspired by the American felons of Shepton Mallet prison and the 'Filthy Thirteen' of the 506th Airborne Infantry Regiment, who demolished targets behind enemy lines, the American author E. M. Nathanson was inspired to write the bestselling novel *The Dirty Dozen*. When later adapted for the 1967 film of the same name – which starred Second World War veterans Ernest Borgnine, Charles Bronson, Lee Marvin and Telly Savalas – filming took place at Beechwood Park. In the Hertfordshire countryside, to the north of Hemel Hempstead new town, the manor house and estate had been bought in 1961 by Group Captain Peter Stewart OBE. With his customary zeal and enthusiasm, Stewart had renovated the dilapidated property and by 1964 had established Beechwood Park School, an act of philanthropy that ensured Old Beechwoodians would recognise him as the school's true 'onlie begetter'.

*

Following such a succession of discoveries at Peenemunde, on 3 July 1943 Churchill had directed that the 'maximum possible contribution to the investigation was to be made by the CIU, who were to be given any necessary facilities and manpower required to meet the commitments'. By 26 July Sandys summarised the current state of the investigation, and observed that since the latter part of April 1943, the accumulation of intelligence reports indicated the development of very long-range rocket projectiles at Peenemunde:

- Frequent photographic reconnaissance of Peenemunde has tended to confirm this condition;
- An object having the appearance of a long-range rocket 38 feet long with three fins has been seen in the open;
- Intelligence indicates range in the order of 130 miles;
- Intelligence indicates propulsion by some new fuel of very high calorific value;
- Having regard the size of the projectile, it should be taken as certain that the projector sites will be rail served. Certain unexplained installations, rail served, have been observed in north France, notably at Wissant, Marquise and Watten. Intelligence indicates that the rocket is directed by radio.

The same day the German Long-Range Rocket Detection investigation became known by the code name 'Bodyline', the reference being to the cricketing tactic considered by many to be intimidatory and physically threatening to the point of being unfair in a game – like warfare – that was once supposed to have gentlemanly traditions.

On 7 July 1943 Arthur Harris had held a meeting at the headquarters of Bomber Command to consider the best method of attacking Peenemunde. In his later book *Bomber Offensive*, he records that the only successful attacks on single factories up to this point in the war had been made by small forces of exceptionally experienced crews during the day, or at night when there had been an unusually good chance of identifying the target. Destroying such a large and challenging target as Peenemunde would mean flying at night. Since there could be no question of relying *completely* on the H2S radar system, and with the area concerned being beyond the range of the Oboe radar set-up, a large bomber force and a full moon would be required. Peenemunde was such a large experimental site, with buildings scattered in a narrow strip on the peninsula in a remote part of the Baltic, that there was a great risk of bombs being wasted. The challenge was further complicated by the fact that Peenemunde was known to have smoke generators that could effectively mask the site with a giant smokescreen devised specifically to shield the site from prying eyes.

Ground Intelligence and Target Intelligence at Medmenham had been run as separate sections until September 1942 when Squadron Leader David Linton took command of a combined Intelligence Section. After stints as a first-phase interpreter at Wembley, St Eval, Wick and Leuchars, the former Edinburgh University geographer was posted to Medmenham. As an outsider it was hoped he could amalgamate the two sections, which had grown organically since their creation at Wembley, and provide direction to the different personalities working in them. Although reconnaissance pilots would visit Medmenham on occasion, Linton came to miss the daily contact with them and the camaraderie of having them around. He was now responsible for the collation of the information contained within the aerial photography, obtained at great risk by the reconnaissance pilots, that ensured targets could be attacked in the most effective way and gave aircrew the best chance of surviving. The section prepared detailed Target Folders for every enemy objective selected by the Air Ministry, which had standing committees deciding on the attack priority for enemy-controlled oil, industry and transportation facilities as well as other targets – such as Peenemunde.

Intelligence from Medmenham was supplied to the Air Ministry Target Mapping Centre (AI3c) at Hughenden Manor – code-named 'Hillside' – a red-brick Victorian mansion located to the north of High Wycombe, Buckinghamshire, and the former home of Benjamin Disraeli. It was here that Air Ministry cartographers drew their detailed target maps, essential guides for the Allied bomber crews in their raids over Europe. It meant that in 1943 they were conveniently located two miles south of the Bomber Command headquarters at Naphill, less than ten miles from Medmenham and only three miles from the American Eighth Air Force headquarters (code-named Pinetree) at High Wycombe. The target maps were created from information provided by Medmenham, including annotated aerial photographs and associated interpretation reports. On arrival at Hughenden, the information was checked for accuracy before being passed to the cartographers who drew target maps using three colours that could be more easily seen by moonlight: purple for built-up areas and roads, black for railways and magenta for the actual targets. Rivers were shown in white and targets were drawn in the centre of concentric rings, which helped ensure better navigation to them. Each draft map went to the Checking Office before a definitive copy was produced, from which a photo-plate copy was made and hundreds of copies printed.

By mid-1943 Bomber Command had been transformed from the force much maligned in the 1941 Butt Report initiated by Cherwell and written by the economist David Bensusan-Butt. The Area Bombing Directive of 14 February 1942 followed, authorising Bomber Command to use area-bombardment methods, a policy strengthened a month later by a memorandum from Lord Cherwell that became known as the 'dehousing' paper. By the summer of 1943, Bomber Command had passed a number of mileposts including the first thousand-bomber raid on Cologne on the night of 30/31 May 1942, and was just coming to the end of a five-month-long campaign of strategic bombing against the Ruhr area. It had a force made up in the main of twin-engined Vickers Wellington medium bombers and the four-engined heavies, the Short Stirling, Handley Page Halifax and Avro Lancaster. Bomber Command had the type of heavy and long-range capability that the Germans simply did not have – a fact which must have given German hopes for the V-weapon programme an element of urgency, if not desperation. British bomber crews had greatly improved navigational aids, had shown astonishing ingenuity in the Dambusters raid and were about to embark on Operation Gomorrah, the Battle of Hamburg.

W Section at Medmenham had been closely involved in the planning of the Dambusters raid, when the specially formed 617 Squadron flew nineteen modified Lancaster bombers on the mission against the German dams. In the basement of Danesfield House the Wild Machines had been used to calculate the volume and depth of water in the dams, information that was to prove as essential to Barnes Wallis as the height and angle at which the bombs would need to be dropped, and the speed the aircraft would need to travel at in order for the bombs to bounce across the water to the target and, on

Soon after the outbreak of hostilities, the innocuously named No. 2 Camouflage Unit at Heston, under honorary Wing Commander Sidney Cotton set about modifying a Supermarine Spitfire. Flying solo with cameras in place of armaments, and with a seemingly bizarre colour scheme that reduced its visibility against the sky, Robert 'Bob' Niven and Maurice 'Shorty' Longbottom relied on the fighter's speed, climb, and ceiling.

Group Captain Frederick Winterbotham – Head of Air Intelligence – was responsible for the organisation, distribution and security of Ultra intelligence throughout the war.

Involved in all manner of combat and test flying, Longbottom would work with Barnes Wallis on the first 'live drop' of the bouncing bomb, before losing his life in January 1945, aged twenty-nine, during a test flight.

Right: A colourful character with a love of aviation, money and women, Cotton masterminded aerial espionage during the Gathering Storm before developing RAF Photographic Reconnaissance (PR) into a war-winning capability.

Below: Back from one of the early photographic reconnaissance sorties, pilot Gordon Hughes, later awarded the Distinguished Service Order, is debriefed by Intelligence Officer Edward Hornby.

Above: Robert 'Bob' Niven became co-pilot and navigator for Sidney Cotton. The Canadian Airman is standing beside G-AFKR, the luxurious Lockheed-12A 'Electra Junior'. When later flying on a strike mission, Wing Commander Niven DFC was fatally hit by flak. His body was never recovered.

Right: Major Harold 'Lemnos' Hemming (left), managing-director of the Aircraft Operating Company, immediately recognisable with his trademark black eyepatch. Francis Wills (right) founding-father and first managing-director of Aerofilms Limited, the first British company to specialise in aerial photography

Left: Twenty-six year old Cambridge don Dr Glyn Daniel was posted to Wembley in September 1940 for PI training. Transferring to India in 1942, he was responsible for establishing the PI Unit that supported the military campaign in the Far East.

Above: The colossal Swiss-built Wild – pronounced Vilt – Machine. This Autograph A5 model (Instrument No. 50) was purchased by the Wembley-based Aircraft Operating Company in 1938 for the mass production of maps. Its powerful optics allowed the precise measurement of three-dimensional data contained within aerial photography.

Left: Arriving at the Photographic Development Unit, Wembley, in the summer of 1940 Peter Riddell developed and formalised the three-phase approach to photographic interpretation. He recognised that additional phases of interpretation would provide data that would have long-term intelligence value.

The partnership of Squadron Leader Claude Wavell, head of G Section at Medmenham, and the Scientific Intelligence Unit headed by Dr R. V. Jones played a major part in unearthing the German's offensive and defensive radio systems.

Wing Commander Douglas Kendall, Technical Control Officer at Medmemham, was responsible for the operational direction of the Unit. The only Ultra-cleared interpreter, from mid-1943 he co-ordinated the investigation into the V Weapons.

Michael Spender operated the Wild Machine at Aerofilms and would pioneer first-phase photographic interpretation during the war. A few days before VE Day, whilst in an aircraft flown by Gordon Hughes, Spender was fatally wounded during a crash landing.

Returning to Heston Aerodrome on 30 September 1938, Neville Chamberlain addressed the gathered crowds, holding aloft the Munich Agreement signed by the German Chancellor, Adolf Hitler. Whilst the British Prime Minister placed his hope in a policy of appeasement, MI6 planned aerial espionage from the same west London aerodrome.

The impact from two high-explosive bombs dropped by the German Luftwaffe on 17 October 1940 rendered the main building structurally unsafe and destroyed the drawing office. With photographic interpretation increasingly recognised as a war-winning intelligence capability, there was now added impetus on the Air Ministry to secure alternative accommodation.

660   N/183. I.PRU. 21·5·41. F/20" P ↑

It was from Wick, on 21 May 1941, that one of the most famous early photographic reconnaissance sorties – Sortie N/183 – was flown. Flying over the Dobric Fjord, Norway, twenty-year-old Pilot Officer Michael 'Babe' Suckling (*above right*) spotted and photographed the battleship Bismarck. Six days later the German battleship would be destroyed.

Assistant Section Officer Ursula Powys-Lybbe, had a life-long love of all things photographic, and served in the WAAF at Medmenham where she ran the Airfields Section.

Flight Lieutenant Villiers David was the Medmenham go-between with MI6 and maintained a vast card-index-system to manage intelligence sourced from secret agents and POW interrogations.

The purpose-built Beresford Avenue home of the Aircraft Operating Company – and its subsidiary company Aerofilms Limited – off the North Circular road in north-west London. Once the vital importance of the company to the war effort was recognised, sandbags were hastily filled in an effort to protect the newly established Photographic Development Unit (PDU).

On 7 January 1941, the Wembley-based Photographic Interpretation Unit became the Central Interpretation Unit (CIU) and the following April Fools Day moved to Danesfield House, near the Buckinghamshire village of Medmenham.

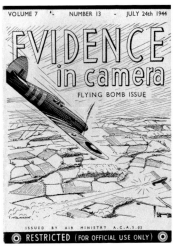

Special issue of the in-house Medmenham Magazine, which chronicled the war as it happened.

The Palladian villa of Nuneham House in south Oxfordshire – with landscaped grounds by Capability Brown – was transformed into RAF Nuneham Park during the war. A satellite station of RAF Medmenham, it would accommodate the PI school and a second of the colossal Wild Autograph A5 Machines. In a plot worthy of a James Bond novel, the machine was secretly smuggled into the country from the factory in Switzerland, and ensured that vital interpretation would continue even if RAF Medmenham was attacked.

The labyrinth of Nissen huts constructed by the Air Ministry Directorate of Works in the grounds of Danesfield House, highlight the vast scale of this wartime intelligence unit.

The Würzburg radar installation at Bruneval near Cap d'Antifer on the northern French coast. Spectacularly captured during Operation Biting on the night of 27/28 February 1942. When preparing for their mission the commandos of British Combined Operations carefully studied a scale-model created by V Section at Medmenham.

Above and right: At RAF Medmenham and RAF Nuneham Park the Photographic Sections were equipped with cutting-edge automatic printing and developing facilities. With an average of 250 photographers employed in the two sections, over 1 million prints were produced per month. The responsibility of allotting priority fell on Progress Section, whose tally board gave an immediate indication of workload.

contact with the dam, drop below the surface and explode. From aerial photography of the Möhne, Eder and Sorpe dams in the Ruhr, Target Folders were created for Wallis and Bomber Command – detailing where power lines, natural features and anti-aircraft guns were – and, along with detailed models from V Section, meant that Medmenham played a key part in planning the operation.

On the morning of 17 May 1943 a photographic-reconnaissance mission was flown from RAF Benson and provided evidence of the successful outcome of Operation Chastise. Led by twenty-four-year-old Wing Commander Guy Gibson, 617 Squadron had succeeded in making a breach about 240 feet wide in the Möhne Dam, 180 feet wide in the Eder Dam, and damage to a stretch of about 200 feet on the crown of the Sorpe Dam. Aerial photographs showing the breached Möhne and Sorpe Dams and the destruction brought about by the flooding downstream in the Ruhr Valley were quickly distributed by the Medmenham Press Section. The King, Prime Minister and Air Staff were quick to congratulate those responsible and leaflets were shortly dropped in occupied Europe showing before-and-after aerial photographs of the sites, along with an explanation of the operation. Whatever the debates about the impact and significance of the raids, the technical wizardry displayed and the propaganda value of breaching dams in the industrial heartland of Germany were significant.

In addition to the oblique aerial photography carried out by the No. 1 Camouflage Unit, aerial photography was also conducted over Britain during training missions and for bomb-damage assessment. This led to the creation of M Section at Medmenham. At Wembley in late 1940, after the Coventry and Southampton Blitzes, Hamshaw Thomas was asked by Air Marshal Sir Philip Joubert de la Ferté – the soon-to-be Commander-in-Chief of Coastal Command who was then on special duties investigating the potential of radio navigation – to write a report comparing the relative effectiveness of German and British bombs. The two men had known each other since the First World War, and de la Ferté had re-established contact with Hamshaw Thomas earlier in the year to ask his expert opinion on the quality of photography being created by Sidney Cotton. On this occasion Hamshaw Thomas used aerial photographs taken after the devastating German raid of 14 November 1940 on Coventry, code-named Operation Moonlight

Sonata by the Germans. This attack had prompted Hitler to invent a new verb, *coventriren* – to coventrate, meaning to destroy. Little was known at the time about the effects of blast, and the fact that British bombs created holes in the roofs of buildings that they hit but no other damage – compared with the relative mass destruction of Coventry – was apparent in the aerial photography. This failed to dismay Bomber Command, who unconvincingly explained that all the damage was hidden from view inside the buildings. Hamshaw Thomas's report 'sowed the seeds of doubt' about British bombing strategy.

The need to develop more effective bombing techniques became the responsibility of Section 8 at the Research and Experiment Department (RE8), which had originally been created by the Ministry of Home Security to work on the problems of firefighting during the Blitz. Based at the Ministry's research unit in Princes Risborough, Buckinghamshire, RE8 used statistical analysis and mathematical modelling to develop a new bombing strategy. This required the recruitment of a group of outstanding academics including the Cambridge statistician and geneticist Professor Ronald Fisher, the chemist Professor George Finch of Imperial College London and Dr Jacob Bronowski, the brilliant mathematician, biologist and historian of science. At the end of the war the Polish-Jewish polymath Bronowski was part of the British team who visited Hiroshima and Nagasaki to document the effectiveness of the atomic bombings. This experience that raised moral issues about science and warfare was something he later addressed in his 1973 television series *The Ascent of Man*. For one programme this involved visiting Auschwitz – where members of his own family had been murdered, and where he movingly analysed the importance of scientific responsibility in bringing about the defeat of Nazi Germany.

In 1943 and despite a groundless MI5 warning that anything secret Bronowski saw would go straight to the 'Communist Party Headquarters' he was appointed to RE8 and left University College Hull for Buckinghamshire. The work of RE8 was dominated by the debate about the relative destructive advantages of incendiary and high-explosive bombs. The chief protagonists were Fisher who favoured the theory that incendiary bombs should be dropped in concentrations, and Finch whose theory was that incendiaries distributed in a pattern tended to create fires that coalesced. To help answer their theoretical questions they turned to Medmenham. Throughout the war, various approaches to the systematic bombing of targets were developed and throughout periods of night-time raids were assessed by the specialist third-phase interpreters in Night Section. As the pioneer of interpreting night photography in Bomber Command and later as officer in charge of Medmenham N Section, Bernard Babington Smith – the older brother of Constance – had witnessed the evolution of the bomber war through the photography taken by the open-shutter cameras mounted in every bomber. A senior lecturer in experimental psychology and fellow of Pembroke College, Oxford, Bernard was a statistician who grappled with risk, data and probability. His understanding of random phenomena allowed him to analyse patterns of bombing and the destructive impacts of attacks throughout the war.

Bernard believed in learning through action, and with Bomber Command flying mostly at night he faced the seemingly impossible challenge of developing techniques to interpret photographs taken at night. Experimentation led to the use of photographic flash-bombs that, dropped from the same aircraft that carried the camera, were synchronised to detonate and illuminate the area where the high-explosive bombs landed. So much light was produced by the flash-bomb that it was possible to recognise ground features, and the impact points of bombs could be plotted on a map. Bernard Babington Smith discovered that, in addition to flash photography, a lot of information could be obtained from open-shutter cameras. They were subsequently carried by all the bombers. In an open-shutter camera, the film would slowly advance through the camera and be exposed to light sources below to create an image that was often likened to spaghetti. Learning the language of night photography was considered something of a black art among most of the interpreters, but since every light source could be identifiable it was possible to accurately plot the intensity of flak, the spacing and method of operation of spotlights, the position of anti-aircraft batteries and the spread and efficiency of fire as a destructive force. This information helped to illustrate how a particular 'cocktail' of incendiary and high-explosive bombs performed and allowed for the better planning of future attacks.

The early inaccuracy of British bombing distressed many of the interpreters, who knew how poor the bombers' navigation systems were and what little impact their bombs were having, all of which helped fuel a sense of futility that so many airmen were losing their lives needlessly. With flak forcing them to fly at up to 15,000 feet, and their training having been at only 8,000 feet, their view of the landscape below through their bombsights was minute in scale. The interpreters could sympathise with the pilots' reactions to the photographic evidence, which ranged from the disbelief of those who thought they knew where they had been and who simply could not accept the photographic evidence that they hadn't been, to those who were more accepting and interested but nevertheless shaken and frustrated and intent on doing better next time. They could hardly be expected to do better until the development of better navigational aids, something that was outside their control, but photographic intelligence had at least allowed for the systematic and unemotional assessment of the facts.

The development by Bomber Command of the Pathfinder technique, in particular the marking of targets by special 'target indicators', had immediate repercussions for N Section. Since aircraft using H2S or Oboe radar systems could identify a target that was within range, even if it was covered by cloud or a smokescreen, the Pathfinder Squadrons dropped target-indicator flares suspended below parachutes in order that the bomb aimers should have something to aim at. If the target was clear there would be no requirement for parachute flares to illuminate the objective but since target indicators were either red, yellow or green – with each colour used to identify primary and secondary targets for the waves of bombers that followed – there was a need for colour film to identify these flares in order to assess the accuracy and effects of the bombing. Experiments were carried out with Dufaycolor – in which Sidney Cotton still had a commercial interest – before Kodachrome and later Kodacolor film was used.

In time, 80 per cent of the films used were colour and the information they contained, alongside an increasing quantity of cine film, allowed N Section to piece together the destruction caused sequentially during a raid by using the information from the films created in all the bombers over the target, from the first to the last. An illustration of the dramatic growth of the bomber war is given by the average monthly intake of night films, which went from one thousand in 1942 to four thousand in 1943, and at its peak in 1944 and 1945 this grew to ten thousand per month. The total number of night films supplied to N Section by Bomber Command amounted to 211,000, while another 56,000 films which were also interpreted by the section were added from bombing operations during daylight hours.

In June 1943 the Target Section at Medmenham compiled a Target Folder on the 'experimental establishment at Peenemunde, situated on a tongue of land on the Baltic Coast, about sixty miles north-west of Stettin'. Target 3/AIR/389 was broken down into seven distinct areas: Experimental Station, Factory Workshops, Power Plant, Unidentified Apparatus, Experimental Establishment, Sleeping and Living Quarters and the Experimental Airfield. Protectively marked as SECRET and NOT TO BE TAKEN INTO THE AIR, the folder contained a detailed site plan that identified the location of flak guns and searchlights and noted the absence of barrage balloons. There were also annotated aerial photographs showing each of the seven areas, along with the Target Information Sheet. Careful not to explain the true purpose of the raid to the bomber crews, the sheet mentioned that Peenemunde was serving a number of experimental purposes, and its importance at the present time concerned its being used for the production of high-concentration hydrogen peroxide. The target area measured some 8,000 × 2,750 yards and was shaped roughly like a pear. Detailed descriptions for each of the seven areas highlighted specific landmarks, including the large circular earthworks, particular buildings, and the location of residential quarters, including the horseshoe-shaped Karlshagen holiday camp.

With detailed target information from Medmenham, target mapping from Hughenden Manor, and details of specific targets contained within Bomber Command Operation Order 176, the night of 17/18 August 1943 was chosen for Operation Hydra, the attack on Peenemunde. The true nature of the target was never revealed to the bomber crews, but the importance of destroying it was emphasised by the words: 'If the attack fails . . . it will be repeated the next night and on ensuing nights regardless, within practicable limits, of casualties.' Following shortly after the Battle of Hamburg – which began on 24 July 1943 and became the heaviest aerial assault in the history of warfare to date – Operation Hydra benefited from Bomber Command's successful deployment of 'Window' during Operation Gomorrah.

On examining the Würzburg radar equipment brought back to Britain from Bruneval, British Scientific Intelligence had discovered that German radar was vulnerable to jamming, and this was to prove crucial to the success of Operation Hydra. It had been discovered during Operation Gomorrah – when the large industrial city of Hamburg experienced one of the largest firestorms of the war – that when several hundred metal strips – known as Window – equal in length to half the wavelength used by the German radar stations, were released from an aeroplane, they reflected as much energy as a bomber.

The night chosen to attack Peenemunde was the earliest occasion when there was bright moonlight, reasonable weather and sufficient hours of darkness to provide the 596 bombers deployed – 324 Lancasters, 218 Halifaxes, 54 Stirlings – with maximum protection. To improve their chances of success, and while the heavy bombers were flying over the comparative safety of Denmark and the Baltic, Operation Whitebait involved eight Mosquito fighter-bombers from 139 Squadron crossing the German coast at Rostock, from where they headed south towards Berlin. Being only eighty miles from Peenemunde, they managed to sound the alarm there but continued on their way towards Berlin. With the first Mosquito arriving at 11 p.m. local time, the plan was for them to fly at different heights around the city and release their modest payload of three 500-pound bombs, target indicators and Window. Since indicator flares were normally the forerunner of a heavy raid and Window gave the Germans the impression that the main bomber force had arrived, the Germans reacted as hoped. Whilst one of the Mosquitos was lost and two men were killed, in little over thirty minutes 139 Squadron had managed to focus the attention of the German night fighters on Berlin, and while they waited for the main bomber force to arrive searchlights swept the sky and anti-aircraft guns blazed away noisily.

As the first bombers came within sight of Peenemunde, they could see in the bright moonlit sky that the Germans had activated their smokescreen but their searchlights and flak could not be seen. Flying at only 8,000 feet to ensure better accuracy, the Pathfinders established their position by H2S radar and released their coloured target indicators, which were soon followed by high-explosive and incendiary bombs. There were three aiming points – the scientists' living quarters, the experimental station and the rocket factory – that were monitored throughout the attack by Master Bomber Group Captain J. H. Searby, the Commanding Officer of 83 Squadron. With Peenemunde under attack the local anti-aircraft guns opened up and since the raid had taken longer than anticipated the last wave of bombers was attacked by night fighters who had seen the flames from over Berlin, more than a hundred miles south, and had made for the target area. That night nearly 1,800 tons of bombs were dropped and, whilst forty bombers were lost at a cost of 215 lives, the casualty rate could easily have been so much greater if the Germans had not fallen for the diversionary tactics.

On 18 August 1943, and while N Section began the task of evaluating the success of the mission using 450 photographs taken on the night, Squadron Leader Gordon Hughes flew sortie N/902 from RAF Leuchars in a PR Mosquito, and at 10 a.m., from 29,000 feet and in daylight, took the post-strike reconnaissance photographs which recorded the impact of the attack. For the most part the photographs were good and meant that, along with the results of a subsequent sortie N/904 flown on 20 August, André Kenny and the interpreters in D Section could issue Interpretation Report DS34 on 4 September 1943. With damage plotted building by building on an attached plan, the photographs recorded no damage to the elliptical earthworks but extensive damage to buildings throughout Peenemunde, including the design offices, scientists' quarters and the Peememunde South Labour Camp. The tragic consequence of a target indicator being dropped accidentally on what Churchill later described as a KdF 'Strength Through Joy' establishment – (*Kraft durch Freude*) – meant the very prisoners of war and foreign labourers who supplied the British with vital intelligence on the installation would tragically suffer from the attack.

Of the 12,000 people living at Peenemunde, 732 were killed. Since one of the key objectives of the raid had been to kill as many of the scientists and engineers as possible, the fact that only 120 were killed – and of that number the only notable death was that of Dr Walter Thiel, the head of engine development – meant it had not been as successful as hoped. An extra casualty of the raid was Generaloberst Hans Jeschonnek, Chief of the General Staff of Nazi Germany's Luftwaffe and a devotee of Hitler. Jeschonnek committed suicide on the morning of 18 August after hearing about the extent of the damage at Peenemunde. As his suicide note revealed, 'I can no longer work together with the Reichsmarschall. Long live the Führer'. He was a broken man after bitter battles with Hermann Goering, and Operation Hydra was to prove the final straw. Estimates of the extent to which Bomber Command slowed Germany's rocket programme vary from four weeks to six months, and while Peenemunde would continue to be closely monitored the interpreters were mindful that new facilities would probably be set up elsewhere. The bombing raid signalled the fact that the hunt for Hitler's V weapons had only just begun.

The intelligence summary issued by the Bodyline Committee at the Ministry of Supply on 26 July 1943 concluded that rocket-firing sites would have to be rail served and required the interpreters to study all new railway constructions in northern France, particularly new sites within a 130-mile radius of London. The area of France north of the Seine and within 130 miles of London would ultimately be photographed five times throughout the investigation, each time requiring around 140 sorties. By October 1943 the area under investigation had increased to a radius of 150 miles from the British capital, to include Normandy and the Cherbourg peninsula where it was feared that installations might have been targeted against areas vital to the Normandy landings. These could have included Portsmouth, Southampton and other obvious targets, including Bristol. This involved photographing an area of approximately 7,000 square miles and the detailed interpretation of thousands upon thousands of images. But the first suspicious rail-served site had already been spotted by the Army Section in the Pas-de-Calais region on 17 May 1943, when a large clearing in the Forêt d'Eperlecques a few miles from the village of Watten was spotted and identified as 'a large rail- and canal-served clearing in the woods'. Many such sites were originally reported in the course of routine interpretation work carried out by the Army interpreters who were closely watching northern France in their preparations for Operation Overlord. The secret-weapon sites were recorded by B Section in the BS series of reports.

On 6 July 1943 a ground report was received at Medmenham from secret agent Michel Hollard, a French wartime Resistance operator who founded the espionage group the Réseau AGIR. Hollard's report indicated possible rocket activity at the same geographical pinpoint near Watten. Further photo-reconnaissance sorties were flown and resulted in Interpretation Report BS4 – dated 15 July 1943 – which described the construction work that was taking place. Increased activity was shown, with a new railway line leading to 'heavy constructions' which had developed considerably from the formless excavations shown in the aerial photography of 17 May 1943, into a site with scaffolding and wooden shuttering for the pouring of concrete. This ensured that from this point photographic-reconnaissance sorties were flown from Benson on a weekly basis to monitor closely the development of the site at Watten. These missions led to the discovery of all the enormous reinforced concrete structures in the Pas-de-Calais and Cherbourg peninsula of northern France that were suspected of being linked to the German rocket programme and that became known to the interpreters as the 'Heavy Sites'.

Planning for the construction of launch bunkers had begun at Peenemunde in the autumn of 1942, when Reich officers and engineers recced possible sites in northern France over Christmas 1942 and chose locations including Watten. Under the code name Kraftwerk Nord West – 'North-Western Power Plant' – construction work at Watten was begun by Organisation Todt, the Third Reich civil and military engineering group that was responsible for building large parts of Hitler's Atlantic Wall using 6,000 labourers from the Nazi-occupied countries. By 17 August 1943, when Bomber Command attacked Peenemunde, the main outline of the Watten project could be seen from aerial photography and this resulted in a detailed plan being issued with 'Interpretation Report DS23', which had been written by WAAF Section Officer Winifred Bartindale of Medmenham's Industry Section. Although the true purpose of the site was unknown, the enormous construction effort in a forest clearing was suspected of having 'nefarious intent' and of being connected with the German secret-weapons programme.

Advice was sought from Sir (Thomas) Malcolm McAlpine, the leading construction expert of Sir Robert McAlpine Limited, the civil engineering and building company established by his father, 'Concrete Bob'. (McAlpine chaired the Contractors Committee on the project to construct the Mullberry Harbours that would be towed across the Channel and sunk into position to form the three temporary 'ports' used to supply the Allied invasion force.) He recommended that any attack on Watten should be carried out once concrete had been poured but not fully hardened. On 21 August the interpreters reported that more wooden shutters were in position, ready for the concrete to be poured. Watten was attacked six days later by over 180 Flying Fortresses, in the first assault by the US Eighth Air Force on a rocket site. The Medmenham damage-assessment report written by K Section that followed the attack described how

> the site and buildings in course of construction within the excavations have been heavily hit, particularly on the N. side. The general mass of construction on that side appears to have been destroyed for nearly half the depth it was seen on 19.8.43, and it is possible that several bombs have penetrated nearly to the foundations. The deep excavation at the W. end has suffered a considerable collapse of the S.W. wall from a direct hit.

When copies of the damage-assessment photographs were examined by Malcolm McAlpine they showed a jumble of steel reinforcing girders and cement which had set into a rigid mass. McAlpine estimated this would set the Germans back about three months and that in effect it would be easier for them to begin construction again from scratch. In the weeks that followed the raid photographic reconnaissance continued and, whilst additional anti-aircraft batteries were spotted, construction work was observed to be at a standstill. By October the decision was made to reduce photographic reconnaissance to one sortie a month: by early November, new construction work was spotted and recorded in Report BS77, dated 5 November 1943, and reconfirmed in Report BS209, dated 27 December 1943. McAlpine's hunch was correct: the Germans had cleared up the site and begun building south of it.

Through September to November 1943, the remaining Heavy Sites were identified and a careful watch was kept on a number of suspected areas. Watten had been first reported in May 1943, then in September came Marquise-Mimoyecques and Siracourt in the Pas-de-Calais, Martinvast and Sottevast in Normandy and in October Lottinghem and Wizernes, also in the Pas-de Calais. Just as the interpreters had been perplexed by the activities at Peenemunde, now they faced the challenge of identifying the network of manufacturing plants that the Germans must have, understanding how the rockets were transported, where they were stored and assembled, and the location of any more Heavy Sites.

At the beginning of November 1943 a message was received from a French secret agent that a French construction firm had been contracted by the Germans to build eight unusual structures in northern France. This so-called 'ground intelligence' was vital to the interpreters, but on its own one source of information was never enough. For the interpreters it was essential to have access to as many sources of intelligence as possible so that facts could be checked and double-checked. Intelligence was rated with the letters A to E according to the perceived reliability of the source, whilst the numbers 1 to 5 rated the probability of the information itself being true. The amount of A1 information available during the war was never sufficient and meant that photographic intelligence became prized on the basis that photographic reconnaissance was reliable. Photographic intelligence was regarded as the perfect means by which to corroborate intelligence gathered from other sources and could be used as a means of determining the reliability of information received from an agent and, in turn, of evaluating how trustworthy and valuable the agent was. Even though there was scope for PIs to misinterpret what they saw through their stereoscopes, their training instilled in them the importance of only reporting known facts, which meant the reliability of their information was high.

Intelligence from 'All Sources' was to prove vital in the hunt for the secret weapons. In March 1943, after von Thoma and Crüwell were secretly recorded at the Trent Park Military prison, a steady stream of German POWs – mainly Luftwaffe aircrews – routinely supplied information, often unwittingly through bugging and interrogations. A common interrogation technique was to ensure that several prisoners from the same unit who had been captured separately were kept apart. Information gleaned from the interrogation of Prisoner A would then be revealed to Prisoner B ,who would often come to the conclusion that since the British seemed to know everything already there was little point in holding back. The prisoners would often gossip happily about new types of aircraft being developed by the Luftwaffe and would speculate about when the planes concerned would be introduced. Whilst much of this information had to be treated with caution, it did help enhance the general intelligence picture. The division of the

British Directorate of Military Intelligence responsible for obtaining information from enemy prisoners of war was MI19, who developed Combined Services Detailed Interrogation Centres at Latimer House and Wilton Park in Buckinghamshire and Trent Park in North London. Every three to six months, Douglas Kendall visited the Palladian mansion at Wilton Park, ten miles to the east of Medmenham in the market town of Beaconsfield, where senior German officers were secretly eavesdropped on and their conversations recorded on wax discs for later transcription. Kendall ensured the interchange of information between interpreters and interrogators to the benefit of the quality of the intelligence supplied by both.

Personnel in the Ground Intelligence section were responsible for the compilation of a vast card-index system, gazeteers, and writing the Medmenham Operations Record Book (ORB). Also known as RAF Form 540, the ORB is written to provide a complete historical account of an RAF unit and requires a report of each operation carried out to be kept for the historical record and for the benefit of future operations. The form was carefully written while the Unit was based both at Wembley and at Medmenham, and records various key mileposts in the war – such as the breaching of the Ruhr dams – and the involvement of the Unit in them alongside day-to-day information such as when personnel arrived, or left the unit, when people were ill, when they were promoted and demoted or even court-martialled. Whilst it is a unique social document that records the activities of Medmenham, it is also nevertheless notable for the absence of the mention of many secret operations which, for security reasons, were never recorded in it.

The Ground Intelligence Section received, registered and created card-index entries for all reports from 'ground sources' that were supplied to Medmenham. This intelligence would come from various sources, including R. V. Jones. Whilst some of this intelligence was sanitised, and some would not even be supplied to Medmenham, such was the flood of ground intelligence available that from the end of 1941 until the beginning of 1944 a team of five Area Officers – who were responsible for Scandinavia, Germany, the Low Countries, France and the Mediterranean – was required to manage the mass of information. They were also responsible for maintenance of a library of foreign guidebooks, technical books and dictionaries, plus the filing and indexing of all the photographic-interpretation reports that were created at Medmenham. This reflected the same philosophy behind the creation of the Print Library: that all intelligence should be catalogued – whether it was an aerial photograph, an interpretation of an aerial photograph, or a ground report – in order that all the information could be fully exploited. This approach meant that they could produce from early 1942 a Monthly Interpretation Review that provided a general summary, country by country, of all photographic reconnaissance undertaken and special technical articles. With their specialist knowledge of particular geographical areas, the Area Officers also doubled as liaison officers and would regularly advise the Admiralty, War Office, Air Ministry and in particular the Ministry of Economic Warfare, which was responsible for the activities of the clandestine Special Operations Executive (SOE), created by Churchill in 1940 with the order to 'Set Europe Ablaze'.

Whilst fifty-eight-year-old Wing Commander Hamshaw Thomas was leading the investigation into the V weapons, he also commanded C Squadron and from early July 1943 was responsible for the work of the newly established X Section that supplied vital intelligence to the SOE. Section Officer Irene Marsingall-Thomas was in charge, although no space was allocated to the section until November 1943 when they took up residence in Room 168. Such requests for information had initially been dealt with by R Section – Combined Operations – but after they became preoccupied with preparations for the 1942 Allied landings in North Africa – Operation Torch – the work was transferred initially to senior second-phase interpreters in Z Section. Although these interpreters were cleared to 'Most Secret' level, they were often in the dark about what the information was required for. They were often asked to provide aerial photography of a particular geographical area and provide interpretation using the 'latest and best' photography available in the Print Library. Often, however, when information was requested for a pinpointed location within a two-kilometre radius a new photographic-reconnaissance mission was necessary. Between 1 January 1943 and 8 May 1943, 67 reports were issued providing interpretation of 134 areas but, as the pace of the war quickened, from 9 May to 7 August 1943 another 61 reports were issued describing 133 areas, including those originating in extensive requests for mosaics and topographical descriptions of areas in Norway for the Royal Norwegian Forces.

These requests concerned operations involving the landing and take-off of two types of aeroplane and the dropping of agents and supplies by parachute over occupied France and Norway. There were six interpreters involved in this work and by mid-1943 they worked in three pairs working continuous shifts of twelve hours on and twenty-four off, with continuity of their work ensured by a logbook and the Medmenham Duty Intelligence Officer. The Duty Intelligence Officer was responsible for all requests from external sources whether they arrived by telephone, signal or letter and provided the link between Medmenham, those requesting aerial photography and interpretation, and the photographic-reconnaissance squadrons. In view of the operational character of the work and its high degree of secrecy, as small a number of people were involved as possible, and although other personnel were used to search for relevant photography, reproduce photographs and annotate them, they never knew any other details of the task in hand. This detachment from the reality of what they were doing is exemplified in a statement by WAAF typist Myra Nora Collyer who worked in K Section. She typed reports on the destruction of German cities but later recalled that as a touch typist the words went from her brain, down her arm and then appeared on the page, and that often she would never know what she had typed. Her primary motivation was to finish her shift and go to the local dance, which by then was frequented by many of the Americans stationed nearby. As big spenders the Americans were well liked by the landlords of Marlow, who were even known to charge them a pound for an omelette.

As the demand for special reports increased, a standardised form was designed and convenient code words were agreed with the Air Ministry in order that the interpreters could know the type of operation contemplated and the size of area required. The code words used included QH which meant a Lockheed Hudson aircraft, QL which represented a Lysander and QX a parachute-drop operation. The use of a Lockheed Hudson was a reminder of how Sidney Cotton and MI6 had also made use of Lockheeds on their clandestine missions. The QL and QH missions in France, when Lysander and Hudson aeroplanes were landed to take in secret agents and to bring out others with information or for training, or for the evacuation of important people or refugees, called for extreme accuracy on the part of the interpreters. When making QL and QH reports on the suitability of a proposed landing ground the interpreters had a lot to bear in mind. There had to be good cover in the form or woods or thick scrub immediately adjacent to the dropping point, with the ideal site being a large clearing in a wood or a field with woods on both sides. It was always necessary to keep as clear of villages, hamlets or even isolated farms and houses as possible. Pylons and power lines, isolated trees or tall trees proved to be a danger not just to those landing by aeroplane but to those landing by parachute, as was nearby military activity, under which heading the interpreters' reports considered the danger of flak, searchlights, hutted camps, ammunition dumps, or indeed any other relevant evidence. If a proposed landing ground still looked promising it was necessary to undertake a detailed interpretation of the area.

The challenges encountered while landing an aeroplane in a clearing, given that the pilots often had little more than a map, a compass and possibly members of the local resistance shining torches from the ground, were considerable. When reporting on potential QL and QH landing grounds, and in order to see objects such as wire fences or telegraph poles, it was necessary to have large-scale aerial photographs. Nevertheless, if the shadows of objects happened to fall on dark ground they could be easily hidden, which meant it was essential to assess the landing ground with photographs taken at different times of the day. There were many instances where telegraph poles were quite invisible in one sortie and very apparent in the next. Since so many of the flights involved landing on farm fields, many photographs were taken of ploughed fields where the light was cast in such a way that the furrows became indistinguishable and the ground appeared even. Conversely, when photographing during times of long morning or evening shadows, the depth of furrows could be easily exaggerated. This highlighted the essential need for comparative photography if an accurate picture of an area was to be obtained.

In January 1944 Medmenham were asked to provide a detailed report and model of Amiens prison where many high-ranking French Resistance prisoners were being held. The interpreters' ability to provide especially detailed information from aerial photographs taken on specially flown missions meant the heights of walls and security arrangements at the prison could be determined. This information was used in the planning of Operation Jericho, during which twenty-eight-year-old Group Captain Percy Charles 'Pick' Pickard commanded the legendary raid by Mosquito bombers that blew a hole in the prison wall – on 18 February 1944 – through which many prisoners escaped. As a reminder of the dangers facing so many of those featured in *Target for Tonight*, Pickard, who had also been the first officer to have been awarded a Distinguished Service Order and two bars in the war, was killed on the raid. His earlier work flying on Lysander missions provided a lifeline to the French Resistance and had proven critical to the establishment of a network in France that provided vital intelligence in the hunt for the secret weapons.

One of the most important people in the hunt would be Frenchman Michel Hollard, who would not only provide British intelligence with engineers' drawings of a launch site but would follow this up with the actual measurements of this weapon of the future. The month after Operation Hydra, Bodyline became the responsibility at Medmenham of Wing Commander Douglas Kendall, and by the end of 1943 Wing Commander Hugh 'Ham' Hamshaw Thomas, the pioneer of photographic intelligence since the First World War, was feeling the strain and increasingly unwell. He resigned from the RAF on health grounds on 22 December 1943. As an indication of the esteem in which he was regarded he was presented with an album, signed by his fellow PIs, which showed vividly how much Hamshaw Thomas was liked.

CHAPTER 6
The Bois Carré Sites

In September 1943, unbeknown to the Medmenham interpreters who were busily hunting for rocket sites in northern France, the Réseau AGIR espionage network – named after the French word for 'to act' – were about to make a series of breakthroughs that would prove critical to Operation Bodyline. The French patriots were led by forty-five-year-old businessman Michel Hollard, who had run away from home aged sixteen to join the French Army during the First World War and was now a sales representative for Maison Gazogène Autobloc – a manufacturer of gas generators for motor cars – an occupation that provided the means and justification to travel around France. In January 1942, Hollard secretly crossed the Swiss frontier, visited the British Legation in Berne and offered his services to the Assistant Military Attaché. Undeterred when his offer was rejected out of hand, he began building a resistance network throughout France that included hotel owners, dockyard workers and notably railway workers. The British were understandably pleased when Hollard reappeared in May 1942, by which time they had read a report that he had passed to them on the French motor and aeronautical industries and that was graded as highly valuable. He was immediately recruited and became 'Source Z165'.

Hollard set about developing his network of agents who gathered information that he smuggled to the MI6 station in Berne every couple of weeks. Berne in turn forwarded it in secret cypher telegrams to MI6 Headquarters in Broadway Buildings, London. At its peak, the Réseau AGIR would have over one hundred agents, strategically located throughout the country, who provided information on the

German order of battle, and the disposition of military personnel and equipment, throughout France. It was in August 1943 that Hollard received an intriguing message from one of his agents, Jean-Henri Daudemard, a railway engineer, who had overheard a café conversation about unusual construction projects in Normandy. Hollard went to investigate and on arrival in Rouen went straight to the local labour office and asked for the welfare officer. With the cover story that he represented a Protestant welfare organisation, and with a ready supply of Bibles and religious literature in his briefcase, the seemingly devout Christian expressed his concern for the spiritual welfare of itinerant construction workers who might have been uprooted from their families and could be exposed to moral dangers.

The French official was won over. Hollard needed to visit the men at their workplaces, he explained, and so the Resistance agent was duly provided with a list of local construction sites. Taking a train to the village of Auffay, between Dieppe and Beauvais and 24 miles north of Rouen, Hollard changed clothes en route and emerged from the train masquerading as a labourer. He had chosen Auffay on the basis that it appeared to be the most accessible place on his list, but when enquiries in the village drew a blank he began to explore the roads leading out of the place. Walking westward for three miles, he came across a noisy construction site at La Ferme de Bonnetot near the small hamlet of Beuville where several hundred labourers were hard at work. Anxious to learn more about this construction site set amid the apple orchards on this farm in the Seine-Maritime *département*, and having spotted a wheelbarrow in a nearby ditch and 'commandeered' it, Hollard boldly entered the site, steering straight past the German guard, and headed towards a group of French labourers. After making some small talk, he filled his wheelbarrow with building materials and went on a tour of the site which consisted of ten individual buildings, none of them more than one storey high, all connected by a winding concrete track. What particularly intrigued him was a wide strip of concrete about fifty yards long – like 'a miniature runway' – with wooden posts down the middle. Since Hollard always carried a pocket compass with him he leant down and, pretending to adjust his shoelaces, read the magnetic bearing of the concrete strip.

Despite the token presence of the German guard, Hollard exited the place as brazenly as he had entered it. To all intents and purposes the site was civilian in nature, with German engineers working alongside French construction labourers. Returning to Auffay, he travelled by train on to Rouen and was in Paris by nightfall. Staying in a hotel near the Gare de Lyon, he spread a map of northern France (which also covered southern England) on his bedroom floor and, plotting a line from the construction site, discovered that the concrete strip pointed directly towards London. Convinced that he had discovered something of importance, Hollard passed the information to the Berne station who are recorded as being singularly unimpressed but nonetheless passed it on to London because of the reliable status that Hollard had by then earned. For Operation Bodyline this intelligence was to prove a sensation. Just as the Réseau AGIR had previously pinpointed the enormous construction site at Watten in early July and thereby managed to increase suspicion about the Heavy Sites that were being developed in northern France, so the discovery of the development at La Ferme de Bonnetot would soon result in the investigations at Medmenham dramatically expanding in scale and importance.

Following instructions from London, Hollard searched throughout October for similar sites in the region and attempted to secure detailed blueprints of one of them. To achieve this he gathered a team of his most reliable agents, allocating each of them a zone covering a stretch of coastline and giving them each a bicycle. With a brief about what they were looking for and a map to record any sites' locations, the agents began searching and discovered more construction zones, all at various stages of completion. Some, it would turn out, were related to secret weapons and others weren't. Many of them had miniature runways aligned towards London. As the month progressed, and with a number of the sites looking as though they were nearing completion, Hollard was painfully aware he had not yet secured detailed blueprints. But the Réseau AGIR were to have a lucky break when Hollard was introduced to André Camp, a young graduate engineer who worked at one of the most completed sites as an architectural draughtsman. Just under one mile east of Yvrench village, 40 miles south of Calais, like many of the construction zones it was located in a wood for the purposes of camouflage. In this case the area concerned was square in shape and was thus named the Bois Carré – or square wood.

On first meeting André Camp, Michel was not particularly impressed with the young man who appeared more interested in his first graduate job than in helping the Resistance. But when André explained that his task was to draw detailed plans for the site, Michel knew that the chances of finding anyone else with such privileged access to the information were limited. With time running out, Michel convinced André of the need for the information and gave him two weeks to create copies. With his privileged access, André was able to secure copies of all the plans, except for the critical one detailing what would be constructed on the concrete strip that was aimed at London. André knew that a plan existed because he had seen a German engineer inspecting that part of the site with a blueprint in his hand. But the man always returned it to the inside pocket of his overcoat. As far as André could see, the German was meticulous in always having his overcoat with him – except when he punctiliously visited the latrine at nine o'clock every morning. From his adjoining office, André could see the engineer walking to the latrine and began timing his visits, which varied between three and five minutes (longer when he had a newspaper). As his deadline got closer, two days before he had agreed to rendezvous with Michel in Paris André spotted the German walking towards the latrine, newspaper in hand. Heading to the man's office next door, he quickly grabbed the blueprint from his overcoat, returned to his own office and began tracing the plan and copying essential measurements. Completing his task within three minutes, André returned the blueprint before the German returned, none the wiser about what had just happened.

On the earlier advice of Hollard, and in order that he could keep their rendezvous, the following day André claimed to be ill. Despite the scepticism of the German doctor in Yvrench he was given four days' leave and permission to be treated at home. Returning to Paris he met a justifiably pleased Michel, who insisted that André's presence would be required for several days to ensure that the plans, assortment of sketches and notes were pieced together in as coherent a way as possible. While a friendly doctor, Raoul Monod of the Rue du Sergent Bauchat in Paris, sent the German doctor in Yvrench a certificate stating that André required complete rest and requested an extension of his leave, the two men went to work. With Michel also a trained engineer, they set about the task of analysing the site. Thanks to the blueprint, they now understood that the purpose of the concrete strip was to support an inclined ramp about 150 feet long that rose at an angle of fifteen degrees and carried two metal rails. With the track pointing towards London, 130 miles away, the Resistance men were in no doubt that the purpose of the site was to launch missiles against England.

As soon as the plans were completed, Hollard travelled to Switzerland. After crossing the frontier on 20 October 1943 he telephoned to arrange a rendezvous with his handler who went by the code name 'OP' and who knew something interesting was coming when Michel said he had 'bought the tickets for the performance' and they were 'very good seats'. The plans were despatched in the diplomatic bag – which ensured immunity from search or seizure – to London. They arrived with MI6 on 28 October, by which time André Camp had returned to work at the Bois Carré site from where he continued to report on developments and where he pensively awaited the inevitable bombing raid. In London the information from the Réseau AGIR was to prove a decisive influence on Operation Bodyline.

In London, MI6 had already been supplied with related ground intelligence courtesy of the naval officer-in-charge on the Danish island of Bornholm in the Baltic Sea, 80 miles to the north-east of Peenemunde. During the early part of the German occupation of Denmark, which began with Operation Weserübung on 9 April 1940, the Danish government was considered the model Protectorate Government by Hitler – largely because it was considered to be a Nazi-friendly puppet regime. This ensured that many of the country's institutions – including the Danish Royal Navy – continued to function with a surprising degree of peacetime-like normality. It was for this reason that Lieutenant Commander Hasager Christiansen was able to gather and circulate intelligence on 22 August 1943, after an object resembling an aeroplane approached Bornholm from Peenemunde at considerable speed. Eyewitnesses later described hearing a strange 'humming sound' when a low-flying projectile approached, before it narrowly missed some trees and crashed onto a meadow, from where it bounced over a road, smashed through a telegraph wire, and came to rest in several pieces in a field of mangel-wurzels. It landed in countryside in the south-east of the island at Bodilsker, a little over a mile from the thirteenth-century St Bodil's Church. A local policeman was the first person on the scene where he discovered a strange yellow aircraft-like object and reported the incident to his HQ.

The police alerted Hasager Christiansen who quickly drove the sixteen miles from Rønne, the largest town on the island and where the Germans were also garrisoned. At the crash site he found a pilotless aircraft and noted that it seemed to have no motor or propeller, nor room to carry anyone, and that where the pilot's seat would normally be was 'a complicated apparatus which resembled the inside of a radio receiver'. Christiansen quickly set about taking detailed measurements, noted that the code word V-83 was painted on the machine in several places, and even managed to take photographs that recorded the main body of the projectile and steering device. Then two Wehrmacht soldiers appeared. Clearly perplexed by the contraption, they set about salvaging the wreckage, hiding it from prying eyes with a tarpaulin. Christiansen must then have made duplicate copies of his photographs and report, since multiple copies of the information made it to Broadway Buildings from different sources – including Aage Andreasen, otherwise known as Agent Elgar.

Opinion differed among the British scientists and military thinkers about what had been found, with some believing it was a larger version of the Henschel Hs 293 glide bomb that used a radio-controlled guidance system to devastating effect against Allied warships in the Mediterranean. The Bornholm machine was destined to reignite a debate that was becoming increasingly confused, a debate about the differences between pilotless aircraft and long-range rockets. Along with the discovery of the construction sites in northern France with their 'miniature runways' aimed at London, the Bornholm aircraft proved to be a lucky break for Operation Bodyline. The projectile had dramatically overshot the Greifswalder Oie – the small island seven miles from Peenemunde in the Baltic Sea that was used for target practice – and within days of it landing on Bornholm the Germans declared martial law in Denmark. As the war had progressed, and particularly after decisive Allied victories – notably the Second Battle of El Alamein in late 1942 and the Battle of Stalingrad in 1943 – acts of sabotage and strikes had become increasingly common in Denmark. While the Danish resistance movement grew, the already uneasy relationship between the Danes and the occupying Germans became increasingly strained and now reached breaking point.

The impact of martial rule on 29 August 1943 was felt immediately by the Royal Danish Navy. Whilst many ships were scuttled to prevent them falling into German hands, Paul Mørch, the Chief of Danish Naval Intelligence, was aboard one that successfully eluded German capture and made it to neutral Sweden. Mørch had earlier passed copies of Christiansen's report on the pilotless aircraft to Captain Henry Denham, the British Naval Attaché in Stockholm, and the Swedish intelligence agency – C Bureau – who both independently forwarded the information to MI6. On arrival in Sweden, Mørch visited the British Legation – where Denham was based – to discuss the recent happenings on Bornholm. A vital source of intelligence for the British in the Scandinavian region, Denham was a thorn in the flesh of the Germans, who suspected that he had discovered the deployment of the *Bismarck* and *Prinz Eugen* in May 1941, courtesy of his network of ship-spotting agents in the neutral country.

The Germans were right to assume that Denham had a valuable network of agents at his disposal in Sweden and that the British claim that spotting the battleship *Bismarck* – which now languished on the floor of the North Atlantic Ocean – had been solely the consequence of photographic reconnaissance was untrue. They desperately wanted the Swedes to declare Denham *persona non grata* and put pressure on the Swedish Naval Attaché in Berlin in the summer of 1941 for him to be put on round-the-clock surveillance to prove that he was illegally intelligence gathering. When the Swedish secret police went through the motions of installing a microphone in the chimney of his flat, Denham – who was famed for his good humour – took great delight in taking a group of party guests upstairs to say hello. Since late March 1941, C Bureau had been actively sharing their naval intelligence in exchange for British intelligence, including aerial photography taken by the PRU. C Bureau systematically interviewed Swedish ship captains who could navigate around the ports of the Baltic with relative impunity and frequently saw U-boats and shipbuilding in German ports and the movements of military ships. At Medmenham the veracity of the Swedish intelligence was checked using aerial photography and was found to be generally accurate, but by early 1944 the Germans had grown wise and screened the seaward sides of their shipyards.

In London the evidence from Bornholm tallied with a separate report from a disgruntled officer in the German Army Weapons Office who had earlier leaked information about plans for 'winged rockets'. This new report was much more specific, and revealed that a pilotless aircraft known as the *Phi 7* was being tested at Peenemunde and that the Germans had set 20 October 1944 as Zero Day for the deployment of the A4 (V2) rocket against London. A third report, from Agent Amniarix of Resistance agent George Lamarque's Druids Network, even connected activity at Peenemunde with sites in northern France. The report from Amniarix contained intelligence on the research being undertaken at the experimental site on bombs and shells guided independently of the traditional laws of ballistics, on stratospheric shells, and on the use of bacteriological weapons. For the first time it revealed details about Colonel Max Wachtel, who was reportedly in the process of creating a regiment that would have its headquarters in Amiens and would operate 108 catapults located in the Pas-de-Calais region. It was reported that reinforced concrete platforms were already under construction, and the sites would be fully operational in November when the catapults would be capable of firing a bomb every twenty minutes. Amniarix went on to record the plan to have 400 catapults that would be sited from Brittany to the Netherlands.

The report from Amniarix – designated the Wachtel Report by MI6 – provided a wealth of information even down to the colour of the security passes required to enter the Peenemunde facility, and prompted R. V. Jones to enquire more about the source. All he could discover at the time was that it came from 'une jeune fille la plus remarquable de sa generation' and he would only discover later that she was twenty-three-year-old Jeannie Rousseau (who spoke five languages and worked as a translator for the Germans) when Marie Madeleine Fourcade published her 1973 book *Noah's Ark: The Secret Underground*. Her life as a secret agent began on the night train from Paris to Vichy – the spa town in the Auvergne, then home to Marshal Phillipe Pétain and his Nazi-collaborating Vichy Government – after a chance meeting with George Lamarque, who remembered her as a student at the University of Paris, and recruited her to his Druids Network. In 1976, during the making of their documentary *The Secret War of Dr Jones*, Yorkshire Television tracked down Amniarix, who was by then the Vicomtesse de Clarens. Meeting for the documentary, Jones discovered that Agent Amniarix had used her language skills and personable manner to get her job with the Germans. But she was eventually captured by the Gestapo and from 1944 was imprisoned in Ravensbruck, Konigsberg and Torgau concentration camps. After writing the foreword for his book *Most Secret War* in 1978, they were reunited again in 1993 for a ceremony at the Central Intelligence Agency Headquarters in Langley, Virginia, when she was awarded the Agency Seal Medallion for her contribution to human intelligence – HUMINT – and he became the first recipient of the R. V. Jones Intelligence Award created in his honour.

One of the many problems now facing British Intelligence was discovering a coherent thread that ran through all the information they received, particularly in view of the fact that intelligence about the long-range rocket was still frequently being confused with that concerning pilotless aircraft. When Michel Hollard supplied MI6 with intelligence in September 1943 about the mysterious construction site at La Ferme de Bonnetot, the Medmenham interpreters involved in the hunt for secret weapons were told nothing about it, nor about the feared immediacy of the threat presented by the 'pilotless aircraft' that propaganda chief Goebbels would brand the V1. In B Section, Captain Robert Rowell's investigation was still focused on the 38-foot-long rockets that had been spotted at Peenemunde and were working on the same basis they had been briefed on: that the rockets were so large and heavy that rail access to the deployment sites would be essential. But although construction sites in northern France connected to the rail network received special attention in their BS Reports, the Army interpreters had nevertheless independently noted the existence of smaller and suspicious sites. For the time being, though, they had just noted them as having 'no suspicious railway spurs or anything that suggests a Bodyline installation'.

This was to change partially when Medmenham was supplied with an anonymised copy of the 'Pingpong' report – the term MI6 used for agent-derived intelligence on secret weapons – dated 1 November 1943 that contained some of the information recently supplied by the Réseau AGIR. Report CX/12421/A was received at Medmenham from MI6 with a CX reference – the reference assigned to all agent-derived intelligence – and was held by the Ground Intelligence Section who ensured a reciprocal flow of intelligence by compiling monthly summaries of work being undertaken at Medmenham for inclusion in the Air Ministry Weekly Intelligence Summary. This CX report provided Medmenham with information on a network of construction sites near Yvrench's Bois Carré – where the young draughtsman André Camp still worked – in the neighbouring villages of Saint-Josse-au-Bois, Gueschart, Maison-Ponthieu, Ray-sur-Authie, Cramont, Campagne-lè-Hesdin and Yvrencheux in the Pas-de-Calais region. This resulted in sortie E/463 being flown on Wednesday, 3 November 1943 from RAF Benson by Flight Sergeant James McGinn Aitken of 541 Squadron who took 'A Quality' photographs of the sites at 10:45 in the morning from 26,500 feet in a PR Spitfire.

On the same day when this aerial photography was being taken over northern France, Colonel Terence Sanders, the Director of Technical Development at the Ministry of Supply, visited Medmenham to brief B Section on the methods that the Germans might use to launch missiles. This Old Etonian and Cambridge Blue had won a Gold Medal at the 1924 Paris Olympics and before the war had been a lecturer in engineering and a Fellow of Corpus Christi College, Cambridge. Sanders revealed to the Army interpreters that the 'Peenemunde rocket' was probably not yet ready for large-scale production and the Germans might try a lighter projectile first, one that would probably be launched from two rails inclined at a steep angle. This was news indeed to the interpreters, who had been working tirelessly to a different brief: that the projectile was so heavy that any firing site would need to be rail-served. At the time they were busily analysing the photographs taken during the second systematic survey of all areas in northern France within 130 miles of London, in order that they should have a complete overview of any new constructions which had been started since the first comprehensive photographic reconnaissance of summer 1943.

With each sortie involving hundreds and sometimes upwards of a thousand photographs Captain Rowell and his small team had been overwhelmed with the deluge of demands. The Medmenham investigators had been focusing on new railway extensions or loops off the main railway network, which had resulted in the discovery and detailed analysis of the Heavy Sites. In Report BS76, the information derived from the 1/9,000-scale photographs taken during sortie E/463 confirmed that the eight places referred to in the text of the CX report 'appear to be amply confirmed by the photographs and the sites are such that a Bodyline significance is not improbable'. But as Rowell noted: 'At each place some activity is seen, but, if these are projection sites, they at present bear little or no resemblance to any installations which have hitherto come under observation in connection with the investigation. There are, however, a number of features which make it clear that all these sites are similar.' The report goes on to point out that there were no railway lines anywhere near any of the locations, which were only served by roads, and that like Yvrench's Bois Carré the sites were being constructed in woods, presumably for camouflage purposes. Included in the report was a drawing of a distinctive ski-shaped building – one that looked like a ski lying on its side – which had walls approximately three feet thick. The appearance of standard structures suggested these might be 'associated features' of the firing installations and henceforth they would be reported whenever seen. Since Bois Carré was the first of these new sites to be examined – all of which featured three of the distinctive ski-shaped buildings – all such sites thereafter became known as 'Bois Carré sites'.

Unbeknown to the interpreters at the time, the ski-shaped buildings were built for the storage of the pilotless aircrafts' fuselages and wings that were to be assembled on-site. They were designed with a distinctive curved entrance to provide the maximum protection for the explosive contents inside the buildings from any Allied air attacks. Since these structures were quickly identified as a common feature of the Bois Carré sites and provided a useful signature on the aerial photography, the interpreters nicknamed them the 'ski sites'. In a footnote to Report BS76 of Friday, 5 November 1943, it is recorded that four similar sites had already been photographed – one of which was on the Cherbourg peninsula – and that detailed reports would be issued shortly. In order that the original Bois Carré site – which appeared to be the most complete – could be examined in greater detail, the RAF's 170 Squadron, a fighter-reconnaissance squadron that flew the North American P-51 Mustang, were tasked with taking low-level oblique photographs of the site near the village of Yvrench at the earliest opportunity.

The intelligence from the Réseau AGIR ensured that Operation Bodyline became an even higher priority, and would come to highlight more than ever the split amongst Churchill's scientific and technical advisers: one side, including R. V. Jones, argued that a rocket projectile was quite possible; the other side, led by Lord Cherwell, considered the whole thing a hoax aimed at drawing the bombing effort away from current priorities and onto what they considered useless targets. The impasse prompted Churchill to ask the Minister of Aircraft Production – Sir Stafford Cripps – to hold a short inquiry into the evidence and to assemble the arguments for and against the pilotless aircraft being a glide bomb operated by controlling aircraft from a distance. The urgency of the situation meant that a meeting was convened for Monday, 8 November, only five days after Medmenham had learned that a projectile lighter than the long-range rocket was likely to be a more immediate threat. Rather than go over the ground already covered by Duncan Sandys, it was proposed to discuss evidence about rockets, pilotless aeroplanes and glide bombs: evidence of the method of projection, evidence of manufacture, evidence of preparation for use and any evidence that refuted their existence. For each of these subjects it was noted that there was photographic evidence, intelligence from other sources and deductions to be made from German propaganda. All invited to contribute were asked to consider whether the rocket was merely a creation of the German Propaganda Ministry, whether the rocket story had been deliberately planted as part of a cover plan to conceal something else, whether the construction sites in northern France were in reality intended for a quite different purpose, and whether the Germans had not yet solved the technical problems associated with the development of a long-range rocket.

Wing Commander Douglas Kendall later recalled attending the first meeting of the Stafford Cripps Enquiry with Captain Neil Simon and Flight Lieutenant André Kenny in Conference Room D at the War Cabinet Offices, Great George Street, Whitehall at 10 a.m. on Monday, 8 November 1943. With the Minister of Aircraft Production at the head of a U-shaped table, flanked by Duncan Sandys and Lord Cherwell, the interpreters found themselves in the bottom seats whilst all the others present were senior-ranking officers from the Navy, Army and Air Force, as well as senior scientists including R. V. Jones. The meeting began with a review of non-photographic intelligence, including an analysis of German propaganda about the secret weapons. One line of argument followed was that Goebbels never made a major persistent claim without there being some truth to it because if the German leadership built up an expectation amongst their people and then subsequently failed to deliver on their promises, the consequence from a propaganda point of view could be a potentially dangerous drop in morale. For this reason a representative of the Ministry of Information argued that the claims should be taken seriously: the threat was real.

The remainder of the meeting was given over to a detailed discussion about photographic evidence. Douglas Kendall later recalled that when Cripps raised questions all the senior officers and scientists looked at one another up and down the table until the questions came to rest with the interpreters, who had all the answers. Kendall began by presenting the available evidence about the Peenemunde experimental site before turning to the Heavy Sites in northern France. It was concluded that they were connected to rocket deployment on the basis of ground intelligence, their lack of resemblance to, or connection with, known military or any other installations, the fact that work had started on all the sites at around the same time, the fact that they were all served by main railway lines, that with the sole exception of Watten they were targeted on or at right angles to London or Bristol, and that the scale and intensity of the work was an obvious sign that the enemy set great store by their rapid completion.

At this point Cripps asked the meeting whether there was any further evidence of secret-weapons activity in northern France, a question in reply to which Douglas Kendall later observed he 'had the unpleasant task of telling him that during the previous four days we had located nineteen sites in the early stages of construction'. When asked what evidence connected these sites with the Bodyline investigation, Douglas and the Army interpreters explained that whilst there was no photographic evidence of a positive nature there was a considerable amount of negative evidence: for example, the sites did not appear to have anything to do with the Atlantic Wall defences – the extensive system of coastal fortifications built by the Germans along the western coast of Europe to defend against the anticipated Allied invasion – but they did all share common features, including three ski-shaped buildings and a platform that always appeared to be facing towards London. Shocked reactions among the group prompted Stafford Cripps to adjourn the two-and-a-half-hour-long meeting for two days to allow the interpreters time to examine more areas to see whether any further installations of this type were under construction.

Kendall and the others had less than forty-eight hours to prepare the next stage, working late into the night after finishing their normal shifts: Captain Neil Simon analysed the photography and calculated the measurements while Kendall made the deductions about what each site was and how it operated, after consulting other Medmenham specialists. The Réseau AGIR had shown the way, but now it was the interpreters who faced the challenge of precisely pinpointing locations in northern France. This was now a race against time to locate all the sites. It prompted Kendall to reorganise B Section overnight, and with a need for a much larger investigation team the B2 Sub-section was created under Captain Robert Rowell, and Operation Bodyline was immediately expanded to include officers and other ranks from across the Army, RAF, WAAF and American forces. This resulted in Neil Simon working under Rowell alongside a growing team that included Flight Officer Nora Littlejohn, Barbara Slade and Humphrey Spender. Barbara Slade apparently became an artist with the stereoscope and could smell out the launching sites. Once she reportedly had a stereo pair and told her colleagues that she knew there was a site there, which nobody could see – only for one to appear days later, on the next cover.

Humphrey Spender, brother of Michael and Stephen, was an Army interpreter and had been conscripted as an official war photographer before becoming a PI. After Charlcote Boarding School, Humphrey had studied art history at the University of Freiburg-im-Breisgau and spent time in Weimar Germany in the company of his brother Stephen Spender, W. H. Auden and Christopher Isherwood. He was inspired by the avant-garde style and took photographs of the Jarrow Marchers and the British Union of Fascists for the *Left Review* and as 'Lensman' for the *Daily Mirror*. Travelling extensively around Europe for *Picture Post*, Humphrey spent time with the British-American novelist Christopher Isherwood in Berlin. Drawn there by its reputation for sexual freedom Isherwood was inspired to write his 1939 novel *Goodbye to Berlin* – featuring a cover designed by Humphrey – which inspired the later musical and film *Cabaret*. Focusing on the nightlife at the seedy Kit Kat Club and set against the backdrop of the violent rise of the National Socialist movement, the impact of Nazism on German society is chronicled. Arriving at Medmenham in 1941 Humphrey met dancer and choreographer Frederick Ashton, with whom he is recorded as having had an affair.

Interpretation Report BS89 was completed in the early hours of 10 November by Neil Simon, in time for the second Stafford Cripps meeting later that morning. The eight-page digest outlined the layout of the Bois Carré site, contained detailed measurements, plans, diagrams and data on the orientation of the platform – courtesy of W Section and the Wild Machines – and featured descriptions of the square building with non-magentic Zinquat nails that perplexed Michel Hollard, two rectangular buildings, a small hut with blast walls, a pyramidal excavation, the ski-shaped buildings that were still under construction, pipelines and overhead cables, the current absence of any flak defences, and a speculation about the date of construction. Appended to the report was an annotated enlargement of vertical photograph frame 4089 from sortie E/463 taken on 3 November 1943, and corresponding photographs of the Bois Carré – frames 3025–3026 – taken on 15 August 1943 during sortie D/988 which showed that only preliminary work was then in progress. The report benefited from dramatic low-level oblique photographs taken the previous day by 170 Squadron – sorties R/134/170 and R/135/170 – in which considerable activity could be observed, including quickly fleeing construction workers – not altogether surprisingly given the extremely low level at which the photographs were taken from the two P-51 Mustangs.

The second Stafford Cripps meeting met at 11 a.m. on the morning of 10 November 1943. Kendall and Simon had worked on the Bois Carré interpretation report throughout the night; Simon had only managed to get one and a half hours of sleep before going into the 'witness box'. Operation Bodyline was becoming a particularly intense investigation requiring long hours of sustained work, and compelled Robert Rowell to take sleeping tablets to get any rest. The interpreters were amazed at how quickly Stafford Cripps absorbed their report: they found his questions masterly and completely to the point. Twenty-six Bois Carré-type sites had by then been located and all of them pointed towards England. But there did not appear to be any indication that the sites were intended for the firing of a rocket that was supposed to weigh 45 tonnes and, based on the interpretation of Peenemunde which had been carried out, would require heavy-lifting gear. Sir Stafford Cripps felt able to report to the War Cabinet that in his opinion pilotless aircraft and rockets existed, principally because the amount of activity being devoted to the projects was too great to be a mere diversionary hoax. Kendall's point was that since there was no evidence of heavy-lifting equipment, there was no reason to assume the sites could be used to fire the rockets that had been spotted at Peenemunde. This refutes a later assertion by R. V. Jones that he had to convince the interpreters before they would disallow the hypothesis that the Bois Carré sites were designed to fire A4/V2 rockets.

By this time in mid-November MI6 was being inundated with intelligence about secret weapons and the network of associated manufacturing sites. This included one report from August 1943 that recorded 'a very well-informed source' known as Ishmael as having provided information on the two most important sites associated with the manufacture of the liquid propellant used in the rockets. These were a factory near the southern Bavarian town of Altötting, Germany, and another at Auschwitz in Upper Silesia, Poland. The source explained that alongside Auschwitz was the I. G. Farben plant, one of Europe's largest wartime factories, built by the German chemical conglomerate who had requested the creation of the neighbouring concentration camp to provide the slave labour that manufactured the high-explosive content of the rocket. The source explained that no 'free' workers were employed in the plant that also made synthetic oils and rubbers, only 'concentration camp, politicals and prisoners of war' who totalled '65,000 workers', of whom 32,000 were Jews.

In another report, dated 10 November 1943, a source who described himself as a 'good German' was convinced that the Allies would win the war and wanted to warn them about the deployment of the rocket on the grounds that if Hitler was to 'wipe out southern England' it would create so much hatred against the German people that the 'last hope of any future for Germany would be lost'.

After supplying blueprints for the site at Yvrench's Bois Carré, and information on the eight construction sites in the Pas-de-Calais, Michel Hollard and his agents continued to supply the Berne station with more information on the network of construction sites in northern France. On 2 December 1943, Pingpong No. 53 provided information on construction sites at Guechard, Ailley-le-Haut-Clocher, Bois de Roche Earl, Yvrencheux, Bois de Grambus and Maison Ponthieu, plus a further eighteen in the wider region. In this report, Hollard provided information on the construction materials used, dimensions of buildings, and he promised to get a sample of the non-magnetic Zinquat nails used in the square building that had prompted speculation among the scientists at the Stafford Cripps meeting that it might be connected with the installation of an atomic warhead. This was followed by further Pingpong reports on suspected sites, and one report listing all the civilian French companies then involved in developing sites for the bombardment of Britain.

For Michel Hollard, the Normandy village of Auffay had been the starting point for his most productive initiative to date. So, when he was informed by Pierre Bouget, the Chef de Gare in Rouen – with whom he had an 'arrangement' – about unusual crates being transported to Auffay, he decided to investigate with one of his agents, Pierre Carteron. From the Chef de Gare in Auffay they discovered that the Germans had shunted the railway wagons concerned into a railway shed belonging to the local sugar factory, where they were under guard. One night Hollard managed to sneak past the sentry and in the darkness discovered that the contents of the wagons were in three parts: a large cigar-shaped body and two flatter parts. Measuring each of them in turn, he supplied the information to London, who classed it a Pingpong report.

Despite having safely crossed the Swiss frontier ninety-eight times, Hollard was ultimately betrayed and caught by the Gestapo on 5 February 1944 when he was arrested in the café des Chasseurs at 176 Rue de Faubourg, near the Gare du Nord, in the Saint-Denis quarter of Paris – along with fellow members of the Réseau AGIR, including Jules Mailly who would die in Mauthausen concentration camp. After suffering prolonged torture in the Centre Pénitentiaire de Fresnes, to the south of Paris, in the prison favoured by the Gestapo – and the Vichy Regimes paramilitary force the Milice Français – for captured SOE and Resistance agents, Hollard was sent to the Neuengamme concentration camp near Hamburg. Lucky to survive the war, Hollard was decorated with the Rosette de la Résistance by the French and with the Distinguished Service Order by the British, but his remarkable wartime achievements would be largely forgotten. When preparing for his 1959 BBC documentary series *Men of Action*, Lieutenant General Sir Brian Horrocks researched the work of Hollard and came to describe him as 'the man who saved London'. The epitaph became the title of George Martelli's 1960 book in which he observed that many statues had been unveiled in 'London – the city he saved – to less deserving people'.

The strategy of the Réseau AGIR operating as a freelance group outside the mainstream of the French Resistance movements, coupled with the reserved nature of its leader Michel Hollard, perhaps contributed to this apparent neglect. But nevertheless, eleven years after his death in 1993 at the age of eighty-nine another Anglo-French project – Eurostar – named one of their high-speed trains that travel between northern France and London after him. Hollard would prove vital in focusing the discussions about how Britain, in particular London, would shortly come under attack not just from the long-range rocket but from the pilotless aircraft code-named by the Germans the *Vergeltungswaffe 1* or Revenge Weapon 1: the V1.

Despite this information, the debates continued. R. V. Jones was requested by the Joint Intelligence Committee to provide them with raw information in order that they could pass their judgement. This prompted a letter to the chairman of the committee in which he outlined that it had been his duty since the beginning of the war to foresee the application of science to warfare. Jones observed that pilotless aircraft and long-range rockets were two such weapons that he had been studying throughout the war, and that unless he had failed in his duty there was no need for the committee to cover the same ground, boldly adding that the chairman had no experience of this work anyway. In his view the most important challenge was investigating the scientific and technical facts, and he claimed that when these were established it would be relatively easy to determine where the weapons were manufactured and how they were transported, stored and deployed. On the basis that any judgement by the committee would perform the same function as his office, he wrote that it would indicate a lack of confidence in his methods and if he was compelled to share them it would lead to his resignation. Citing Churchill's belief that committees could easily become functionless bodies, he ended by reporting that his section would continue in their work regardless of the opinion of any such committee and trusted that 'we shall not be hindered'. He did not receive a reply.

CHAPTER 7
The V1 Flying Bomb

While André Camp continued his work as a draughtsman at Yvrench's Bois Carré in December 1943, under the direction of Air Commodore Claude Pelly MI6 briefed its agents to look for locations linked to secret weapons and soon began receiving similar confirmation from agents in northern France who had infiltrated other sites, some of which had progressed to the stage of being camouflaged. Throughout this whole process everyone involved remained mystified by the fact that they had never found a ski-site in Germany. It was considered implausible that the Germans would not have completed and tested a prototype at one of their experimental stations before building so many sites in northern France. When briefed, the agents were warned to be mindful of the considerable volume of enemy-inspired propaganda and were told that they must be careful to scrutinise closely any information couched in general terms. Reports that failed to mention a specific place, whether a firing site or a factory, or which talked about a weapon in vague terms – such as that the Germans were developing a weapon that would lead to the total destruction of London or southern England – were considered to have no value and were a misuse of the agent's wireless-transmitter time.

Information was wanted about the purpose of any building on a ski-site and the contents of any such building, plus knowledge of any other activity that was invisible from the air and carried out underground or under camouflage. Agents were told that on sites that were still under construction the most important place to watch was the presumed firing point. Information was requested about the arrival at a ski-site of pipes, cranes or winches, or any bulky objects such as tubes of exceptionally large diameter, and any radio or radar installations. It was assumed that once the initial work had been completed the Germans would remove all the French and foreign workers, and this would indicate that the site was approaching operational use. Agents were instructed to look for, and treat as first priority, any indications that the site was becoming operational such as the arrival of German engineering or military parties, the arrival of fluids in bulk, special vehicles, anti-aircraft batteries, or the evacuation of the local population near the site.

When briefing Agent Alibi (MI6 source 99567), a French diplomat in Madrid who had established effective wireless communications with resistance movements in France, particular emphasis was placed on the Heavy Sites. His network was told that these sites were small in number, were characterised by the fact they were all rail-served and contained excavations and underground works. Since information on them was not as complete as for the ski-sites, the agents were posed a series of questions. What is the general purpose of the sites? What is the purpose and construction of any buildings, excavations or tunnels? What connection do they have to the ski-sites? They were asked to provide detailed descriptions about equipment, materials or supplies arriving at a site, and in the case of equipment, since such items were clearly important to the operation of the secret weapon itself, they were told that every effort should be made to ascertain where the deliveries came from. This was particularly the case for agents who had contacts in the railway industry.

The importance of censorship to ensure that negative morale among the public was minimised was considered paramount by the British government throughout the battle against the secret weapons, and the reasons why were spelled out to newspaper editors invited to a secret conference on 12 October 1943. This followed a report by Herbert Morrison, the Home Secretary, to the Prime Minister on 5 October 1943 that Fleet Street was in possession of a good deal of information on German secret weapons: that articles touching on them had already been published, that recent developments would inevitably come to the attention of the public and would invariably lead to difficult questions being asked. At the conference Herbert Morrison and Brendan Bracken, the Minister of Information, met with the editors of the major London, weekend and provincial newspapers and explained that they had been invited for two reasons: firstly to be given 'in-confidence information' about an important possible development in the air war, and secondly to obtain their advice on the best method of handling publicity on the matter if and when the time came.

Acknowledging that Hitler and Goebbels had repeatedly promised the German people reprisals that would destroy London, the politicians reflected on how the previous month the German Führer had said: 'It is only from the air that the enemy is able to terrorise the German homeland. But here, too, technical and organisational preparations are in hand which will enable us not only finally to break his terror tactics, but also by other and more effective means to retaliate'. Observing that 'for once Hitler may have been speaking the truth', the editors were reminded of a speech made by Churchill to the House of Commons on 21 September 1943 which described a rocket-assisted glider that could be released from a considerable height and then be guided towards its target – a reference to the Henschel 293 glide bomb – and that the Prime Minister had gone on to observe that: 'the Germans are developing other weapons on novel lines with which they hope to do us damage and compensate to some extent for the injury which they are receiving daily from us'.

The editors were told that for some time a wealth of intelligence, which frequently conflicted in matters of detail, about a long-range missile the Germans had been experimenting with, had been received from a variety of sources. Unclear how true or untrue much of this information was, the politicans were at pains to stress that the enemy would certainly like the British to believe all the propaganda and become 'terror stricken'. Whilst some sources believed the threat to be from a glorified Big Bertha – the super-heavy howitzer, developed by the armaments manufacturer Krupp, that bombarded Paris during the First World War – other accounts told of an enormous rocket with an explosive charge. The importance placed by Hitler on the development and manufacture of secret weapons was emphasised by revealing the existence of the Peenemunde experimental station and the fact that the Germans were then busily constructing a network of bombproof installations in northern France from where London and other cities within range might soon be bombarded.

In spelling out the editors' responsibilities, Morrison and Bracken told them that their business was to ensure that while the enemy continued its experiments with secret weapons no information should be released that could improve the morale of the Germans, either by revealing that the Allies were making preparations at the cost of other war activities or by providing details of what those preparations might be. They were reminded of their responsibilities under existing censorship rules and the most important issue was identified as their duty not to give the enemy any technical details of defensive preparations. Since existing air-raid warning systems were likely to be completely inadequate against these new weapons, which would be launched from so far away and would travel so quickly, for the time being the only warning would come from the firing of anti-aircraft guns. They were advised that sirens would be replaced with the firing of maroons – explosive flares – but for the time being nothing should be communicated to the public: they would be told in good time. Any premature alert might cause unnecessary panic and the enemy must not learn anything about British preparations.

For this reason it was explained why a recent newspaper article mentioning that 20,000 tons of steel had been reserved for additional Morrison shelters in London and the strengthening of existing shelters had been a mistake. Since heavy bombardment akin to that experienced during the Blitz was considered a very real possibility, civil defence reinforcements and labour would need to be brought in to rescue trapped people, clear debris and demolish damaged buildings. To accommodate this army of workers temporary camps would need to be created now and that required the clearing and levelling of bomb sites. The politicians were keenly aware that much was being made in the country, at this stage in the war, of the aggressive will to victory, and they realised that any mention in the newspapers of new shelters in London or of labour being used on purely defensive measures could easily lead people in Britain, and indeed the enemy, to draw their own conclusions. At the same time as they were briefed on the necessity of controlling the British media, the newspaper editors were told that, providing it didn't disclose anything to the enemy, they were to continue publishing tales from neutral sources since this would have the effect of familiarising the British people with the possibility and would lessen the shock if the secret weapons ever came.

*

From being created overnight on 8/9 November 1943, B2 Section steadily grew in size throughout the remainder of the year and in 1944, at the height of the investigation at Medmenham, 2 Army officers, 6 RAF officers, 11 WAAF officers, 2 American officers and 5 other ranks from the Army – a total of 26 people – were dedicated to the hunt for secret weapons. On discovering a Bois Carré site, the interpreters used comparative cover – recently flown photography alongside historical photography from the Medmenham Print Library – in order to understand the stages and speed of development of the sites. In doing this they discovered that as early as 24 September 1943 they had recorded tree clearance and dumped building materials at a ski-site that was later confirmed. As the interpreters continued their hunt for the ski-sites in late November 1943 – in all locations within 130 miles of London – and with frequent rumours from ground-intelligence sources prompting the reinterpretation of entire areas, it was soon observed that, as yet, no small pilotless aircraft had been spotted in any of the aerial photographs.

To help find this evidence, on 13 November 1943 Douglas Kendall turned to third-phase interpreter Constance Babington Smith and asked the Aircraft Section to look again at Peenemunde. During the investigation earlier that year, the aircraft specialists had been asked to watch Peenemunde for 'anything queer' and in the process had discovered the jet-powered fighter, the Messerschmitt Me 163. Kendall now hoped they would have similar success in finding a photograph of the pilotless aircraft at the experimental station. Such was the level of secrecy surrounding Operation Bodyline that until then nobody in L Section knew anything of the rocket investigation, the discovery of rockets, the Stafford Cripps meetings, or the existence of Bois Carré sites in northern France. With the briefing that they should look for something smaller than a fighter aeroplane, the most detailed prints of Peenemunde were retrieved from the Print Library, on which the Messerschmitt Me 163 – provisionally dubbed the 'Peenemunde 30' – had been spotted.

These same photographs had been viewed by Frank Whittle, the pioneer of jet propulsion, who visited Medmenham in the summer of 1943 while attending a three-month War Course at the RAF Staff College. Whittle later reflected on the fascinating visit to Medmenham, during which Constance virtually detached him from the rest of the visiting party on discovering his identity. Captivated by the attentions of the attractive young lady who shared his fascination with flight they spent most of his visit head-to-head, poring over photography of Peenemunde and other airfields that showed indications of German jet activity. Both were fascinated by the telltale signatures – double scorch marks in the grass – which indicated, even though none appeared in any photographs, that the Germans must be flying a twin-engined jet aircraft. Whittle received many envious glances from the other students on the course, and on the return journey to the RAF Staff College – then based in Bulstrode Park to the north-west of Gerrards Cross, Buckinghamshire – he was the butt of friendly banter from his fellow students. When one of his tutors, Air Commodore Arthur 'Fido' Fiddament, the Assistant Commandant of the College, asked what perfume Constance wore Whittle replied, much to Fiddament's amusement: 'Judging from its effect on you, sir, it must have been Air Commodore's ruin'.

Revisiting these photographs of Peenemunde again in November 1943, L section spotted a small object, not beside the airfield but in a small enclosure behind the aircraft hangars, immediately adjoining a building that the interpreters suspected was being used for jet-engine testing. With a wingspan of only twenty feet it was christened the 'Peenemunde 20' and captured the interest of both Douglas Kendall and Squadron Leader Michael Nikolaevich Golovine. Born in Petrograd, Russia, Golovine had worked for the Aero Engine Department of Rolls-Royce before recruitment by the Technical Intelligence Branch at the Air Ministry, where he specialised in the evaluation of enemy aircraft. Golovine was involved throughout in the hunt for the secret weapons and after the war would become an authority on military aspects of space flight, a powerful advocate of European participation in astronautics and President of the British Interplanetary Society. In his 1962 publication *Conflict in Space* he advanced the hypothesis that developments in military spacecraft could lead to active conflict between the Soviet Union and United States of America beyond the Earth's atmosphere. He considered the scenarios should one of the superpowers destroy the other's space fleet and questioned whether this would lead to a mutually destructive nuclear war on Earth or conditional surrender with the victor then able to control the world.

Whilst it would be accepted that the newly discovered jet-propelled 'Peenemunde 20' had such unusually small dimensions that it could conceivably be an expendable aircraft, as yet there was no evidence regarding the quantity of production. While the Aircraft Section continued searching for more evidence, they had to depend on their experience of recognising the unusual among the normal activities of enemy aircraft factories and airfields. With Constance recalling an earlier rumour that circulated at Medmenham about launching rails being used for secret weapons – which doubtless followed the visit by Terence Sanders of the Ministry of Supply on 3 November 1943 – they spotted what looked very much like a ramp pointing out to sea at the Peenemunde airfield. Reckoning that this could very easily have been a catapult for launching pilotless aircraft, Constance records taking the photographs to the Industry Section who told her that the area had already been studied and interpreted as having something to do with land reclamation. Being perhaps more aware of the possibilities of pilotless aeroplanes, she was not altogether convinced and resolved to show the photographs to Douglas Kendall. At the time of this discovery on 1 December 1943, Kendall was attending a meeting of the Crossbow Committee – which reported to the Joint Intelligence Committee – to explain the theories developed by the Medmenham interpreters from their studies of the Bois Carré sites. In recognition of the fact that the British and Americans were now actively planning countermeasures against the Heavy Sites and the Bois Carré network, all of which had been assigned 'NO-BALL' references during the British-led Operation Bodyline – the term referring to the ungentlemanly delivery of a ball in the game of cricket – on 27 November 1943 the investigation had been renamed 'Operation Crossbow'.

At the Crossbow Committee meeting on 1 December 1943, and while Captain Neil Simon dealt with the network of associated supply sites, Wing Commander Douglas Kendall revealed a dramatic development in their understanding of how the Bois Carré sites worked, thanks to information captured during sortie N/980. Flown on 28 November 1943 by Squadron Leader John Merifield in a PR Mosquito with his navigator Flying Officer William Whalley, the 540 Squadron ORB records them as having left RAF Leuchars at 09:55, on a flight of six hours and ten minutes, to take post-strike photographs of Berlin. This followed the Bomber Command attack on the city the previous night, when 443 Lancasters and 7 Mosquitos attacked an area marked by the Pathfinders 6–7 miles north-west of the city centre. Whilst most of the British bombs fell on the semi-industrial suburb of Reinickendorf, Berlin Zoo was also heavily bombed and this resulted in several large and dangerous animals escaping, including apes, leopards and panthers which roamed the streets until they were shot on grounds of public safety. On reaching the German capital that morning, and discovering their target was obscured by 10/10 cloud cover, Merifield photographed alternative high-priority targets in surrounding areas. These included Stettin, where the Kriegsmarine's one and only aircraft carrier the *Graf Zeppelin* was docked; Swinnemunde, where Merifield and Whalley photographed the German crusier *Admiral Scheer*; airfields at Anklam, Greifswald and Stralsund; plus vertical and oblique photography of the Peenemunde area, including the town of Zinnowitz.

Eight miles south-east of Peenemunde on the island of Usedom, the Hotel Preussenhof in the popular spa town of Zinnowitz had been home to Colonel Max Wachtel since 12 May 1943. While staying in the hotel beside the white sandy beaches of the Baltic Sea, the Luftwaffe officer had been recruiting personnel for Flakregiment 155(W), and would soon be commanding the network of Bois Carré sites then being developed throughout northern France. Between Zinnowitz and Peenemunde, three experimental Bois Carré sites had been built, hidden among woodland on the coastline of the Baltic Sea, to both test the V1 and train its operating personnel who were then being recruited throughout the Luftwaffe. The V1 had been designed under the code name *Kirschkern* (Cherrystone) by Fritz Gosslau and Robert Lusser, chief designer and technical director at Heinkel. The fuselage was mainly welded sheet metal and the wings plywood, and the sound of its simple pulse-jet engine was responsible for its later nickname of 'doodlebug' or 'buzzbomb'. Known for a while in Germany, on Hitler's orders, as the *Maikäfer* (mayfly) and *Krähe* (crow), the pulse-jet engine was started with a current from a portable starting unit. Acetylene gas was commonly used for starting, with a panel of wood held across the end of the tailpipe to prevent the fuel from diffusing and escaping before ignition. Once the engine was started and its minimum operating level reached, the hoses and connectors were disconnected. The catapult system was then triggered and the missile launched. Take-off speed was about 360 mph.

Squadron Leader John Merifield had first photographed Peenemunde on 19 January 1943 – sortie N/709 – after aborting another damage-assessment sortie over Berlin when his PR Mosquito left a telltale contrail in the sky to the north of Berlin and attracted the attention of enemy fighters. Whilst the pilot later reflected how much he enjoyed flying solo in PR Spitfires, since poor visibility meant he had to bank the aircraft to check his position he couldn't be as accurate as he could in a PR Mosquito where he had the advantage of flying alongside his navigator, which both increased the accuracy of his flying and that of the aerial photography and also improved the odds of survival because two pairs of eyes were keeping watch for enemy fighters. Merifield would become closely identified with the de Havilland Mosquito and on 6 September 1945 he achieved what was then the fastest Atlantic crossing when he flew a PR Mosquito from RAF St Mawgan, Cornwall, to St John's in Newfoundland, Canada – a distance of 2,300 miles – in 7 hours and 2 minutes. The high performance of 'The Wooden Wonder' infuriated Hermann Goering who lamented in January 1943 that:

In 1940 I could at least fly as far as Glasgow in most of my aircraft, but not now! It makes me furious when I see the Mosquito. I turn green and yellow with envy. The British, who can afford aluminium better than we can, knock together a beautiful wooden aircraft that every piano factory over there is building, and they give it a speed which they have now increased yet again. There is nothing the British do not have. They have the geniuses and we have the nincompoops.

Following the briefing from Colonel Terence Sanders on 3 November, that the Germans might be planning to deploy lighter projectiles – possibly launched from two rails inclined at a steep angle – Kendall had been puzzling over how the Bois Carré sites could be intended for the rockets seen at Peenemunde. Perplexed at how anything as large as the rocket, could possibly be stored in the ski-buildings or even moved to the sites – given the tight corners on many of the rural roads in northern France – Kendall's suspicions were only strengthened when the aircraft specialists discovered the 'Peenemunde 20'. By analysing every Bois Carré building in turn the alternative hypothesis was developed that the two longer ski-buildings were used to store the main body of the projectile and the shorter structure housed the wings. Since the platforms pointed towards London – the direction became known as the London Line – Southampton, Bristol and Portsmouth, the likelihood of this being the launch site of some form of projectile was high. Desperate to confirm and develop his understanding of how the Bois Carré sites worked, and with the second meeting of the Crossbow Committee scheduled for 1 December 1943, the photographs taken over Zinnowitz on 28 November 1943 allowed Kendall to analyse and clarify the operation of an active Bois Carré site for the first time. The photographs of Zinnowitz had been taken to record a wireless station near the town and were brought to the attention of the Crossbow interpreters on 30 November 1943, in time for copies to be taken to the Crossbow Committee. The photographs of Zinnowitz taken during sortie N/980 provided an important link that had until now been missing: the Luftwaffe had a prototype site in Germany for trials and the training of personnel.

Revealing the aerial photography of Zinnowitz, Kendall explained that the Bois Carré sites in northern Germany matched those in northern France in every way, except there were no ski-buildings – which tended to confirm that they were used for storage – and at Zinnowitz ramps could be seen inclined at about ten degrees, pointing out to sea. Kendall explained how he had analysed these new photographs alongside photography of the sites in northern France, and advanced his theory that the weapons deployed from the Bois Carré sites would be pilotless aircraft or glide bombs. He explained that each operational site had three ski-shaped buildings, with two of equal length for the storage of fuselages whilst the shorter one was used for the storage of wings. Making reference to a diagram prepared by W Section – 'CIU Plan R12: Layout of Typical Ski Site' – Kendall observed that access to these buildings could only be achieved through the large rectangular one, which consisted of a chamber running end to end and containing five small rooms. Suggesting that the weapon could arrive by lorry, already mounted on an undercarriage, and be unloaded into the rectangular building, Kendall pointed out that on every site, close to this building, was a smaller rectangular building that was divided into three rooms. A buried pipeline ran from the centre room of that building to the large rectangular building, and since several CX reports had indicated that the small building contained air compressors it was proposed that the projectile would be charged with compressed air in the large rectangular building on arrival. The bomb would then be taken on its undercarriage along the winding concrete pathways and be stored in a ski-building. Since there was no indication of fuel storage on any site, it was suggested that the projectile would arrive with its warhead complete and its fuel tank already full.

Being able to look at the internal layout of walls inside a building, before the roof went on, was to prove critical in the construction of scale models that helped in the long discussions about the different hypotheses that various people had. One high-ranking Army officer, on a visit to Medmenham, examined one of the models, lifted its roof off and enquired, 'How on Earth could you possibly know what was inside from just staring at photographs?'

With the use of aerial photography of Zinnowitz from sortie N/980 and Yvrench's Bois Carré site shots, Kendall showed that another square building, 23 feet wide, was always orientated parallel to the firing site, towards England. Since agent intelligence – from the Réseau AGIR – indicated that no metal, except non-magnetic metal, was used in that building it was suggested that the control mechanism for the weapon was a magnetic automatic pilot. The evidence leading to this came from the wireless experts who worked under Squadron Leader Claude Wavell, who had been tasked with identifying any evidence of new wireless transmitters in northern France that might be used to guide a pilotless aircraft. For some time in the later part of 1943 G Section had been spotting lattice structures, 30 to 35 metres in height, built on the highest points in northern France. This evidence included photographs taken on sortie E/463 flown by Flight Sergeant James McGinn Aitken on Wednesday, 3 November 1943, which had recorded the busy construction site at Yvrench's Bois Carré for the first time. Whilst it was apparent that some structures were built over trigonometrical stations operated by the Institut Géographique National, some were not and therefore their purpose remained unclear. Nicknamed the 'spires' by the PIs, their connection with the ski-sites became clear when the bearing of each firing ramp was tied into the new geodetic points with great precision. Interpretation Report G1R was completed for Kendall on 29 November 1943 in time for his meeting with the Crossbow Committee on 1 December. The report drew attention to the significant efforts being made by Wehrmacht cartographers resurveying large parts of northern France – something that would have been unnecessary if the V1 could be steered – and highlighted not only the relative inaccuracy of the French maps but the connection between the 'spires' and the long-range bombardment of England.

Kendall then contested the assumption by which the interpreters had been advised that a rocket would be fired (wrongly, it would prove) at a steep angle of approximately seventy degrees rather than at a low angle up a ramp, such as was implied by the platform. The lack of any heavy equipment to handle a large projectile and the tight angles of the path network indicated more some form of aircraft than a rocket. A study of the path layout revealed important evidence about the method of operation of the site. Since every location had been physically arranged dependent on the local topography, it was possible to study all the sites together and see what buildings could provide access to other buildings. A process of elimination therefore allowed a reasonable firing sequence to be deduced. It was suggested that the fuselage on its undercarriage would be brought from one of the two larger ski-buildings to the large square building, the wings would be brought from the shorter ski-building, and the bomb would be assembled and mounted on a base while the automatic pilot was set in operation. The projectile would then be taken to the firing ramp and mounted on the ramp where a rocket would then be placed under each wing. The firing point itself consisted of a rectangular concrete platform, a small concrete shelter for the person operating the firing control, and six pairs of posts leading from the platform in the direction of London and a growing number of targets in southern England. The small rectangular building with three small rooms, one of which was believed to contain an air compressor, was also where the electrical power serving the site arrived. From analysis of the low-level oblique photography, taken by 170 Squadron, a 15-kilovolt transformer could be seen, which meant the amount of electrical power entering the site – which would do little more than provide lighting – was small. Instructions to fire would be received by telephone at the firing point. Immediately before and after firing, water would be sprayed at the bottom of the ramp to prevent the surrounding trees catching fire.

After Kendall, Captain Neil Simon revealed the network of Bois

Carré Supply Sites to the Crossbow Committee. They were at six locations, equally spaced in the firing belt, which implied that each of them served eighteen firing sites and that, if this theory was correct, a maximum of 108 firing sites could be expected.

By this point in early December 1943, sixty-four Bois Carré-type sites had been pinpointed through aerial photography. But since the Réseau AGIR had identified approximately one hundred, the search continued in earnest. It was estimated that twenty-six of these sites were 50 per cent or more structurally complete and it was becoming increasingly evident to many that if the threat that they posed was to be kept under control the destruction of the more advanced of them had to begin without delay. But this was not an opinion yet shared by the Joint Intelligence Committee, whose views had not changed since their meeting on 21 November 1943 when they had considered the first report of the Crossbow Committee, which had reviewed the threat to Britain from attack by long-range rockets, pilotless aircraft and glide bombs, concentrating on new evidence, and modifications to conclusions or deductions expressed by Duncan Sandys or Sir Stafford Cripps. The site at Bois Carré was recorded as being the most advanced, with an estimate that, at the current rate of construction, it could be completed in one month.

The least advanced sites would probably take another three months to complete, based on figures prepared by a specialist from Sir Robert McAlpine Limited who estimated that, based on the number of men seen to be working at the Bois Carré site, construction would require approximately '450 man months' from start to finish. But with a far larger labour force, if the task was considered urgent enough by the Germans, then it would take two months or around 300 man months from start to finish. The committee did not share the view of the Cripps Enquiry that the sites should be attacked immediately. The careful planning of countermeasures was favoured over the diverting of large-scale bombing to targets that were considered not yet 'ripe'. It was accepted that no appreciable threat from long-range rockets could materialise before the middle of February 1944, and ongoing information-gathering on the movement of personnel and heavy railway traffic, alongside photographic intelligence, would provide around one month's warning. At the end of the meeting, the committee were recorded to have been somewhat sceptical about the flying bomb theory, but nevertheless asked for the theories to be submitted as a report.

Returning to Medmenham late on 1 December from the Crossbow Committee, pale-faced and exhausted, still clutching his jacket and briefcase, Douglas Kendall rushed to see Constance Babington Smith in the Aircraft Section to ask L Section to begin searching for more intelligence from Peenemunde on 'the aircraft that could be launched from the ramps'. When Constance swiftly showed Kendall the ramp found at Peenemunde airfield earlier in the day, he instantly confirmed their suspicion, and while he began writing Interpretation Report BS164 for the Crossbow Committee – which he worked on through the night – the third-phase interpreters began searching the photography from the Print Library to understand the chronological development of the Peenmunde airfield experimental site. Photography from January 1943 revealed one completed firing site and a second under construction, while photography of May 1943 showed that the second site had been completed.

The most recent photography of Peenemunde, taken by Merifield during sortie N/980 on 28 November, had been doing the rounds of the specialist third-phase sections. When print copies arrived in L Section a few hours after Kendall's visit, the aircraft specialists looked immediately at the photography of Peenemunde airfield and on examining frame 4031 which covered the ramps at the start of a run over Peenemunde – unfortunately no stereo pair had been taken – they could see a cruciform object at the bottom of the ramp. Finally Medmenham had the irrefutable evidence of a 'Peenemunde 20' sitting on a ramp, ready to be fired. In Interpretation Report BS164, Kendall was able to record not only two prototype ramps additional to those at Zinnowitz but the signature marks left by pilotless aircraft that had crash-landed into the estuary sands. To reinforce the fact that Crossbow was now a hunt for both rockets and flying bombs, rockets could even be seen next to the elliptical embankment at Test Stand Seven.

Turning to the installations at Peenemunde on the reclaimed land to the north-east of the airfield, the most important fact highlighted was 'the identification of a Peenemunde 20 glide bomb on one of the two prototype firing points'. When comparing the sites at Zinnowitz, which had been photographed in stereo, with the prototype site at

Peenemunde the firing points were seen to be identical and, based on quick measurements, the ramps were 130 feet in length, the ramp end was twenty-one feet above ground level, and the gradient was around nine degrees. Kendall ended his Most Secret report with several key conclusions: that the Bois Carré sites were designed for the projection of glide bombs of the Peenemunde 20 type; that the site at Zinnowitz provided confirmation that the body of the bomb must pass through the larger rectangular building; that the absence of any ski-buildings at the experimental Zinnowitz site tended to confirm that these were for the storage of the bombs; and that since the firing sites in northern France had already developed on similar lines to those at Zinnowitz and Peenemunde it seemed probable that they would follow the same later stages as their construction neared completion.

A copy of Frame 4031 from sortie N/980 was sent to Squadron Leader Michael Golovine at the Air Ministry, who visited the Royal College of Science to use their high-precision microscope to measure the cruciform shape. Alongside him was Peter Endsleigh Castle, a gifted artist who had been recruited by Air Intelligence at the beginning of the war. One of his first tasks – given that detailed illustrations of aircraft were frequently limited to those contained in magazines such as *Flight* and *The Aeroplane* – had been to illustrate aircraft-recognition booklets about foreign aircraft. Much of his time was spent with No. 1426 (Enemy Aircraft) Flight RAF – nicknamed the 'Rafwaffe' – who were responsible for evaluating captured hostile planes. This involved inspecting crashed enemy aircraft that would be carefully photographed, sketched and measured. Alongside information from espionage sources detailed cutaway drawings would be created. With the wingspan of the pilotless aircraft scaled, using the microscope, to be 16 feet 9 inches and with the photographs and sketches supplied earlier by Hasager Christiansen of the pilotless aircraft that had crashed into a field of mangel-wurzels on the island of Bornholm on 22 August 1943, they could begin to build a picture of the secret weapon that would soon be christened the V1.

All these discoveries at Medmenham happened on the last day of the Tehran Conference, when the Big Three – Prime Minister Winston Churchill, Marshal Joseph Stalin and President Franklin Roosevelt – met and agreed that the Western Allies would open a second front against Nazi Germany. Alongside Winston Churchill, acting as a family aide-de-camp because of the Prime Minister's frequent ill-health, was his daughter Sarah who had been given special leave from her duties at Medmenham. During the trip she was taken on a flight over Sicily in a Mosquito, and immediately recognised places that she knew so well from scrutinising them at her desk at Medmenham. With no need for a map, the only unfamiliar aspect of the experience was looking at places from above in colour rather than black and white. Also attending the conference was the United States Ambassador to Great Britain, John Gilbert Winant, who had a wartime affair with Sarah Churchill much to the alarm of some in Washington D.C. who were concerned at the extent to which senior Americans fell for the charms of the British.

In February 1943 the US Eighth Air Force had established their British photographic reconnaissance base at RAF Mount Farm, five miles north-west of Benson, and gave it the USAAF designation 'Station 234 (MF)'. The first Americans to occupy Mount Farm were 13th Photographic Squadron, who flew the Lockheed F5 (P38) Lightning, and as their films were initially processed and first-phase interpreted at Benson, American PIs became increasingly frequent visitors there. On 7 July 1943 the US 7th Photographic Group (Reconnaissance) was established at Mount Farm and Squadrons 13, 14, 22 and 27 flew a combination of Lightnings, North American P-51 Mustangs and, in time, Supermarine Spitfires Mark IX. And whilst the Americans established their own film-processing facilities on the base, all the first-phase interpretation was undertaken by shifts of fourteen interpreters – recruited from throughout the Allies – at Benson.

In his 1978 book *Most Secret War*, R. V. Jones records that Zinnowitz was identified by the French agent Jeannie Rousseau (code-named Amniarix) as the location of a facility used by Luftwaffe Colonel Max Wachtel. Jones claims that it was this revelation which prompted him to request photographic-reconnaissance mission N/980 in the hope that it would establish a definitive connection with the Bois Carré sites in northern France. Furthermore, by intercepting German radar tracks, which indicated when the Germans were firing projectiles, he could recommend the optimum time of day for the sortie to be flown to stand a chance of photographing a pilotless aircraft. But the Zinnowitz radar station was in any case a priority target, which explains why Merifield photographed it after discovering his primary target in Berlin was obscured by 10/10 cloud cover. The suspicion that Jones had a tendency to embroider the facts is further reinforced by his observations on the achievements of L Section – whom he failed to acknowledge in his book – when he fleetingly described their breakthrough detection of the pilotless aircraft on the ramp at Peenemunde as nothing more than an accidental discovery. Their masterful photographic interpretations, particularly given how little information was available to them, proved to be one of the most important and dramatic intelligence discoveries of the war.

The relationship between R. V. Jones and Medmenham was frequently strained but never more so than during the hunt for the V weapons. On one occasion, on 18 June 1943, Jones took credit for spotting a rocket on a Peenemunde railway truck in photographs produced by sortie N/853 on 12 June, and later had the audacity to criticise Medmenham for having 'missed it'. In the detailed third-phase interpretation report DS9 of 16 June 1943, André Kenny makes reference to an object measuring 30 feet by 8 feet. He later claimed never to have doubted that it was a rocket but, not being a scientist, he was constrained by the fact that interpreters were forbidden to speculate about objects until their existence had been officially accepted. Whilst Kenny could perhaps have been more forthright in his report, it describes in considerable detail the existence of a column about 40 feet in height on the foreshore, which was missed by everyone at the time – even R. V. Jones – as being a rocket in the vertical position. The discovery of a rocket in close proximity to the elliptical embankments and moveable cranes that everyone assumed were required to handle the bulk and weight of the rocket had the effect of focusing the investigation on this part of the site. What Kenny had noticed – and if investigated further this could have been realised earlier – was that the rockets were ultimately more mobile than the pilotless aircraft, and in effect did not require a specific launching site.

Jones would also lambast André Kenny for identifying the ramp at the Peenemunde airfield as 'sludge-pumping equipment', something that Kenny later asserted had been misconstrued from his early reports. Certainly his pre-war experience of land drainage in the English Fens, combined with his academic studies into hydraulic engineering in Greek and Roman civilisation, made Kenny an easy target for the 'father of scientific intelligence'. But Douglas Kendall certainly had sufficient concerns about Kenny – considering him to have a 'closed mind' – which prompted his putting him firmly onto the industrial aspects of the investigation when he took over from Hamshaw Thomas. Kendall considered Kenny to have misrepresented information during some of the early Cabinet meetings, and reckoned that whilst Kenny might have spotted rockets lying on the ground at Peenemunde and reported in great detail on the large experimental testing installations, he had entirely missed the significance of the asphalt-covered area near Test Stand Seven. This area was located near the water's edge and was accessed by a road leading from the main experimental site away from the elliptical earthworks and was actually the location from where the test launches were staged. Some photographs had been taken of rockets in the upright position for launching, which had been reported by Kenny as towers. This led to the failure to fully interpret the site, delaying the realisation that the rockets ultimately did not need a launching site.

CHAPTER 8
No-Ball Targets, Hottot and Belhamelin Sites

After writing Interpretation Report BS164 overnight – in which he meticulously detailed the unmistakable links between the Bois Carré sites on the island of Usedom and those in northern France – Douglas Kendall sent his report to the Joint Intelligence Committee on 2 December, which resulted the following day in Air Ministry Signal AX549 and the order that attacks on Crossbow targets should 'be carried out forthwith'. With each site believed to house twenty flying bombs, and with ground intelligence indicating there could be upwards of a hundred sites, the fear was that 2,000 flying bombs could be fired on England every day. In deciding the order in which the growing network of Crossbow targets should be attacked, the military planners turned once again to Medmenham whose interpreters had been studiously preparing Target Folders on each Bois Carré site since the end of November, consisting of a map pinpointing the various features of each site and aerial photography of the target and surrounding area. With the reconnaissance programme for northern France providing detailed vertical coverage of each known Bois Carré site on a monthly basis, and low-level oblique photography where particularly valuable information could be obtained, both the growth and the destruction of the network would be recorded in the Medmenham Print Library.

When the mission planners debated the best way to destroy ski-sites they considered using a limited number of heavy bombers that used radio-navigational aids. Deciding they should experiment with different bomb loads – so that they could determine the best composition for this type of target – they requested that nobody else attacked these sites in order that the damage assessment from their attacks could be accurately carried out. On learning about these proposed raids, RAF Station Benson sent a secret cypher message to the headquarters of the Allied Expeditionary Air Force, US Ninth Air Force and Second Tactical Air Force to inform them that they were constantly involved in photographing Crossbow targets in northern France and requested information about the approximate times, flying routes, targets and altitudes of the bombers concerned. In finishing the cypher they politely requested that they would 'much appreciate if all your aircrew be informed that lone Spitfires or possibly Mosquitos might be operating in areas at same time at heights between 30,000 and 40,000 feet'.

With so many competing demands on the Allied Air Forces meaning that there were no resources to waste on attacking sites prematurely, it was imperative that Medmenham worked quickly to locate all the sites and identify those nearest completion. But with so many different contractors using different construction methods in the network of Bois Carré sites, a scoring system was developed by the Crossbow team to assess the relative state of completion of each site. With the main components of a ski-site awarded points according to their relative importance and each site totalling one hundred points, as new aerial photography was taken each component was reassessed with points added according to progress. When the total reached fifty or more, a sufficient amount of civil engineering work was deemed to have been completed and the site was recommended for attack. Points were allocated on the basis of:

Ski-buildings	30
Half-ski	5
Square building	10
Rectangular buildings (8 points each)	16
Launching point	10
Blast-wall building	5
Pyramidal hole	5
Concrete paths	15
Other installations	4
GRAND TOTAL	100

Based on the assessment that each site would be ready after 133 days of construction, and since weather conditions often prevented photographic-reconnaissance missions taking place, each site was automatically allocated three-quarters of a point per day. This meant that sites were often added to the attack list due to accumulated points even without photography having been possible because of bad weather. Once a site was attacked, the damage was reflected in the points system and meant that, if the Germans rebuilt a site, it would reclimb the chart. On 1 December 1943, while Douglas Kendall was at the Crossbow Committee, work was already under way planning attacks on the Crossbow sites in northern France. That day the Deputy Chief of the Air Staff (Operations), Air Marshal Norman Bottomley, wrote to the Commander-in-Chief of the Allied Expeditionary Air Force – Air Chief Marshal Sir Trafford Leigh-Mallory – with information about twenty-four sites where the civil engineering was more than 50 per cent completed. Spelling out the importance of Operation Crossbow, Bottomley explained that:

It is desired to make it quite obvious to the enemy that we are making a concerted attack against these 'ski sites'. The initial attack by the Tactical Airforce should, therefore, include as many of the 24 sites as the force available will allow. Thereafter the Tactical Air Force and 9th Air Force should proceed with the methodical destruction of individual sites.

This was followed on 5 December 1943 by the Air Ministry circulating a 'Summary of Information' on Operation Crossbow to all the Allied Air Force headquarters, a report that highlighted the threat presented by these new weapons. Now that the operation had advanced to the stage of countermeasures, the existence of German pilotless aircraft, the connection between Peenemunde and ski-sites in the Pas-de-Calais and Cherbourg peninsula, and the threat they represented, could be more openly discussed. It was explained that with the estimated number of ski-sites the enemy could achieve the equivalent effect of at least a 2,000-ton bombing attack in a period of twenty-four hours and that in northern France there were also four or five large construction sites which had not been proved to have a connection with the ski-sites but could possibly be connected with a long-range attack of a different kind. That same day attacks were made by Lancaster bombers carrying 1,000-pound bombs on three sites in the Pas-de-Calais region including Ligescourt (No-Ball Target XI/A/40) and Saint-Josse au Bois (No-Ball Target XI/A/19).

By 9 December Bottomley showed his considerable discontent at the lack of progress in a letter to Leigh-Mallory, when he observed that since the date of the Air Ministry signal only three attacks had been made and no effective results had been achieved. Whilst accepting the fact that weather conditions had restricted offensive operations, in view of the extreme importance of destroying Crossbow targets, given the threat they posed to the cities of southern England and ports essential to Operation Overlord, he stressed that every practicable opportunity was to be taken. In reply Leigh-Mallory accepted the need for urgency but, reflecting a degree of the indifference felt by many who believed that Crossbow operations represented a diversion of resources from preparations for Operation Overlord, he stressed that operations to support the strategic bomber had overriding importance. But with thirty-five of the ski-sites more than 50 per cent completed by 15 December, and bad weather continuing to hamper the Tactical Air Forces, it was agreed that the American Eighth Air Force would play a greater part in the offensive countermeasures once the weather improved.

The lack of success had also prompted Air Vice-Marshal Horace Wigglesworth, at the headquarters of the Allied Expeditionary Air Force, to question Air. Officer Commanding No. 11 Group RAF – Air Vice-Marshal Hugh Saunders – whether they understood what was meant by 'maximum concentrated effort'. Replying on 15 December, Saunders expressed full appreciation and confidently observed that now the Eighth Air Force was also involved, as soon as they had favourable weather, he had 'little doubt that a fair percentage of the "Crossbow" Targets will be knocked out in one day'. The Americans had their pick of the thirty-five ski-sites with several notable exceptions, including the site at Yvrench's Bois Carré that was being monitored for intelligence purposes and was reserved for tactical trials by Bomber Command. For the avoidance of doubt, the importance of Crossbow targets was spelled out to them with the words 'this operation is to have overriding priority over any other operation when weather conditions over northern France allow of the operation'.

*

During the first half of December key information derived from aerial photography of the Bois Carré sites was consolidated into Interpretation Report BS194. Issued on 14 December, it was accompanied by a list showing the state of completion of the sixty-four known sites, plus information about an additional twelve newly located sites. By this date the site at Yvrench's Bois Carré (No-Ball Target: XI/A/39) is recorded to have been 99 per cent complete, a level of completion that was doubtless making the draughtsman André Camp – who had supplied Michel Hollard with the detailed plans of the site – increasingly concerned. By 16 December, and with the Crossbow operations well under way, the site that had been the focus of the investigation from the start was added to the 'No-Ball Target List'. Until that day, ski-sites had been classed as 'Rhubarbs' – the code name given to opportunity targets – but now they were strategically important targets, each of which had to be destroyed.

On 19 December the Crossbow Section at Medmenham was expanded and became operational around the clock – in advance of large-scale attacks beginning on the ski-sites from 21 December, after which date the Crossbow interpreters were fully occupied on damage assessment and the rapid provision of target information, which included spotting the appearance of anti-aircraft batteries at the Bois Carré sites. After particularly heavy raids on Christmas Eve 1943, on 29 December a memo was sent from Medmenham to the Assistant Director of Intelligence (Photography) – Group Captain 'Daddy' Laws – about damage-assessment policy. In reviewing operations taking place against No-Ball Targets over the previous ten days it was emphasised that the points system needed improvement, as information was required more quickly on whether particular Bois Carré sites needed to be bombed again. The procedures adopted involved the duty controller at RAF Benson being telephoned with a list of targets and the attack timetable so that post-strike sorties could be despatched to arrive over target as soon as possible after the attack. Daily at 18:00 hours, Benson telephoned Laws at the Air Ministry on extension 8787 to provide a verbal interpretation report while the first-phase interpretation report was circulated.

In a memorandum of 19 January 1944 – 'Operation Crossbow: Attacks on Targets in northern France' – Leigh-Mallory lamented the dramatic rise in Bois Carré sites more than 50 per cent complete and 'ripe' for attack, and that from there only being twenty-four on 1 December 1943 the figure had worryingly jumped to eighty-nine by 12 January 1944. With the main factors preventing the destruction of the targets cited as bad weather and difficulties of target identification, he noted that little marked improvement in bombing results could be expected from medium and light bombers until some navigational aid was provided. With neither sufficient time nor resources available to equip or train units in the Allied Expeditionary Air Force to become Pathfinder squadrons, three Mosquitos were requested from RAF Bomber Command that were already equipped with the Oboe navigation system for the accurate dropping of Target Indicators. Leigh-Mallory also reviewed the possibility of using a small number of Hawker Typhoon fighter-bombers, given how fast and manoeuvrable they were, as an alternative means of marking the target. In conclusion, he emphasised that the rate of destruction by his forces was barely equalling the rate of construction and that the number of targets 'ripe' for attack at any one time could only effectively be reduced by the efforts of the heavy day-bomber forces of the US Eighth Air Force. Noting that a greater effort was planned by the RAF heavy night-bomber force against Crossbow targets, he voiced his hope that the Americans could increase their daytime efforts too.

In deciding how best to destroy the No-Ball Targets, the Americans reckoned that low-altitude attacks could achieve better results than heavy bombers and they set about proving this by means of a practical test. On the instructions of General Henry 'Hap' Arnold, Commanding General of the Army Air Forces, a complete Bois Carré site was constructed at the Air Force Proving Ground at Eglin Field, Florida, in early 1944. A series of experimental attacks followed, with various types of aircraft attacking from low, medium and high altitudes, with various weights of bombs from 150 to 2,000 pounds, and with rockets, glide bombs and 75mm cannon. Norman Bottomley visited Florida on 19 February to observe the attacks, later detailed in a report by General Arnold that presented the most effective and economical aerial countermeasures against ski-sites as those using low-altitude attacks by fighters. Much to the apparent annoyance of Air Chief Marshal Arthur Harris – who shared the American position that diverting heavy bombers onto No-Ball Targets at the expense of others was folly – Bomber Command were increasingly deployed against those ski-sites and Heavy Sites which were also classed as No-Ball Targets. As evidence of how robustly constructed the Bois Carré sites were, and as with so many of the sites in northern France, remains of the Eglin Field site in Florida survive. They can be found on the coast of the Gulf of Mexico, listed on the United States National Register of Historical Places.

Despite the involvement of the bomber forces, the most dangerous high-mortality attacks continued to be flown by fighters, fighter-bombers and rocket-equipped fighters. The meteorological conditions required for Typhoons and Hurricanes was either clear weather, or when the cloud base was not below 4,000 feet and visibility was at least three miles. Except in an emergency, strict radio-transmitter silence was to be observed on the outward flight. On reaching the target pilots were to use the square and rectangular buildings at the centre of the site as aiming points and they were required to study carefully aerial target photographs, plans and models supplied by Medmenham. When returning from a sortie, aircraft were instructed to recross the English Channel at between 1,000 and 2,000 feet in order to allow recognition.

Despite the flippant observation by Hugh Saunders, Air Officer Commanding No. 11 Group RAF in December 1943 – that a fair percentage of the Bois Carré, sites would probably be taken out in one day – the ski-sites were proving difficult targets to destroy. According to an official briefing by Wing Commander Denys Gillam of 197 Squadron, who flew Hawker Typhoon single-seat fighter-bombers from RAF Tangmere, near Chichester in West Sussex, against the Bois Carré sites. The first challenge was actually finding the right wood; and whilst they could guarantee hitting their target this depended to a large extent on first dropping a marker bomb. The anti-aircraft fire was recorded as varying considerably from day to day and site to site. Targets under construction were heavily camouflaged with netting if they had never been attacked, but once a target had received a severe attack no effort was made to conceal the damage. Both enemy and friendly aircraft were an issue, with one pilot referring to the airspace as being 'like Clapham Junction'. While pilots were busy attacking one target, bomb bursts could be seen at targets in nearby woods and above them all the PR pilots flew in Mosquitos and Spitfires at over 30,000 feet.

On 6 January 1944, Captain Neil Simon had written a supplement to Kendall's Interpretation Report BS194 about the Bois Carré-type sites.

This updated report included information, for the first time, from the detailed official German plans copied by André Camp at Yvrench's Bois Carré, which had been supplied to MI6 in October 1943. By careful comparison with aerial photography the plans allowed all their earlier findings to be corroborated and additional information discovered; the report tabulated all measurements obtained from the photography and the German blueprints and in most cases they conformed very closely. By 23 January, ninety-six Bois Carré sites had been located by the interpreters, but puzzlement why only that number of sites had been found, compounded by frequent reports of additional sites from secret agents, prompted the Air Ministry to authorise complete photographic reconnaissance of northern France again. The Crossbow Section examined every photograph and created a set of annotated 1/50,000-scale maps of the region, with areas marked up when it could be taken as certain that no activity existed – with woods left unshaded where activity might be concealed by trees. No additional sites were revealed. It would ultimately transpire that, since the Bois Carré sites were the responsibility of Luftwaffe Colonel Max Wachtel, for operational purposes they had been arranged as if they were a standard German Air Force squadron. This equated to six squadrons each with sixteen Bois Carré sites, giving a grand total of ninety-six.

The Boeing B-17 Flying Fortress bombers of the US Eighth Air Force were particularly effective in the No-Ball attacks, since they operated by day and were better equipped for attacking small targets, whilst the RAF night-flying tactics were better suited for attacks on the larger No-Ball Targets. In total the number of attacks on Bois Carré sites in northern France was:

RAF Bomber Command	26
US Eighth Air Force	203
US Ninth Air Force	201
RAF Tactical Air Force	623
TOTAL	**1,053**

The monthly attacks totalled:

December 1943	72
January 1944	313

February 1944	196
March 1944	127
April 1944	181
May 1944	123
June 1944	40
July 1944	1
TOTAL	**1,053**

By early March 1944, when the Germans virtually gave up trying to repair the bombed sites, many at Medmenham believed the battle had been won when progressively every single site was reduced to a 'safe' category. But on 4 March 1944 Medmenham received a letter from the Air Ministry with a list of several hundred pinpoints reported by secret agents as being connected with Bois Carré sites. Instructed to inspect good-quality 1/10,000-scale photographs of these locations the pinpoints were photographed immediately. This task was to take priority over all others. The fear at the Air Ministry was that some Bois Carré sites had been missed, and the Chiefs of Staff were therefore unable to calculate the amount of bombing effort that should be dedicated to No-Ball Targets in the coming months. Whilst Medmenham were convinced they had located all the Bois Carré sites, and claimed that from March onwards no launching rails could be seen at ninety-two of the ninety-six sites, evidence from agents on the ground cast doubt on the assessment that all the sites had been located. It would shortly transpire that the success of the photographic interpreters in locating the Bois Carré sites, and the subsequent No-Ball attacks, had made the Germans particularly aware of the efficiency of photographic intelligence, the benefits of camouflage and the danger of using foreign labour. Unbeknown to Medmenham, the Germans had devised an alternative approach to the deployment of the pilotless aircraft.

On 28 April 1944, by when countermeasures against the No-Ball Targets were thought to have neutralised the threat from pilotless aircraft, the Crossbow Section were alarmed to discover a new kind of deployment site – they duly logged it in Report BS372 – at Le Bel Hamelin, in the commune of Nouainville to the south of Cherbourg in the Manche *département* of Lower Normandy. From photography taken during two recent sorties – 16/44 on 23 April 1944 and 106W/135 on 26 April 1944 – a launching point 130 feet in length could be seen running between two farm buildings, the smaller of which had been partially demolished in order to accommodate the ramp, in due course. In a field to the south of the farm, ground scarring indicated the imminent construction of the square building and, whilst a rectangular structure was spotted, it was so well camouflaged that it could not be measured accurately. It was reckoned that the site might well be a modified and simplified form of a Bois Carré one since none of the distinctive ski-buildings could be seen; the buildings here were dispersed haphazardly throughout the area and extensive camouflaging had clearly been carried out from the early stages of construction. When a second site in the Cherbourg peninsula, at L'Epiney Ferme, was located on earlier photography, orientated on the Bristol Line, these new sites were christened the Belhamelin Sites. With twelve more of the heavily camouflaged sites located by 3 May 1944, the Medmenham Technical Control Officer, Douglas Kendall, tasked the Crossbow Section with providing answers to the following key questions:

- How long do the simplified sites take to construct?
- How many are there and where?
- What is the method of construction?
- What is the exact orientation of each firing ramp?
- Is there any new construction at the Supply Sites?

To answer these questions, and with preparations for Operation Overlord at their peak, in early May complete photographic coverage of northern France – this time of areas within 150 miles of London – was flown for the fourth time. A team of around fifteen interpreters was assembled under the Crossbow Section to hunt for launch sites. They discovered quickly that the ramps were aimed not only at London, Plymouth, Bristol and Southampton but at a whole host of additional targets. The detailed calculations were made by W Section, using the Wild Machines, who measured precisely the projection and bearing of the launch ramps as well as providing all the diagrams and plans for the BS Interpretation Reports. Three WAAF officers each worked twelve hours on duty, followed by twenty-four off, and were responsible for a first-phase signal called the BOD Teleprint that was issued on every known site every time it was photographed. Most of the Crossbow interpreters involved in hunting for possible Crossbow sites, and those involved in answering the questions posed by Kendall, worked the same shift pattern and monitored all the known Crossbow sites. The majority of new sites were discovered by the interpreters but every week up to one hundred pinpoints were reported by ground-intelligence sources and received in the increasingly familiar 'Pingpong' reports, all of which had to be investigated. At the height of the investigation three hundred sites were under routine reconnaissance, which meant that at any one time about four interpreters were involved in writing detailed BS Interpretation Reports.

The development of the Belhamelin Sites had been a consequence of Adolf Hitler's decision that a joint headquarters should be created to allow maximum coordination between the Luftwaffe-controlled Bois Carré sites and the Army-controlled Heavy Sites. After approving the formation of LXV Armee Korps, on 1 December 1943, its newly appointed commanding officer Lieutenant-General Erich Heinemann, an experienced artillery officer, visited northern France for a tour of inspection. The general was shocked to discover the size of the Heavy Sites, knowing that they could never be concealed from photographic reconnaissance, and correctly assumed the use of so many foreign workers meant the sites were probably swarming with enemy agents. He also held a similarly dim view of the Bois Carré sites, with their French construction workers, and unnecessarily elaborate and distinctive buildings. With the Bois Carré sites being destroyed around him, and the news that the V1 would not be ready for dispersal until May or June 1944, General Heinemann had an opportunity to rethink the deployment of the weapon. It was decided to construct a new kind of launch site – consisting of a metal ramp that could be manufactured off-site and mounted onto previously prepared and camouflaged foundations – with the distinctive ski-buildings replaced by camouflaged huts. Using 10,000 German and conscript workers, to minimise the risk of agent infiltration, it was planned to build 150 of these new sites, with the prefabricated ramp only transported and installed days before the offensive. The ramps would often be ingeniously hidden inside farm buildings, in areas that would frequently be cleared of the local civilian population first. To further confuse British Intelligence it was decided to give the impression that active efforts were being made to reconstruct the bomb-damaged ski-sites.

In early May the detailed investigation into the Belhamelin Sites – their method of construction, physical characteristics and the number of buildings required for their operation – began. Since the site at Le Bel Hamelin – No-Ball Target X1/A/218 – was by far the most advanced of the sites then known to exist, it was considered the prototype and in Report BS398 of 7 May 1944 a plan of the installation – ACIU Plan R38 – clearly highlighted the deliberately irregular layout and highly camouflaged nature of the target. It soon became evident that Belhamelin Sites had virtually no servicing facilities or capacity for storage and led to the hypothesis that the bomb must be prepared elsewhere, making the whole question of supplying the Belhamelin Sites critical. With the Allied invasion of northern France scheduled to happen in a matter of weeks – and with embarkation ports among the targets – yet again this became a race against time, requiring the interpreters to locate all the Belhamelin Sites and associated Supply Sites, and to prepare accurate target material in order that countermeasures could be taken. The camouflaged nature of the Belhamelin Sites was a particular challenge, not only for the interpreters who had to locate them via aerial photography but also for the aircrews who would soon be tasked with destroying them. Throughout the Second World War camouflage was used on a monumental scale and, since this was total warfare, not only military installations and equipment were disguised but factories, airfields and railways. The Camouflage Directorate, based in Royal Leamington Spa, had frequent contact with Medmenham – particularly M Section who were responsible for aerial photography of the United Kingdom and E Section who monitored enemy camouflage and smokescreens. Created in April 1941, following the move from Wembley, the primary function of this third-phase section was to help with bombing accuracy, and all the information they collected on targets which had sometimes been rendered invisible to the bomb aimers was passed on to the Target Section.

Led by Flight Officer Molly 'Tommy' Thompson, E Section regularly supplied information to the designers in the Camouflage Directorate, given their daily analysis of enemy camouflage techniques. Everyone attending the PI course at nearby Nuneham Park was given a camouflage lecture by Molly, who explained some of the techniques used to identify hidden targets. With camouflage designed to deceive human eyes, colour was used to blend an object with its background in the hope that it would be hard to spot by aircrew flying overhead at speed. For the camouflage interpreters at Medmenham however, who painstakingly studied black and white photographs, the absence of colour information meant that the tonal differences caused by different surface materials were revealed in the photography, even though they might have been indistinguishable in reality because of a good use of colour matching. This practice of using monochrome imagery rather than colour would continue throughout the Cold War and beyond, albeit information was also derived from across the electromagnetic spectrum. Whilst shadows and vague outlines would often show through camouflage netting, wind and snow would frequently cause netting to become transparent and then sag against the object it concealed, giving a sort of quilted effect. Camouflage files were grouped into different types of targets, including industrial objectives, communications, naval and military installations, landmarks and airfields. From 1942 Area Officers were created, and with enemy-occupied territory split in four – Germany, the Low Countries, France and Scandinavia – the camouflage interpreters were allocated geographical areas of responsibility.

Elaborate structures were often designed and built on top of camouflage to alter the whole outline. Above the roof of the Fokker Aircraft Works in Amsterdam, netting had been stretched over the main building at a wide angle from the roof to the ground, in order that outlines and shadows should be hidden. On this covering small dummy houses were built and roads painted, while gardens and trees were installed just like those on an adjacent housing estate. A small canal basin near the factory was covered with netting, and more dummy houses, trees and gardens were installed near an anti-aircraft battery. But whilst the camouflage merged well with the surroundings, the effect in this instance was nevertheless flawed since the continued presence of a distinctive system of waterways, which had not been camouflaged, provided a perfect 'watermark' for anyone tasked with attacking the factory. This highlighted the importance not only of camouflaging the factory and its immediate vicinity but also of sanitising surrounding features that could be used by aircrews to pinpoint the target even if they couldn't see it.

It was for this reason that particularly important canals, aqueducts, railway lines and bridges would be covered with extensive netting. Railway stations too received similar treatment to the Amsterdam factory.

On 16 May 1944, less than three weeks after the first Belhamelin Site was discovered in the Cherbourg peninsula, information from the Réseau AGIR, contained in Pingpong Report 109, resulted in the examination of twenty-five of the Belhamelin Sites for evidence of curved excavations in the foundations of the Square Building. With knowledge that this structure was used to set a control compass, Interpretation Report BS430 records this as the final confirmation that the sites were intended for launching pilotless aircraft. This followed the discovery of a suspected site at Hottot-les-Bagues, Normandy, in photography taken on 7 May 1944 during sortie 106W/285, which Interpretation Report BS418 records as having a suspected ramp orientated on Worthing and west London. Consideration was given from the outset to the possibility that it could be a dummy site and although it was some time before its reality became an accepted fact it added a new dimension to the hunt for the Belhamelin Sites. Nine Hottot Sites were discovered and like other German decoys in the war they managed to occupy the attention and time of the British.

In the summer of 1941 – when aircrews reported seeing increasing numbers and types of decoys – a second-phase report was written on the subject, which immediately prompted Bomber Command to request that further attention must be given to the issue. This resulted in Pilot Officer Geoffrey Dimbleby being tasked with creating Q Section, a third-phase section specialising in enemy decoys. Educated at Cheltenham Grammar School, Dimbleby developed a life-long interest in natural history during long walks in the Cotswolds as a schoolboy before winning a scholarship to read botany at Magdalen College, Oxford. On graduation in 1940 he joined the RAF, passed the PI course at Wembley, and served as a first-phase interpreter at St Eval and Benson, before working in Z Section at Medmenham, as a second-phase interpreter, where he would meet his future wife. Dimbleby had an interest in how people were selected for interpretation and came to the view over the course of his wartime career that academic types proved to be excellent second-phase interpreters because their minds were trained to be observant and enquiring. But for third-phase interpretation, where you often needed to have an established special subject, recognised authorities were more appropriate – examples included Constance Babington Smith and railway specialist Captain Moody in F Section – but there were notable exceptions even here, including Bernard Babington Smith who carried out groundbreaking work in Night Section.

For the first few months of its existence Q Section concentrated on locating decoys revealed by aerial photography. When three rectangular structures were discovered in open country to the east of the industrial Ruhr region – near the town of Soest, in modern-day North Rhine-Westphalia – surrounded by bomb craters, the nature of the elaborate scheme to deceive and confuse attacking aircraft became clear. From then onwards, all known types of decoy were pinpointed by working in two directions. One was to examine every new sortie that was flown, while the other involved reviewing all the historic sorties in the Print Library for decoys. Whilst it would prove relatively easy to distinguish dummy structures from genuine ones in aerial photography, identifying by day the structures that were designed to simulate burning buildings during a bomb attack at night would prove a little more problematic. This prompted a close liaison with the Night Photography Section who could identify decoy fires in night photographs and began to study their use. As the raids over Germany became heavier, more and more decoy sites were constructed, all of which were pinpointed and provided to the Target Section. When the Pathfinder Force came into existence, whose task was to drop Target Indicator markers on which the following aircraft would aim their bombs, it was noticed by the decoy team that another form of decoy structure had appeared. They discovered that the Germans had developed a launching device from which appropriately coloured flares could be fired, that on descending, gave the impression of being Target Indicator flares.

In addition to his PI duties Dimbleby had responsibilities for station defence at Medmenham, which included organising Gas Practice, when Medmenham personnel had to quickly put on their gas masks and capes – which would provide protection from blistering gas – and take cover in the shelters. He would later vividly recall during one practice session seeing 'what appeared to be an animated gas cape making its way across the parade ground, when everyone was supposed to be in the shelters'. Perplexed for a while, he was quickly reassured when he saw 'it was carrying a little wicker shopping basket' and knew it was only Lady Charlotte Bonham Carter. Demobilised in 1945, Squadron Leader Dimbleby returned to Oxford where, with a fascination in the ecology of human environments, he made extensive use of aerial photography in his peacetime research as a lecturer in forestry. He later maintained this policy when he became the founding Professor of Human Environment at London University.

By 10 May 1944, the orientation of the eighteen firing sites that had so far been identified had been exactly plotted by W Section using the Wild Machines. This triggered Captain Robert Rowell to write Report BS410 which revealed that the Belhamelin Sites at Fiefs and Mouriez in the Pas-de-Calais were targeted on south-west London; Autheux in the Somme and Beaumetz-lès-Aire in the Pas-de-Calais were targeted on central London; Fleury and Crépy in the Pas-de-Calais and Fienvillers in the Somme were targeted on east London. The other known Belhamelin Site in the region at Enguinegatte, in the Pas-de-Calais, was targeted slightly to the south-west of Dover. The remaining nine sites were all located in the Cherbourg peninsula: Belhamelin itself was targeted on Newport, south Wales; L'Epiney Ferme was targeted on the port of Avonmouth on the Severn Estuary; Margot was targeted on Start Bay in Devon where top-secret preparations were then being made for the Normandy invasion; L'Epinette was targeted on Bristol; Piquet was targeted on the picturesque coastal town and fishing port of Looe in Cornwall; Les Fontaines, Le Quesnoy, Rauville-le-Bois and Grossville were all targeted on the major naval port of Plymouth, where preparations were well in hand for the Normandy landings that would take place in little under four weeks' time. The selection of targets at the extremity of the 130-mile range of the pilotless aircraft, as far north as south Wales, east as Dover and west as Cornwall, many of which had no military value, emphasised the fact that these were revenge weapons. The German aim to wreak havoc across southern Britain – not just on London – was clear.

While the Crossbow Section continued searching for Belhamelin and Hottot Sites, the critical importance of locating the Supply Sites involved them investigating the original eight storage locations constructed for the Bois Carré sites. Since these were the only known storage facilities linked to pilotless weapons, the section had to determine whether they had any connection with the Belhamelin Sites and whether they should therefore be attacked. This involved a detailed study of any work at the sites, and comparing the buildings, water and electrical supplies, and road and rail communications at the different locations. The conclusions reached in Report BS453 on 23 May 1944 revealed there was considerable activity at three sites, one of which had been extensively camouflaged. But in view of the rapidly growing size of the Belhamelin network, the suspicion was that alternative sites must exist. Locating and destroying them was considered key to the 'neutralisation' of the Belhamelin Sites, given how much smaller and how highly camouflaged the modified sites were.

With Belhamelin and Hottot Sites now being discovered on a daily basis, on 26 May 1944 Report BS461 on 'The Launching Point of Belhamelin and Hottot Type Sites' was issued, following an investigation into the launching points at the forty known ones. With eight sites ignored because of poor-quality photography – no comparative cover, shadow or camouflage – the relative stage of ramp construction at the remaining thirty-two sites was assessed. This revealed that construction began with the excavation of a shallow trench, followed by the laying of foundations for the posts that would safely anchor the rails, and that throughout the process building work was hidden by camouflage. Since the Bois Carré site featured substantial blast walls on either side of the ramp it was assumed that the foundations were being built – and concealed – on sloping ground. With sites at varying stages of completion allowing the construction process to be revealed, the interpreters were nevertheless perplexed that no rails could be seen at any of the sites. It was almost as though construction had progressed so far and then, as it were, the sites had been abandoned. This gave the impression to many that the sites did not yet present a credible threat – and since the destruction of the Supply Sites had already been identified as key to defeating the offensive from the Belhamelin Sites they were generally considered to be of less importance, given other demands on the Allied Air Forces. But this would not prevent later questioning of the extent to which a bomber assault on the Belhamelin Sites – comparable to that wrought on the Bois Carré sites – would have reduced the destructive impact of pilotless aircraft on Britain.

At this stage in the war Operation Overlord had overriding priority and the Allied air campaign was focused on systematically attacking the railway network in northern France in advance of the Normandy landings. Although it was impossible to conceal the fact that the Allied invasion of Europe was imminent, 'distraction' bombing raids were carried out to confuse the Germans as much as possible about exactly where the landings would take place. One of the preconditions of a successful invasion of Normandy was considered to be the creation of a 'railway desert' in an area of 150 miles around Caen. The idea originated from the realisation that it would be essential to delay the arrival in Normandy of the German Panzer divisions that were scattered throughout north-west Europe. The Italian campaign was now turning in the Allies' favour, and since the bombing of the Italian railway system – on which the Germans relied for reinforcements and supplies – was considered by many of the military planners to have been a key factor, it was arranged for the author of the Italian rail plan to be attached to Air Marshal Leigh-Mallory's team who had responsibility for producing the overall air plan for Operation Overlord.

Professor Solly Zuckerman arrived in Britain from South Africa in 1926 to complete his medical studies, but on graduating divided his time between a research post at University College London and London Zoo, where he pursued his studies of and fascination with primate behaviour. Moving to the Department of Human Anatomy at Oxford University, he was drawn into a wartime career that began in 1939 with him working, alongside his friend the scientist Dr Desmond Bernal, on the biological aspects of the anticipated air raids. Working for the Ministry of Home Security in Princes Risborough, Buckinghamshire, they directed experiments to measure the effect of bombing on people and buildings, which included detonating explosives in close proximity to monkeys that were placed in trenches and shelters on Salisbury Plain. After a year of experimental studies during the Phoney War, the effects of blast and high-velocity fragments were readily available in Britain and meant that Zuckerman and Bernal could work instead on the Bomb Census Survey which had been established by the government to collect and collate information about damage sustained during bombing raids. Initially only information about London, Birmingham and Liverpool was collected but by September 1941 the Bomb Census Survey had been extended to cover the whole country. With the Ministry of Home Security dividing the country into twelve civil defence regions, information was gathered locally by the police, air-raid wardens and military personnel who noted where, when and what types of bombs had been dropped, and this was then passed on to the Ministry who plotted the information onto Bomb Census Maps. With the maps and BC4 Report Forms, a picture emerged of the date and time bombs fell, their type and size, whether they exploded, the damage they caused, whether there had been any air-raid warnings, and casualty statistics.

The findings of a report by Zuckerman and Bernal on the effects of German raids on Hull and Birmingham suggested to Lord Cherwell that different approaches could be taken in the air war against Germany. Their work also came to the attention of the Chief of Combined Operations, Lord Louis Mountbatten, who recruited them to his somewhat eccentric group of advisers. In the summer of 1943, after working on the planning of military operations including the Allied invasion of Sicily – Operation Husky – Solly Zuckerman set up the Bombing Survey Unit to analyse air operations over the Italian mainland. In the Mediterranean arena he found a powerful sponsor in the form of Air Chief Marshal Arthur Tedder and through him gained a degree of support for his views on the role of air power in combined operations from General Dwight Eisenhower. When Eisenhower was appointed Supreme Commander of the Allied Forces in Europe in advance of Operation Overlord and Tedder was appointed his deputy, Zuckerman was brought into the planning as Scientific Adviser to the Allied Expeditionary Air Force where he would enjoy the confidence and support of its commander, Trafford Leigh-Mallory. Zuckerman's promotion of the programme of strategic air attacks on rail centres in the preparatory phase of Operation Overlord in the lead-up to the Normandy landings, although supported by Eisenhower and Tedder, was to prove controversial.

Opponents of the Transportation Plan were known to refer to it privately as Zuckerman's Folly. Its primary aim was to limit the quantity of German reinforcements during and after the invasion and restrict their movement within the area of combat. Since rail centres were large targets it required not only the tactical air forces but the heavy bombers of the RAF and American Eighth Air Force. Zuckerman believed that attacks on railway locomotives, rolling stock and repair facilities would not only create a greater problem for the Germans but would force them onto the roads, where they were suffering from fuel shortages and could be attacked by the Tactical Air Forces. But two important factors made France different from the Italian campaign: firstly the close proximity of French residential areas and therefore the risk of civilian casualties; and secondly the need to avoid giving the Germans any indication of the area chosen for the landing. Opponents of Zuckerman's plan argued that marshalling yards were difficult to destroy and railway lines could be quickly laid around a damaged area after an attack, that this meant trains could move quickly again, and instead fighter-bombers should be used to destroy railway lines at multiple key points *between* marshalling yards. The disagreements took place at a high level over a considerable period, with Prime Minister Churchill, Air Chief Marshal Harris and General Carl Spaatz against the bombing of marshalling yards, and Eisenhower, Tedder and Leigh-Mallory in favour. At a crucial meeting of the Allied commanders on 27 March 1944, Spaatz recommended as Commander of Strategic Air Forces that they should be allowed to continue to destroy the German Air Force and supporting industry, they should attack oil production and they should produce a plan for the direct tactical support of Operation Overlord. The desire to continue with the strategic bombing offensive had made securing resources to destroy the Bois Carré sites difficult. This would prove even harder during Operation Overlord which would severely limit the resources available to Operation Crossbow.

In late May 1944 the Crossbow interpreters were unaware that an experimental Belhamelin Site had already been photographed during a mission over Germany. The photographs revealed that the ramp was constructed from prefabricated sections about six metres long. The suspicion was that the near-complete sites did not feature metal rails because they could be quickly installed when required and would signal that an attack was imminent. The threat presented by the Belhamelin Sites resulted in Douglas Kendall being summoned to a meeting of the Crossbow Committee on 5 June 1944 to discuss the position as it was known at that date. This resulted in instructions being given to the Crossbow team at Medmenham the following day, 6 June 1944, when the station was intensely busy monitoring the Normandy landings. Douglas Kendall tasked the Crossbow team with completing an investigation, within forty-eight hours, into the elusive firing ramps at the Belhamelin Sites. An unprecedented number of photographic-reconnaissance sorties were flown that day – very often two reconnaissance aircraft flew together, on paired sorties, improving the odds that every aspect of the landings would be recorded – and resulted in a uniquely detailed record of the landings as they happened and of the Allies' advance inland.

Whilst it was known that the ramps were manufactured in six-metre lengths, they were asked to determine whether the Bois Carré and Belhamelin ramps were of different lengths, and reminded that if they were, the difference would probably be a multiple of six metres. It was noted that, in addition to Belhamelin Sites, a number of Bois Carré sites had now been sufficiently repaired for use as modified launch points and that any flying bomb could therefore come from one of more than a hundred sites. Since attacking such a large number of targets was not feasible, locating the Supply Sites was imperative: if they were destroyed it would, hopefully, reduce the rate of fire. The Crossbow team were informed by Kendall that he had requested the Crossbow Committee to approve detailed photographic reconnaissance of the known Supply Sites on a twice-weekly basis henceforth, and that he needed a special report on them within seven days. At this time the Crossbow team were also informed that the Belhamelin ramps were prefabricated and could be installed in a matter of days, and it was this that accounted for the peculiar way in which the sites had been developed so far, with the foundations being laid and then the sites seemingly abandoned. For this reason the interpreters were told they had to keep a watch for the first indication of a ramp being installed or modifications to nearby buildings.

With the Crossbow Committee concerned that firing ramps might have been linked with anti-invasion defences, W Section were tasked with preparing a detailed map showing all the Bois Carré, Belhamelin, Hottot and Supply Sites, and the projection lines of their ramps along which the pilotless aircraft would travel. ACIU Plan R/41 was completed and circulated on 9 June with Report BS494, and with all 194 sites plotted the immense scale of the network was shown in graphical form. With projection lines converging over London, the danger faced by the city was made disturbingly clear. Whilst the Bristol and Plymouth areas were targeted from sites on the Cherbourg peninsula, the Southampton and Portsmouth areas were targeted from sites south of the Seine. This evidence was followed the next day with Report BS499, which confirmed activity at seven of the eight known Supply Sites, with camouflage concealing the nature of the activity at many of them. The only exception was the site at Beauvoir, where buildings and road and railway communications had been damaged during an earlier attack on a neighbouring Bois Carré site and had never been repaired.

On 10 June the foundations for Belhamelin Sites were discovered not to be built on slopes. Camouflaging could have led to this assumption, which meant more Belhamelin launch sites could be ready than had previously been thought. On 11 June reconnaissance photographs were taken over northern France and arrived at Medmenham later that day from RAF Benson. Analysed by the Crossbow team late in the night, rails were discovered to have been installed on three sites, which prompted an immediate signal to the Air Ministry warning them of this alarming development. On 12 June Douglas Kendall telephoned the Air Ministry to confirm the signal and discussed the proposition that, if their assessment was correct, firing could be expected to begin at any moment.

The characteristics which led to the discovery of Belhamelin Sites before any flying bombs were fired was typically the ramp, the Square Building, the Rectangular Building, track activity and approaches to roads and turnings into woods. On the aerial photography created as a matter of course – typically 1/10,000 scale – a ramp was around 4.8 millimetres long in the photographs. In open woodland or orchards it was normally recognisable through the clearing of the ground, construction of the base platform, the posts on which the rails sat and finally the finished rail. The orientation of the platform towards known targets was useful in doubtful cases, as was its location parallel to a hedge, a location typically chosen by the Germans. Where the ramp was in thick woods, the most that could be seen was a slight gap in the trees, but after the site had been used, scorch and skid marks left telltale signatures that allowed its position to be calculated even though it could not be seen. Whilst the Square Building often had a distinctive S-shaped approach road, its square proportions and large size made it stand out from normal village and farm buildings. The most characteristic sign of all, though, was its alignment parallel with the ramp. No site was identified by the Rectangular Building alone but it provided useful confirmation and was usually sited parallel with a road, and its rectangular dimensions made it much more like a normal village building than the Square Building. Whilst track activity – the signs left by vehicle movements – often drew the attention of an interpreter to a particular place, so too did approach roads and turnings into woods that had been camouflaged by the enemy. This often had the effect of identifying a genuine site, but frequently drew attention to decoy sites that could be quickly dismissed on the basis that there was inadequate road access.

CHAPTER 9
Diver Diver Diver

From his headquarters in the city of Amiens on 12 June 1944, *Oberst* Max Wachtel announced to his Flak-Regiment 155 (W [for West]) that after months of waiting the time had finally arrived for them to attack. The enemy, Wachtel warned, was, in his own words, trying to 'secure at all costs his foothold on the Continent' and, as they launched their revenge weapons today and in the future, they should 'always bear in mind the destruction and the suffering wrought by the enemy's terror bombing' on Germany. From his strategically located headquarters 75 miles north of Paris, in the capital of the Somme *département*, Wachtel's network of Belhamelin and Supply Sites stretched 250 miles from the tip of the Cherbourg peninsula to the Pas-de-Calais. But, for all practical purposes, after the Normandy landings on 6 June 1944 Wachtel was separated from the sites in the Cherbourg peninsula. And when, on D-Day+6, the Americans captured the town of Carentan – closing the gap between the Utah and Omaha beachheads – this helped ensure that by 30 June, when the port of Cherbourg was captured, targets throughout England's south-west including the Plymouth, Torquay and Bristol areas, as well as south Wales, were saved from mass attack by pilotless aircraft. On 6 June 1944, when the Allies were landing on the Normandy beaches, the senior Allied commanders knew that an attack from pilotless aircraft was imminent and there was a vital need to overrun the firing sites in northern France at the earliest opportunity. While the Battle for Caen raged, the proximity of that front line to the Belhamelin Sites in the Normandy *département* helped save Portsmouth and Southampton and ensured that Wachtel focused his efforts on the largest concentration of projection sites at his disposal, those in the Pas-de-Calais and Somme *départements* that were targeted on London and the south-east of England.

On 16 May 1944, less than three weeks after the first Belhamelin Site was discovered on the Cherbourg peninsula – when the evidence of curved excavations in the foundations of the Square Buildings confirmed that Belhamelin Sites were connected with the deployment of pilotless aircraft – Field Marshal Wilhelm Keitel, Supreme Commander of the Wehrmacht, one of Hitler's most loyal yes-men, issued a directive from the Führer ordering the mass bombing of Britain and declared: 'The bombardment will open like a thunderclap at night . . .' The weapons to be deployed included the *Fieseler Fi 103*, which would be known to the German people as the *Vergeltungswaffe 1*, or Revenge Weapon 1 – the V1. For the most part they were to be deployed from the network of sites in northern France, but also from specially modified Heinkel He 111 medium bombers so that the effective range of the weapon could be dramatically increased. The cross-Channel guns in the Pas-de-Calais region, which had been used to target Dover and other towns on the south coast, were to continue their onslaught, along with the bomber forces of *Luftflotte 3* (Air Fleet 3) that operated in German-occupied areas of northern France and the Low Countries. It had been hoped to open the attack on 20 April 1944, the Führer's birthday, but as it had not been possible to manufacture enough flying bombs – due to the disruption caused by the Allied attacks – the long-range bombardment of Britain was delayed until sometime in mid-June.

As *Oberst* Max Wachtel prepared to deploy the V1 against Britain, information about this secret weapon went from being a closely guarded secret to common knowledge among rank-and-file wardens of the Air Raid Precautions (ARP) and members of the Royal Observer Corps in southern England. On 26 April 1944 an official instruction told them about the existence of a German pilotless jet-propelled aircraft that cruised at 330 mph at a height of 6,000 feet, with an approximate wingspan of 20 feet. Since the pilotless aircraft was designed to enter a terminal dive on reaching its target, it became known by the code name DIVER. From more than 1,500 Observer Posts across southern England, in the summer of 1944 the observers waited pensively for the pilotless aircraft to arrive. By 11 p.m. on the night of 12 June 1944, and with much essential equipment still missing, Flak-Regiment 155 (W) was struggling to make even eighteen of the firing sites operational. Finally, in the early hours of Tuesday, 13 June 1944 at 3:30 a.m., ten V1 missiles were launched into the night sky: five malfunctioned and crashed immediately and one landed in the sea while the remaining four continued flying and cruised above the English Channel towards London.

The first pilotless aircraft to approach England was spotted from the Kent Downs, high above the coastal town of Folkestone, by local farmer Edwin Woods who was on observer duty that night at Observer Post Mike 3. He received a telephone call from the Royal Observer Corps Centre in Maidstone at four in the morning, telling him that something was happening near Boulogne and to keep a close watch through his binoculars. He spotted what appeared to be a fighter aircraft on fire. Providing a bearing to Maidstone he handed over to his colleagues at Observer Post Mike 2, based in the nearby Dymchurch Martello tower that had originally been built to resist a threatened Napoleonic invasion fleet. It was now being used to watch an invading force of a very different kind. Keeping watch that night were local greengrocer Mr Ernest (Ern) Woodland and a carpenter in the building trade, Mr Arthur (Archie) Wraight, who spotted a flaming object in the sky that made a peculiar sound. Following the projectile through the sky, Woodland transmitted the message 'Mike 2, Diver, Diver, Diver – on four, north-west one-on-one'. This told the Maidstone Centre that the pilotless aircraft had arrived.

The first V1 crashed onto farmland between Dartford and Gravesend in north Kent and the second landed on farmland near Haywards Heath in Sussex, twelve miles north of Brighton. Another crashed into the back garden of a large house in the parish of Borough Green, west Kent, destroying a row of glasshouses. The one that reached London that morning would cause fatalities when it landed at 4:25 a.m. on the railway bridge over Grove Road, Bethnal Green, in the East End of London, destroying the railway lines, partly collapsing the bridge parapet, demolishing two houses and badly damaging several others. Six people were killed: Connie Day, aged thirty-two, of 61 Grove Road; at number 64 Mrs Ellen Woodcraft, the nineteen-year-old wife of Thomas – a soldier serving with the Royal Corps of Signals – and their eight-month-old son, Thomas; Dora Cohen, aged fifty-five at number 70; twelve-year-old Leonard 'Lennie' Sherman at number 72;

and William 'Willie' Rogers, aged fifty, of 74 Grove Road. Another thirty people suffered blast injuries. A dry communiqué in the press on 14 June recorded simply, and misleadingly, that a single enemy raider had been shot down over the London area. But those involved with Operation Crossbow were under no illusion: this, at last, was the long-expected secret weapon, the flying bomb. However, it was not quite the opening salvo that the British had expected, and prompted Lord Cherwell to quip that 'the mountain hath groaned and brought forth a mouse'.

In late May 1944 the interpreter Squadron Leader Walter Heath had visited Sweden to inspect and photograph two pilotless aircraft, one that had crashed there on 13 May and another that had been salvaged from the sea by the Royal Swedish Navy. When the Chiefs of Staff received his preliminary report on 9 June – by which date the Allies had a foothold in Continental Europe – this evidence was a timely reminder that the Germans had the means to bombard their enemies with an entirely new kind of weapon. From the captured wreckage it could be seen that the missile was a monoplane with a sixteen-foot wingspan; it was mainly constructed from steel; there was evidence in the design that it was mass-produced; and it appeared to have no radio equipment. Using photographs of the Swedish wreckage and of V1 wreckage from the initial attacks on England, technical artist Peter Endsleigh Castle began creating highly detailed three-quarter-view engineering drawings. Castle was attached to Air Intelligence 1(a) and had worked with Squadron Leader Michael Golovine on analysing the photograph identified by L Section of the V1 sitting at the bottom of a ramp at Peenemunde. When a flying bomb was captured intact in July 1944 after its warhead failed to detonate it was duly delivered to Castle and left in the driveway of the Manor in the north-west London suburb of Harrow Weald, a property which had been commandeered by the Air Ministry during the war. From all this information, Castle and his colleagues were able to create accurate drawings and illustrations that were distributed to all those involved in Operation Crossbow, from the Medmenham PIs to the thousands of Allied aircrew involved in attacking the launch and Supply Sites.

On the same day that the first V1s hit England, Interpretation Report BS511 was written at Medmenham by Captain Robert Rowell about the Belhamelin Site at Vignacourt, a small village in the Somme *département*. Taken on 12 June 1944, during sortie 106G/818, the photographs captured at a scale of 1/1,560 a launching point consisting of twin rails – two feet, six inches apart on a light girder structure – and were of such excellent quality that they would be referred to throughout the remainder of the investigation. This was the archetypal Belhamelin Site. Rowell wrote that the platform was at an angle of seven degrees with the forward end about twelve feet above the ground, and that despite being heavily camouflaged the large rectangular building, control building and blast-wall building could be identified. In a disguise attempt, the large rectangular building was partially hidden underneath a farm building masquerading as a shelter for hay, straw or farm implements. More compelling evidence was obtained on 14 and 15 June about the speed with which the new launching points were being made operational. Report BS533 of 18 June 1944 recorded that at 08:30 hours on 14 June the first quarter of a Belhamelin ramp had been erected and was photographed – during sortie 106G/859 – in the hamlet of Coubronne in the Pas-de-Calais. By 07:05 hours the following day photographic reconnaissance – during sortie 106G/880 – recorded that the entire launching point had been constructed and some camouflage started; while an hour later further photography – taken during sortie 106G/884 – revealed that work to camouflage the ramp was progressing rapidly.

Although the first pilotless aircraft fired at Britain on the morning of 13 June 1944 might have gone unreported, it was followed by another much larger onslaught on Thursday, 15 June 1944 – when 120 bombs, flying at five-minute intervals, rained down on the populations of Kent, Surrey and Sussex, the suburbs and the City of London – which proved quite impossible to hush up. It was the moment that had been feared the previous October when the Home Secretary, Herbert Morrison, and the Minister of Information, Brendan Bracken, had held a secret conference with newspaper editors about the threat posed by the German secret weapons. Now, in mid-June, shortly after the D-Day landings, when most people hoped the war was as good as won and the worst of the air raids were behind them, a new battle in the skies of southern Britain had begun. At dusk, unusual aircraft were spotted, flying at altitudes of between 2,000 and 3,000 feet. Onlookers described them as looking like aircraft that had been shot and were on fire. And yet they were all flying quicker than fighters and were on a fixed course from which they never deviated. The objects – for which those on the ground had no name yet – made a strange loud noise, similar to the clattering harshness of a gigantic cheap motorbike. When viewed from beneath they looked like black crosses travelling across the sky – or medieval crossbows. Looked at from the side they had the appearance of giant rifles fitted with telescopic sights.

When they crashed the pilotless aircraft created an earth-shattering explosion. The early victims of these 'flying bombs' had little or no sense of what had befallen them, but people learnt quickly to listen for the trademark noise of the engine: when that stopped there were just a few surreal heart-stopping seconds in which to find some sort of cover that *might* make the difference between life, maiming or death. When 850 kilograms of amatol high explosive exploded on impact, it caused blast damage over an area of around 400–600 yards in each direction and a huge blast wave rippled out from the epicentre. As it did so it left a vacuum, which caused a second rush of air as the vacuum was filled, causing a devastating pushing and pulling effect.

Just days after Lord Cherwell belittled the first V1 attack, flying bombs began falling on London and southern England with unremitting frequency. Although many crashed on take-off, great damage was caused in Lewisham, Westminster and south London boroughs, especially Croydon, Wandsworth and Battersea, which bore the brunt of the early attacks. Within just twenty-four hours of the first heavy attack, an RAF flight lieutenant talking on the radio was said to have coined an expression for the pilotless aircraft. German propaganda called it the *Vergeltungswaffe 1* but in Britain it became known as the doodlebug, a colloquial expression in south-east England for the cockchafer beetle or May bug. In Germany too, on Hitler's orders, the V1 was known for a short time as the *Maikäfer* or May bug. With adult doodlebugs only living for up to five or six weeks in the summer months, when they fly at dusk on warm evenings and make a noisy hum, a degree of similarity must have been noticeable.

Much of the official reluctance to share information on these flying bombs had been due to uncertainty about what the secret weapons really were, and indeed whether they might be carrying chemical warheads. Standing before the House of Commons on the morning of Friday, 16 June 1944, Herbert Morrison admitted that it had been known for some time that the Germans were planning to use 'pilotless aircraft' against Britain and that use of their 'much-vaunted new weapon' had now begun. Explaining that although a small number had been used in an attack early on Tuesday morning, which caused only a few casualties and relatively insignificant damage, the attack last night was far more serious but as yet he had no information about casualty numbers or damage caused. Stressing that the enemy's development of these new weapons had not gone unnoticed – and that countermeasures had already been taken and would continue to be taken – the probability of the attacks continuing was spelled out. The vital importance of the enemy not being given any information about where missiles landed was stressed and the House was told that no information would be published about air raids in southern England – south of a line from The Wash to the Bristol Channel – beyond saying that it had occurred in that part of the country. The hope was that this would prevent the Germans from zeroing in their projectiles as they came to understand better the impact of weather and how the weapon behaved under combat conditions. Although the nation was to be told that all possible anti-missile steps were being taken, in the meantime the public were to carry on with business as usual. The consequence was that a state of heightened tension arose, and in an information vacuum rumour and speculation flourished, that German paratroopers had landed and even that poison gas had been used by the enemy.

Morrison's statement had immediate effects. The British public had been following with a sense of great expectation the Normandy landings and the advance of the Allied armies. Now they found themselves, particularly the people of southern England, caught up in a situation from which there would be no immediate respite. For as long as their resources would allow, and for as long as their launching sites were within range, the Germans could use their robot weapon to attack day or night. As the raids by pilotless aircraft could occur during daylight when the streets were full of people, the public were told that anti-aircraft guns would be used to shoot them down. So when the public heard the guns – and however curious they might be – while the ack-ack was firing they must take cover for their own safety. In his statement the Home Secretary deliberately made no specific reference to London. The target area was vaguely described as southern England and it would not be until some weeks later that the Prime Minister described it more precisely. Although Plymouth, Bristol and Southampton were also targets, the main one was always London, home to nearly 8 million people. The aim of the Germans was to wreak revenge for the destruction of their cities and to provide a morale boost for the population after the string of defeats they'd suffered since El Alamein and Stalingrad. The successful deployment of the V1 became a vital symbol of hope that was skilfully used by Goebbels who, diverting attention from the success so far of the Allied invasion, reported that southern England was covered with so much smoke that it had been impossible to take aerial photographs of the tremendous damage that lay beneath. The roads north of London were reported to be packed with fleeing refugees. Deliberately likening them to the roads of Belgium and France in 1940, it was stated that if one day of bombardment by V1 weapons could achieve this level of destruction and chaos then there was still hope for the German people.

With its headquarters in the commandeered Senate House of the University of London, the Ministry of Information had responsibility for propaganda and publicity during the Second World War. The activities of the propaganda ministry housed in the art deco landmark in Bloomsbury inspired George Orwell's creation of the Ministry of Truth in his 1949 satirical novel *Nineteen Eighty-Four*. Throughout the war the Ministry issued 'Home Intelligence Weekly Reports' that drew on information from its network of regional information officers throughout the country, as well as from secret sources including postal censors, Special Branch reports and Chief Constables. In London special arrangements were established with 'people from all strata of society' who would be prepared to report the feelings of people they came into contact with. These included 'doctors, dentists, parsons, publicans, small shopkeepers, newsagents, trade union officials, factory welfare officers, shop stewards, Citizens' Advice Bureaus, hospital almoners and local authority officials'. In the 192nd Home Intelligence Weekly Report – covering the period 31 May to 7 June 1944 – restrained excitement and sober confidence was recorded from the majority of people about news of the Allied invasion of France. With a marked rise in spirits there was little mention of war-weariness and a feeling that the final phase of the war had now begun, with many people believing that the war could be over that year, perhaps even by the autumn. The most hopeful thought it might be only a matter of a few weeks. Great surprise and relief was recorded at the lack of retaliatory bombing and there was much speculation about the reason.

By the time of the 194th Ministry of Information report, covering the period 13 to 20 June 1944, the arrival of the flying bombs had resulted in a complete change of morale, with the public frequently referring to the weapon as 'Hitler's last kick . . . his last desperate fling . . . the worst he can do . . . his last patch of awfulness'. The London Region Information Officer reported that people confessed to feeling considerably shaken, despite their calm behaviour, but only at the actual scenes of some incidents was there anything approaching panic. A nervous anxiety about the origins of the aerial reprisals in such a 'weird and uncanny' machine coupled with exhaustion from lack of sleep were considered to be the overriding feelings experienced by Londoners. The Ministry of Information challenged the government's tactics and stated that 'in view of the widespread rumours, it is thought that more details should be published . . . People in target areas and elsewhere are critical of official and press accounts which appear to tone down the raids and the damage that they cause. People ask for less secrecy and more true information.' Rumours abounded that Portsmouth, Plymouth, Southampton and Bristol were also badly damaged: the number of obituary notices was seen by people as evidence that casualties must be heavy. It was agreed that more information should be made available and, although a number of people were recorded as accepting that the enemy should know as little as possible about the effects of the raids, many blamed official reticence for leaving the way open for exaggerated accounts that led people to accept German claims. They maintained that this caused uneasiness in the country.

At first the London anti-aircraft guns – which doubled as a warning system – were used to shoot down the flying bombs but this tactic was soon cancelled on the basis that there was little point bringing down a missile onto the very target it was aimed at. This change of policy was welcomed by Londoners for two reasons: firstly, because the guns were disturbing in themselves – they appeared to do very little good and prevented people from sleeping – and secondly, because it was reckoned that fighter aircraft must now be targeting the flying bombs more successfully. This had the effect of making people who were not directly affected to think that many more of the doodlebugs were being brought down before they reached London, either over the English Channel or the countryside of the South-East. People living in Bomb Alley were accustomed to having enemy bombers on their way to and from London fly over them and were also reported to have expected retaliation after the Allied invasion of France. Although, like the Londoners, unsurprised that it had come, they were surprised at the new form it had taken. They too suffered sleeplessness because of the noise of flying bombs passing over and anti-aircraft guns blazing away but were recorded to have greatly appreciated the efforts of the Allied fighters trying to shoot the V1s down, even though this meant that sometimes the action took place right overhead. In the rest of the country, people in Scotland and northern England were reported to feel safe, believing that the bombs could not reach them, but people in Wales and the Midlands were fearful that the doodlebugs' range might be increased enough to affect them. The mention of a line from The Wash to the Bristol Channel – in the Herbert Morrison speech – was taken by many to mean a dividing line. Some people to its north were observed to show a marked lack of interest on the basis that the V1s would not affect them, although sympathy was expressed by northerners for the predicament of southerners.

In September 1943 fear of aerial bombardment from rockets and pilotless aircraft had resulted in Sir (Samuel) Findlater Stewart, Chairman of the Home Defence Executive, being tasked by the Defence Committee of the War Cabinet with checking whether sufficient accommodation existed in London's underground military citadels to ensure the continued work of government in the event of an attack. He was advised that 'Black Move' – the top-secret plans to relocate the British government away from London if the Germans ever invaded, plans developed at the request of Prime Minister Neville Chamberlain in 1939 – would not be revived and instead the citadel accommodation should be assessed and arrangements made to ensure the work of government could continue in London. Black Move would have involved the Prime Minister, his family and Private Office taking up residence at

Spetchley Park, a stately Georgian mansion near Worcester, while the War Cabinet would have been based three miles away in Hindlip Hall. Parliament would have been evacuated to Stratford-upon-Avon.

It was assumed that a steady and sustained bombardment by rocket projectiles would result in the gradual devastation of London. Because the weapon was unlikely to be accurate, it would not be possible to target specific areas of the capital and the hits were likely to be spread randomly across the Greater London area. Under such a bombardment scenario, London would gradually disintegrate. Road and rail transport would be disorganised to such a point that government workers would be unable to commute to work, and this would mean that essential workers would need to stay in their offices day and night. Although it was known that insufficient citadel space existed, solutions would have to be found using existing resources since no extra labour or materials could be spared. The defence of the civilian population was to be considered separately, but the only planned civilian evacuation allowed for under these plans was for schoolchildren and mothers with children, who would be evacuated to the provinces.

Citadel accommodation had been constructed early in the war, after the Cabinet considered it essential to provide accommodation capable of surviving heavy bombardment for the Admiralty, War Office, Air Ministry, Cabinet Office and Ministry of Home Security. The citadels and their network of deep subterranean tunnels also provided security for government telecommunications systems and the military headquarters in London of COSSAC (the Chiefs of Staff to Supreme Allied Commander) and ETOUSA (the European Theater of Operations United States Army). When the Air Staff were given the opportunity to comment on Sir Findlater Stewart's report to the Defence Committee they estimated that 6,000 staff were essential to deal with a two- to three-week period of bombardment, but only having citadel space for 1,700 meant that they were short of space for more than 4,000. In a show of inter-services rivalry, the Air Ministry lamented the favourable position of the War Office and remarked enviously that 'The Admiralty have their Lenin's Tomb'. Winston Churchill would later describe with disdain the Admiralty citadel as being a 'vast monstrosity which weighs upon the Horse Guards Parade' and in a sign of his hope for the future lamented the challenge that future generations would have demolishing twenty-foot-thick steel and concrete walls. (It remains there to this day.) The dilemmas being faced by the government at the time as they responded to the threat from rocket bombardment were the precursors of the problems that would be confronted during the Cold War. In both cases the desire to protect the country's citizens had ultimately to be weighed against the need for government to continue, given the vast costs that were necessary to ensure this alone.

Defence planning against the V1 threat began at RAF Biggin Hill, four months before the first attack, when the senior sector fighter station to the south of London received a visit from Balloon Command. Barrage balloons were a familiar sight in Britain during the war and are frequently cited by American veterans who were stationed in the UK as their first memory of the approach to the country. The huge inflatables were a distinctive feature of the view from many a city window. This was equally the case throughout enemy-occupied Europe and a careful watch was kept by the second-phase interpreters in Z Section at Medmenham on their location. Balloons were frequently relocated in a deliberate game of cat and mouse, and since their cables posed a serious danger to aircraft they were used by both sides as a means to protect targets from attack. The balloons, however, frequently frightened children, which prompted the publication of *Boo-Boo the Barrage Balloon* with drawings that likened the balloons to elephants with friendly faces. By 1944 the threat of German air raids against Britain had largely abated since the Luftwaffe could no longer sustain such heavy losses to its aircraft and even more critically to its aircrews. To provide protection from the doodlebugs, Balloon Command had been tasked with identifying sites for new balloon emplacements in the North Downs. This resulted in the construction of access roads to a network of sites that would be used to combat the weapons.

After the first V1 attack early in the morning of 13 June, followed by the much larger attack consisting of 120 flying bombs arriving at five-minute intervals on the evening of the 15th, balloon squadrons from around the country began to arrive at Biggin Hill. They were to prove a vital part of the defences against the doodlebugs and by the beginning of July around a thousand were in place, a figure that would almost double to cover an area of 160 square miles by mid-July. The first victory was scored on 20 June when a flying bomb became entangled in a balloon wire, causing the projectile to crash into an orchard. It was followed quickly by another two before a Mosquito fighter that had been flying too close tragically hit one of the steel cables. The first line of defence against the flying bombs were the fighter aircraft patrolling the English Channel and the Kent and Sussex coastlines. Next came a chain of anti-aircraft guns, totalling more than 1,000 at its peak, followed by the balloon belt that stretched from the parish of Cobham in north Kent to the parish of Limpsfield in east Surrey. One of the first teams of 'balloon boys' to arrive came from Glasgow. They had travelled down through the night with special police escorts, and set to work on the enormous logistical exercise of creating a web of metal cables intended to protect London from attack.

With barrage balloons and anti-aircraft batteries in place – and attacks starting on 19 June 1944 against the Belhamelin Sites – Allied fighters were also sent up against the flying bombs in the skies over Kent and Sussex. The Spitfires, Mustangs and Thunderbolt fighter squadrons initially deployed were, however, unable to match the speed of the V1, which was capable of flying at 400 mph. This meant that the Hawker Tempest V fighter – which could reach 416 mph – and the Tempest squadrons of 150 Wing found themselves in the front line of Diver missions. Based at RAF Newchurch on Romney Marsh, Wing Commander Roly Beaumont was responsible for developing the tactics that would be used against the flying bomb. Going into action just after dawn on 16 June he recorded overtaking a V1 that was flying over Folkestone and travelling at 370 mph. After missing it completely with his first rounds Beaumont hit its outer wing with another short burst and then, using up all his remaining ammunition, he hit its fuselage and engine. This stopped the doodlebug's engine and he was able to look quickly at the new weapon, with its high-mounted ram-jet engine and jet pipe at the back. Calling in his number two to finish it off, he records Sergeant Robert Cole firing a well-aimed burst that made the V1 roll over onto its back and dive into a field south of Maidstone, where it exploded in a ball of flames and black smoke. On that first day, 150 Wing learned much about the flying bomb and how to destroy it. They discovered that opening fire at a distance of about 200 yards was the best range for effectiveness but there was considerable danger for an attacking fighter in the explosion of the warhead.

Also involved in the DIVER operations was RAF West Malling in

British and American officers at Medmenham with Douglas Kendall on the front row in the centre. On his left is Elliot Roosevelt, who barely spoke to Kendall for six months after Roosevelt's scheming against the continued existence of the Allied Unit was summarily rejected by General Carl Spatz. To the left of Roosevelt sits Hugh Hamshaw Thomas.

Wycombe Abbey School, Buckinghamshire, was expeditiously requisitioned by the Air Ministry in March 1942 and – codenamed Pinetree – became Headquarters of the US Eighth Air Force.

Housed in the basement of Danesfield House, the presence of Army interpreters (B Section) underscores the inter-service nature of Allied photographic interpretation during the war.

WAAF Section Officer Lady Charlotte Bonham Carter in the Target Section, Medmenham. After all the intelligence on a specific target was collated, the Target Map used on a bombing mission was drawn at the Air Ministry Target Mapping Centre (AI3c) – codenamed Hillside – in requisitioned Hughenden Manor, Buckinghamshire.

Z Section was responsible for second-phase interpretation and with sub-sections specialised in particular subjects or geographical areas; enemy activity was monitored on a day-to-day basis. Assembled on the steps of Danesfield, towards the end of the war, the prominent role of women at this inter-service intelligence unit is abundantly clear.

The Coverage sub-section was operational 24hrs a day and examined incoming sorties to decide which targets had been satisfactorily photographed and which must be flown again.

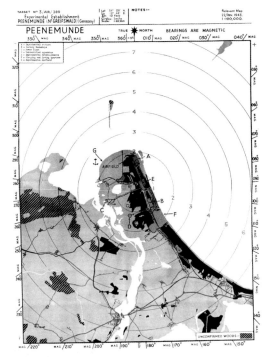

The isolated experimental site at Peenemunde, on the German island of Usedom in the Baltic Sea, was ideally suited to secret rocket and flying bomb testing. Attacked on the night of 17/18 August 1943 by RAF Bomber Command, the Target Map (left) was prepared for Operation Hydra. The photograph below (left) shows Test Stand Seven with V2 rockets lying horizontally.

Scale model of Test Stand Seven at Peenemunde, completed by the model makers on 11 July 1943, when based at the requisitioned Phyllis Court Country Club, Henley-on-Thames. Only three days after the model was completed, a scandalous act by one of the American model makers, resulted in V Section being hastily relocated back to Medmenham.

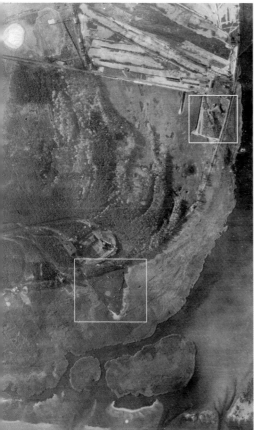

On 28 November 1943 during sortie N/980, Squadron Leader John Merifield photographed targets along the Baltic coastline of northern Germany after he discovered his principal target, Berlin, was covered by 10/10 cloud cover. When photographs of Peenemunde were analysed by the Aircraft Section, a cruciform object at the bottom of a ramp was spotted.

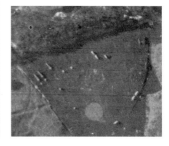

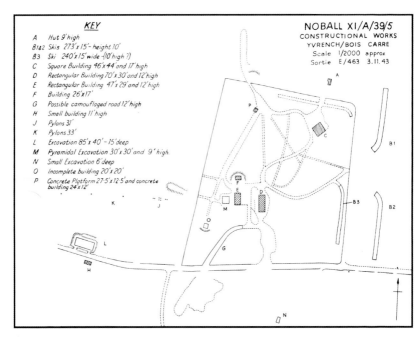

The flying-bomb site in the Bois Carré – or square wood – one mile east of Yvrench village, in the Somme *département* of northern France. When the Réseau AGIR espionage network supplied MI6 with detailed intelligence, it quickly became a key focus of the secret weapons investigation. The dramatic impact of Allied countermeasures against the network of Bois Carré – the first generation of flying bomb sites uncovered in northern France – would prove decisive.

Sortie: E/463, 3 November 1943, Frame 4088.   Sortie: MO/276, 13 April 1944, Frame 5005.

The inimitable Constance Babington Smith – WAAF Flight Officer in charge of L Section. During Operation Crossbow the interpreters in her third-phase section, who specialised in enemy aircraft and aircraft factories were asked to watch Peenemunde for 'anything queer'. In the process they would discover the jet-powered fighter the Messerchmitt Me 163.

Below: The distinctive first generation Bois Carré type sites were comprehensively attacked by the Allied Air Forces from the December of 1943. With their distinctive storage buildings, that resembled skis lying on their side, the interpreters nicknamed them the ski-sites.

TYPICAL SKI SITE
BASED ON PHOTOGRAPHIC AND GROUND INFORMATION

P   LAUNCHING RAMP
K   LAUNCHING CONTROL POST
Rs LAUNCHER SERVICE
C   CISTERN & PUMP STATION
E   WATER COOLING RESERVOIR
Q   FINAL CHECK (NON-MAGNETIC CONSTRUCTION)
R   PRELIMINARY SERVICING
Sus STORAGE BUILDINGS
St PERSONNEL SHELTER or STORAGE
A   STORAGE or DELIVERY BUILDING

BUILDINGS AT TOP OF SKETCH ARE PART OF
EXISTING FRENCH VILLAGE – NOT PART OF SITE.

BASED ON SKETCH DATED 30 JANUARY 1944
PREPARED JOINTLY BY
ASSISTANT CHIEF OF AIR STAFF, INTELLIGENCE
ASSISTANT CHIEF OF STAFF G2 WAR DEPARTMENT
NEW DEVELOPMENTS DIVISION, WAR DEPARTMENT

REVISED BY CROSSBOW COMMITTEE
22 APRIL 1944

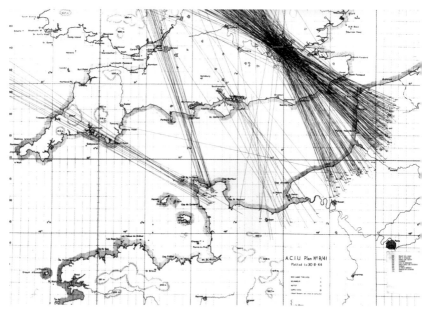

The Operation Crossbow team at Medmenham, plotted the bearing of every firing point throughout the network of flying-bomb sites in enemy-occupied territory. Whilst the destruction of the capital was graphically shown to be the main objective, the German aim to wreak havoc across southern Britain – not just on London – was clear.

Kent, which was on the flying-bomb path to London and at that time home to 91 and 96 Squadrons. Flying Officer Kenneth Collier of 91 Squadron devised a novel method of destroying the V1 while flying a powerful new Griffon-engined Spitfire Mark IX on the evening of Friday, 23 June. Scrambled at 9:50 p.m. to intercept 'Divers' approaching from the Channel, he sighted one flying a steady course at 2,500 feet and dived down to attack it. Despite hitting the bomb, he used all his ammunition without effect and so decided to overtake the missile and fly alongside it. With the tip of his wing he then attempted to flip the doodlebug on its back, which after a number of failed attempts he managed to achieve. This sent the weapon into a spin, as a result of which it crashed into a field and exploded. This wing-tipping method of destroying flying bombs was destined to become part of aviation folklore as the Diver patrols continued.

At 11:20 a.m. during the Sunday-morning service on 18 June 1944, the Royal Military Chapel at Wellington Barracks near Buckingham Palace, opposite St James's Park in central London, was to be on the receiving end of Colonel Wachtel's 500th V1. The chapel was packed with Guardsmen and their families, and eyewitnesses recalled hearing the flying bomb approach and its engine cut out. After gliding downwards it exploded on the chapel roof, covering the congregation with debris. For the following forty-eight hours there was a rescue mission to dig people out of the ruins: 121 people were dead – 58 civilians and 63 service personnel. It was the worst incident yet in the flying bomb campaign. With the Guards Chapel such a potent symbol of the British military establishment, Flak-Regiment 155 (W) could not have hit a more iconic target so early in the flying-bomb campaign.

The next major incident was to follow on 30 June 1944 not long after one bomb exploded in the West End, just after midday, on the roof of a hotel in Brewer Street, Soho, killing a chambermaid in the process. Another followed, less than an hour later, striking Aldwych – the crescent road that connects with the Strand at both ends – in front of Adastral House, the Air Ministry headquarters. Aldwych was packed with people taking their lunch hour, and some WAAF personnel who were based in Adastral House were taking the opportunity to sunbathe on the building's roof. When the bomb struck, its powerful shock wave demolished the ten-foot-high blast wall in front of the Air Ministry building, destroyed buses and other vehicles, and killed the sunbathers who were among the 48 fatalities and 200 seriously injured in what became known as 'the Aldwych massacre'. One of the injured that day was Miss Cecile Daly who worked for the Foreign Office in Ingersoll House, opposite the Air Ministry and next to Bush House. Her job involved preparing propaganda leaflets that were dropped over Europe, and since it had been a sunny day Cecile had opened her office window and was lucky to receive only minor cuts and bruises. But her secret papers were now scattered along the street amidst the scene of devastation. The attack damaged the front of Bush House – the former home of Aerofilms Limited and then home to the BBC World Service – and blew an arm off one of the two entrance statues which symbolised Anglo-American friendship. By this stage a growing fear was being recorded by the Ministry of Information amongst Londoners concerning what Hitler had 'up his sleeve' and whether there was a 'Secret Weapon No. 2'.

With these devastating attacks, and twenty-three days after the first flying bombs landed on England, the Prime Minister broke his silence and stood before the House of Commons on 6 July 1944 to make an official statement. Churchill announced that since the early months of 1943 intelligence had been received from a variety of sources that the Germans were developing a new long-range weapon with which to bombard London. Thanks to photographic reconnaissance and investigations by secret agents, by July 1943 a secret Nazi experimental station at Peenemunde – described by Churchill as a 'Strength-Through-Joy' establishment – had been identified and, thanks to an attack by Bomber Command in August 1943, the onslaughts now being experienced in London had been delayed by many months. The House was told that the following month, in September 1943, large structures at Watten in the Pas-de-Calais that were suspected of being to do with a long-range rocket were attacked and had been repeatedly attacked ever since by the British and American Air Forces. Photographic reconnaissance of northern France was identified as vital in revealing, in addition to Watten-like structures, that around 100 other suspicious sites had been built all along the French coastline (a reference to the 96 Bois Carré sites). It was concluded that these were firing points for a jet-propelled projectile much smaller than the rocket they had first anticipated. Since last December countermeasures had been taken, and all these firing points had been destroyed by the RAF 'with the wholehearted assistance of the growing United States air power' and that without this action the attack on London could have started six months earlier and been on a much heavier scale.

Asking the rhetorical question 'What is the scale of this attack?' Churchill observed that thanks to the destruction of the Bois Carré sites it was not as great as it might have been. Thanks to the efforts of the British and American Air Forces the new firing points had been attacked continuously over the last few months, and attacks were still being carried out. The total tonnage of bombs dropped on France and Germany, including Peenemunde, had now reached around 50,000 tons and the number of photographic-reconnaissance sorties numbered many thousands. With their anonymity ensured, special mention was then made by the Prime Minister about the importance of the photographic-reconnaissance work and the interpretation which was mostly carried out at Medmenham – 'The scrutiny and interpretation of tens of thousands of photographs obtained for this purpose have alone been a stupendous task discharged by the Air Reconnaissance and Photographic Interpretation Unit of the RAF'.

Up to 6 a.m. that morning, 2,754 flying bombs had been launched from the French coast. Of these, a large proportion were reported either to have failed to cross the English Channel, to have been shot down by fighters or to have been brought down by the anti-aircraft batteries and barrage balloons. In providing exact details of the fatalities that had been suffered up to six o'clock that morning, it was described as a remarkable fact that 2,754 flying bombs had been launched and only 2,752 fatal casualties had been sustained. The fact the V1s had killed almost exactly one person per bomb was considered to be a British success, and although it had to be accepted that the figure could increase through any subsequent deaths among the 8,000 people who had been injured and hospitalised, the attrition rate was certainly very small when compared to that experienced by the Germans during Allied bombing raids.

In the country as a whole, flying bombs were now discussed more than anything else. Although it was not generally felt that they could alter the course of the war, people everywhere were now taking them more seriously. Rumours abounded that flying bombs had reached as far north as Kettering, Northamptonshire, and as far west as Penarth in south Wales, that hundreds came in daytime and every two minutes through the night, with some bombs even reported to have been filled with poison gas. Lord Haw-Haw exploited the situation and named places in cities that would be attacked. These rumours were numerous and persistent and were said to circulate most intensely in areas that had already been bombed. Although never hit, the Irish-American fascist warned the 'Yanks' at the headquarters of the US Eighth Air Force (code name Pinetree) in High Wycombe, that they would be 'visited by a buzzbomb at midnight'.

With the Luftwaffe unable to fly photographic-reconnaissance missions over London with the relative impunity that the Allies enjoyed over enemy-occupied territory, the Germans were desperate for any intelligence they could get about where their flying bombs were landing. When the first victims were killed in Bethnal Green on 13 June 1944, Rear Admiral George Thomson – the Chief Press Censor at the Ministry of Information – later reflected that mistakes were made from the outset. In his 1947 book, the self-styled 'Blue-Pencil Admiral' claimed that since the press censors had never been told that the flying-bomb offensive had begun – and however dry the communiqué in the press on 14 June might have appeared – releasing the information that an enemy raider had been shot down across the London & North Eastern Railway line had been a mistake. This was because there had been other enemy aeroplanes over southern England that night, and if they had all returned to base the Germans would have known that one flying bomb had made it to London.

Another source of information that was quickly made unavailable to the Germans after the first major raid on 15 June 1944 was, however, quickly noticed by members of the public. In the 194th Home Intelligence Weekly Report – for the period 13 to 20 June 1944 – the substitution of the live reassuring sound of Big Ben on the radio with a gramophone recording was noted to have been soon spotted by some listeners. This prompted rumours that the Palace of Westminster had been damaged. But in fact it had been arranged for a reason that few would have imagined. Since 10 November 1940, during the Blitz, particular attention had been drawn to Big Ben after a spiritual idea by

Major Wellesley Tudor Pole received the support of King George VI and the Prime Minister.

A First World War veteran, Tudor Pole was an astute and successful businessman who was deeply committed to a spiritual and mystical life. His idea was the 'Silent Minute' – a time when people were encouraged to pray together for peace – in the sixty seconds after Big Ben tolled at 9 p.m. every night. Between 16 June and 8 September 1944, however, a recording synchronised with Big Ben replaced the live broadcast in order that no information would be provided to the enemy during the flying-bomb attacks. The fear was that Flak-Regiment 155 (W) might fire a V1 towards St Stephen's Tower, timing it to arrive and explode during the Silent Minute. Not only would the sound of an explosion cause anxiety among listeners but over successive firings the Germans would be able to zero in their flying bombs on the heart of the government quarter. The ploy did, however, require the programme engineer to start the recording at precisely the right moment, and was recorded on one occasion to have led to a telegram from a listener asking whether Big Ben had taken to drink after a repeating groove in the record made the deception even more obvious.

The press editors had known since October 1943, after their meeting with government ministers, that when the attacks came they would have certain responsibilities and publicity would need to be carefully handled. Following the first major attack on 15 June a meeting of the Newspaper Proprietors' Emergency Committee was held the next day, at which the rules governing the censorship of flying-bomb stories were discussed and it was agreed that no obituary notice for anyone killed by enemy action in southern England should be published. This decision was subsequently revised after the Prime Minister revealed to Parliament that London – rather than places south of a line from The Wash to the Bristol Channel – was the key target. Obituaries would now be permitted providing no more than three from the same postal district appeared in any one issue. But this decision soon prompted Lord Cherwell to write on 20 July 1944 to Sir Findlater Stewart, then Chairman of the Home Defence Committee, about the possible use of newspaper obituary notices by 'the enemy in plotting the fall of his Crossbow shot'. One of his analysts had plotted seventy such obituaries from *The Times* and eighty from the *Daily Telegraph*, and although he admitted that the method used was crude – locating each shot at the centre of a borough or postal district – he observed the results to be 'dangerously near the truth'.

One of the key targets was believed to be Charing Cross – it later turned out that it was actually Tower Bridge – in order that the natural spread of V1 shots would cover the centre of government and the densely populated areas just north and south of the Thames. Lord Cherwell feared that if the enemy obtained a sufficient number of reports of incidents selected in a haphazard way they could plot a mean point of impact close enough to the actual point to provide a good idea of where their shots were falling. This point had been proven by Cherwell whose plots from obituary notices had achieved results only a mile or two off the real ones. In addition to helping the enemy achieve greater accuracy, Lord Cherwell emphasised that the obituary notices could also draw attention to mistakes made by the Germans. One such point of impact was identified as North Dulwich, which was being hit by an unusually large number of flying bombs. As a largely suburban area of south London, the assumption was that the Germans had made a mistake – Cherwell wanted to ensure that they remained in the dark.

CHAPTER 10
Hell's Corner

The first positive photographic evidence of the method of launching a
V1 was recorded during sortie 106G/992, flown on 20 June 1944 over
a Belhamelin site at the village of Hautecôte in the Pas-de-Calais. At
the site named by the interpreters as Haute Cote II, and appearing
since the last photographic coverage of 30 May 1944, the evidence
confirming that projectiles had been fired was the ground-scarring
made by the booster motor and the dolly – the steel piston that travelled
underneath the V1 during its launch up the ramp and that detached
after launch. Starting 350 yards from the railhead and continuing until
422 yards, where they ended in a sharp curve, were a series of skids
where the turf – in a field adjoining the Belhamelin site – had been
thrown to the sides on impact. Other telltale signatures of activity
included vehicle tracks, showing where a lorry had driven to collect
the dolly for its reuse during a subsequent launch.

At the neighbouring site of Haute Cote I, Interpretation Report
BS544 recorded the evidence of bombing and near misses, and noted
that although one bomb had even hit immediately in front of the
rail no damage was visible in the photography. This emphasised
the tremendous difficulty facing the Allied Air Forces in destroying
such a relatively small and skeletal structure as a Belhamelin firing
point. The rails were measured as 160 feet long, and the much-
increased vehicle activity was evidenced by vehicle track marks, so
it was clear that the site had been and could still be used. In the
fields in front of the firing point intermittent ground-scarring was
also visible, with lorry tracks seen leading from the road across the
fields to the ground-scarring. It was even possible to see where a

lorry had backed up for a few yards and reversed when collecting the dolly.

Countermeasures against the Belhamelin firing sites began on 19 June 1944. With a pressing need to destroy the operational sites, they were simply classified as either A, B or C. Sites only reached category A if their launching point, Square Building or Rectangular Building had received a direct hit on their foundations. This evaluation was made on the basis that the light prefabricated structures could be completely blasted away by a near miss but if the foundations were intact they could be quickly rebuilt. In June, nineteen sites were reduced to category A, followed by fifteen in July, ten in August and three in September. But the relative slowness in destroying the Belhamelin firing sites was an indication of how hard it was to locate and demolish them, and reinforced the earlier decision by Medmenham that finding and destroying the Supply Sites would have a far greater impact.

Less than two weeks after the first V1 flying bombs landed on England (on 13 June 1944) Wing Commander Dudley Barker of Air Ministry Branch PR3 – part of Air Ministry Public Relations – wrote to the Ministry of Information with a proposition. In his view the flying-bomb offensive would be short-lived, and since there had been a tendency among the public to criticise the effectiveness of the RAF countermeasures he wondered whether a pamphlet should be issued soon after the attacks finished, as a means to both limit criticism and re-establish whatever respect might have been lost. When the propaganda experts agreed, the Air Ministry immediately sent Squadron Leader H. E. Bates – who was then in Scotland on holiday with his wife – a telegram, ordering him back to work immediately.

When England first came under attack from flying bombs, Herbert Ernest Bates and his family had been at their home, a traditional granary building forming part of a Wealden farmstead which had been converted into a house near the village of Little Chart, north-west of Ashford, Kent. Awoken in the middle of the night by a terrible noise, a mystified and unsettled Herbert Bates spotted what appeared to be burning aircraft in the sky, heading north-west towards London. Little did he know then that two weeks later he would be flown to France to inspect some of the firing sites, which had been overrun by the rapidly advancing Allies in the Cherbourg peninsula and from where Flak-Regiment 155 (W) had planned to bombard south-west England. To provide his writing with the necessary background he was given the first-hand opportunity to inspect the impact of Allied Air Force attacks on the different Crossbow sites, of meeting soldiers on the ground, and of talking with local French people. Throughout his visit he absorbed information about the revolutionary technology that continued to cause so much physical destruction in southern Britain and was having such a negative impact on British morale so soon after the morale-boosting Normandy landings.

While Herbert Bates was still in France, complications emerged as a multitude of government departments on hearing of the project sought assurances that their contribution to the battle against the flying bomb would be covered in the publication. The original plan had been to publish a modest 5,000-word pamphlet akin to the official wartime publications that had been issued by the Ministry of Information and published by His Majesty's Stationery Office – HMSO – on Coastal Command, Bomber Command, the Battle of Britain and other military triumphs throughout the war. As a compromise, Herbert Bates found himself by September in the position of having to write a much-enlarged publication – a 40,000-word book – that would cover not only the flying bomb but all Hitler's revenge weapons.

As the flying-bomb offensive continued throughout the summer of 1944. The Bates family lived in what became nicknamed 'Doodlebug Alley' and Herbert later recalled one beautiful August evening when a V1 flew over his house after being attacked by a fighter and at just before 8 p.m. crashed onto the Little Chart parish church of St Mary the Virgin and the Holy Rood. Only the badly damaged Norman church tower and chancel would survive, and following a decision to rebuild the church in the village, the ivy-covered ruin remains a symbolic memorial to the battle against the flying bomb and the impact of the weapon on Kent during the summer of 1944. In London 111 churches would be damaged by flying bombs, including St Michael Paternoster Royal, in the City of London – a building designed by Sir Christopher Wren. The English Baroque masterpiece had replaced an earlier church destroyed in the Great Fire of London and had already been severely damaged during the Blitz. Then, on 23 July 1944, a flying bomb left only the tower and walls standing.

Before the war, H. E. Bates had been an aspiring author and had written *My Uncle Silas*, a collection of short stories about the bucolic existence of an elderly Bedfordshire countryman and his roguish antics. During the Second World War, using the pseudonym 'Flying Officer X' he wrote a series of stories about RAF pilots that proved hugely popular. A de facto author in residence at the Air Ministry, he secured the position after a friend whom he had approached for career advice introduced him to the Director of Public Relations at the Ministry. Presented with the opportunity of writing morale-boosting stories that would bring the business of war alive in a heroic and exciting way, Herbert Bates accepted on the condition that he was given total freedom of time and movement. This resulted in his becoming a commissioned officer and saw him spending time at airfields discreetly gathering ideas while writing stories on the bravery of RAF pilots and aircrew that initially appeared in the *News Chronicle* and were later published in book form as *How Sleep the Brave* and *The Greatest People in the World*.

With Herbert Bates working on a much-enlarged publication, the writing of the book about the German V weapons continued long after the flying bombs had stopped falling and even after Victory in Europe Day in 1945. With its value as an immediate propaganda tool lost, combined with a War Office reluctance for certain information to be published, there was an increasing risk that the book would never see publication. This prompted Cecil Day-Lewis of the Publications Division at the Ministry of Information – the Anglo-Irish poet who would later become Poet Laureate – to remind those involved that all that the Chiefs of Staff required was a formal security review of the text, but that unless the complete story was told the publication would be cancelled. In an internal Air Ministry communication dated 1 November 1945 it was noted that the position was further complicated by the fact that Squadron Leader Bates was to be demobilised the following week. The net result of this indecision was that the secret manuscript – marked with the caveat *'BIBLE COPY – NOT TO LEAVE PR3'* – remained unpublished and sat in a closed Air Ministry file until 1972, when it became accessible at the Public Record Office. Ultimately, thanks to the efforts of Robert (Bob) Ogley, the editor of the *Sevenoaks Chronicle*, it was finally published on 13 June 1994 to coincide with the fiftieth anniversary of the first V1 attack on Britain. After the war, H. E. Bates wrote screenplays and novels, many of which were adapted for film and television, including his stories of the Larkin family in *The Darling Buds of May*, which charted their exploits in an idyllic rural Kent of the 1950s.

During the battle against the flying bomb in the summer of 1944 many were destroyed by the countermeasures in southern England. Sometimes the unforgettable sound of the motors would change to a stuttering cough and the machines would stall. Either way, many fell on what was customarily referred to as the *open country* of Kent – and although the Allied fighter pilots were careful not to attack doodlebugs near built-up areas there were still people living in that 'open country'. The peak of the attack by V1s came during the first week of July when 800 were launched, and this resulted in bombs landing on a number of Kentish towns in Doodlebug Alley. But a breakthrough came at around the same time after an attack by 617 Squadron – under the command of Wing Commander Leonard Cheshire – against Saint-Leu-d'Esserent, in the Val d'Oise *département* to the north of Paris, on the night of 4/5 July.

After Medmenham received ground intelligence that a number of underground quarries and tunnels were being used for the assembly and storage of Crossbow weapons – mainly in the Oise Valley around 30 miles to the north of Paris – photographic reconnaissance of all the geographical pinpoints supplied was laid on from 14 June 1944. Interpretation Report BS534, issued on 19 June, provided an analysis of activity at eight pinpoints around the town of Creil. Although some of the sites were old and abandoned others showed signs of fresh activity, with new railway construction, excavations, tunnel entrances and mine shafts. The area was observed to have been considerably reinforced with anti-aircraft positions which had appeared since the end of May, with six new heavy batteries seen to arrive on 12 June. It was noted that although there was no evidence of a link between the new activities and Operation Crossbow the fact that flying bombs were now dropping on London with a vengeance ensured the continued interest of the interpreters.

One of the sites observed was near the town of Saint-Leu-d'Esserent on the banks of the river Oise, in a quarry now heavily defended with anti-aircraft batteries and conveniently located approximately three-quarters of a mile from the rail yard at the village of Thiverny.

Saint-Leu-d'Esserent was notable for having vast underground caves, which had been used by French mushroom growers in peacetime that were now used by Flak-Regiment 155 (W) for the storage of flying bombs. In a network of natural underground caves accessed by tunnels strengthened with concrete and sealed with thick steel doors, the Germans hoped they would be immune from bombing attacks. Successfully attacking this site required the taking of oblique photographs that would allow the interpeters to locate the disguised entrances, calculate the amount of overhang over the tunnels and analyse other technical data necessary for target information to be prepared that would ensure the best bombing results.

One particular application of F24 cameras involved using them as forward-facing oblique cameras – which called for cameras to be installed in the wing drop-tanks of Mosquitos and in bulges on Spitfire wings – to make the stereo oblique photography of particularly important targets possible. Known to the PR pilots as dicing sorties – because they were dicing with death – a high degree of skill and daring was required by the pilots as they flew at speed, extremely low and directly towards the targets they were photographing. Another form of dicing photography involved using the F52 camera in the nose of a Mosquito, which meant the pilot could navigate directly on particularly important targets and be more certain of obtaining viable photography. Approaching at speed, the pilot used the element of surprise to photograph the target on his first run and with remarkable skill and daring had to do this, and in the process bank the aircraft to avoid any collision. With a site like Saint-Leu-d'Esserent so heavily defended, his first run would typically awaken the anti-aircraft gunners, which meant that a subsequent run would be near-suicidal.

The bombing of Saint-Leu-d'Esserent began on 27 June 1944 and on 6 July, after twenty direct hits, a large section of earth measuring 200 × 400 feet collapsed into the underground facility. This had the dramatic effect of reducing supplies to the firing sites in the Somme and Seine areas, and within twenty-four hours cut by half the number of flying bombs being fired from there. This proved there were no intermediate sites and highlighted the need to completely destroy the Saint-Leu-d'Esserent location. Destroying all the entrances to the subterranean Supply Sites was proving a difficult challenge but it was clear that supplying the Belhamelin Sites was the bottleneck for the Germans. This meant that twenty-six launching sites in the area north of the Seine were never completed by the Germans and were instead held in reserve in case other firing locations were destroyed or supplies could be increased.

Large-scale vertical photographs of Saint-Leu-d'Esserent were taken on 6 August 1944 – during sortie 106G/1980 – at a scale of 1/3,500. These were followed on 15 August with low oblique photographs taken during sorties 106G/2276 and 106G/2277 with side-facing oblique cameras in two P-51 Mustangs. While one flew close to the tunnel entrances to record them in detail, the second flew at some distance to cover a wider area. This method of photography made the aircraft far less vulnerable and managed to secure just as much information as when the forward-facing oblique camera was used. In a communiqué from Air Commodore D. J. Waghorn – the Air Officer commanding 106 Photographic Reconnaissnce Group at RAF Benson – to Bomber Command headquarters that same day, he noted their interest in the tunnels and drew their attention to these new sorties. While pointing out that this approach was much safer than the forward-facing oblique method, he was careful to emphasise that such sorties would only be sanctioned for special purposes since the aircraft concerned were still extremely vulnerable to attack.

In a report by Douglas Kendall dated 16 August 1944, he studied the vertical and oblique photographs of Saint-Leu-d'Esserent and estimated their relative value for damage assessment. His conclusions were that the large-scale vertical photographs were ideal for interpretation purposes and a full damage-assessment report was possible. The oblique photographs were of excellent quality and as a way of illustrating the target could not have been better, but for damage-assessment purposes had a number of significant disadvantages: they were not in stereo, a large proportion of the areas photographed was 'dead' ground, there was real difficulty in coordinating the photographs with plans and pinpointing damage, and for comparative purposes – upon which all damage assessment depends – the length of time required compared with that needed for assessing vertical photographs was increased enormously. It was recognised that the oblique photographs were ideal for providing information about structures and helped enormously with the planning of bombing missions, but that vertical photography was much preferred by the interpreters for damage-assessment purposes.

On 26 July 1944 a report was issued that gave the results of investigations up to that date. It attempted to identify common features of the Supply Sites and concluded that they were all located near a main road, were frequently near a railway, were usually found a short distance from a town or communication centre, and were well hidden in wooded country. Where the contours of the land allowed tunnelling into a hillside this was taking place, and where it wasn't possible shafts were being sunk into the ground. Many of the sites were seen to be connected with an overhead cable line that did not appear to be part of a telephone system and there was extensive use of camouflage, consisting of netting stretched over supports to conceal the centre of activity from the air. Soon after the report was issued it became apparent that some of the sites that were being discovered across northern France had no connection with the flying bombs being fired from Belhamelin Sites. In late July the heavy bombing attacks on the main supply depots near Paris in the Oise Valley had produced an immediate drop in the number of flying bombs being launched from sites south of the Somme. This led to an assumption that those sites depended on the Paris depots for their supplies and not these underground storage facilities. This assessment was reinforced by the fact that many of them were still under construction, and although some of the tunnels could have been used, the evidence pointed to them having another future function. The obvious inference was to connect them with the A4 rocket.

With the raid on Saint-Leu-d'Esserant using 12,000-pound Tallboy bombs putting the underground storage facility there out of action, the immediate effect was to reduce the number of flying bombs reaching London from around 100 per day to less than 70. As the Allies continued their advance and overran an increasing number of Belhamelin firing sites, the Germans had to change their tactics. They never managed to implement the original plan of Colonel Wachtel to fire the bombs in salvos, which would have meant that while some V1s would be intercepted the maximum number would reach their targets. Despite the continued improvements in countermeasures, which included using newly introduced rocket guns and the deployment of the first British jet-powered combat aircraft, the Gloster Meteor, those living along the projection line of the firing ramps in northern France were vulnerable for much of the battle. Although the Prime Minister had correctly observed that on average one flying bomb killed just one person, and the weapon was clearly never going to affect the outcome of the war, its negative impact on morale was heavily reinforced when one single bomb killed many, such as the tragedy at the Guards Chapel and the later Aldwych massacre. Croydon suffered more than anywhere during the battle with 142 V1s ultimately landing on the south London borough, causing damage that required a makeshift army of 1,500 workers to be brought in from throughout the country. The tradesmen came from as far afield as Scotland and Ireland and were billeted in temporary huts, schools and church halls while they repaired and frequently re-repaired the tens of thousands of houses that had been damaged. In 1944 a map which plotted where each of the 142 flying bombs had fallen on the town would be sold, with the proceeds going to the National Fire Service Benevolent Fund.

At the beginning of July 1944, with the V1 battle at its peak, the miscellany of defensive measures deployed had resulted in a number of friendly-fire incidents with Allied fighters frequently being hit by anti-aircraft fire, on several occasions fatally. This prompted Air Chief Marshal Sir Roderic Hill – Commander-in-Chief, Fighter Command – to fly a number of sorties over southern England to examine the situation. His assessment was that the gun batteries should be relocated from their inland positions to the coast, creating in the process two unambiguous fighter-patrol areas, one over the English Channel and the other between the gun belt and the barrage balloons. This was no mean decision since hundreds of guns had by then been positioned on the North Downs, with thousands of miles of telephone cables laid and accommodation arranged for the gunners. In deciding to relocate such a vital part of the defences, a process that would take many days, the Air Chief Marshal was painfully aware that this would leave London vulnerable.

For several nights during the second week of July no flying bombs were plotted over London and the capital had a welcome respite from attack. This prompted many to think that perhaps the Germans had run out of supplies, but on 11 July twenty-nine flying bombs would land on Hampshire. This was to mark a double departure for Colonel Wachtel, when the ports of Southampton and Portsmouth were attacked by V1s that had been launched from modified Heinkel He 111 medium bombers. Despite the fact that most flying bombs were deployed from fixed sites, trial versions of the V1 had previously been air-launched and as the Allies pushed the Germans back it would increasingly become the only means by which they could be launched against Britain. After the decision about the Diver gun belt had been made, on 13 July the guns began their relocation to the coast. Over the following five days this move was completed and a zone stretching 10,000 yards out to sea and 5,000 yards inland was created.

But the gamble paid off, and from their new positions the gunners shot down 17 per cent in the first week, 24 per cent in the second and by the end of August the figure had jumped to 74 per cent of all flying bombs launched. With the balloon barrage to the south of London increased to 1,000 balloons by 21 July and 1,750 by the end of that month the combined efforts of the anti-aircraft batteries, the fighter pilots and Balloon Command were very effectively protecting London. This was less the case for Kent and Sussex where doodlebugs would still land in open country, but it did result in a more positive sense that the battle against the V1 was being won.

The method of using modified Heinkel He 111 aircraft to deploy flying bombs had actually been introduced on 9 July 1944, when small numbers of V1s arrived to hit London from the Thames estuary, a far more northerly position than on any previous occasion. These attacks came at night and ushered in a completely new phase in the German flying-bomb campaign. Since the maximum range of the V1 was known to be 160 miles, and all known land launch sites were aligned directly on their target, the initial fear was that the Germans had developed a doodlebug with increased range. When the projection lines had been followed to western Belgium and the south-west corner of the Netherlands, photographic reconnaissance was flown but no launch sites were located.

As the bombs from the north continued to come, an investigation of all Dutch ports was carried out to discover whether V1s were being launched from ships or barges that might have been towed by tugboats into the North Sea. With this investigation also drawing a blank the mystery would not be resolved until ground intelligence made the connection between certain airfields and Crossbow activity, with confirmation also coming from radar stations that had observed one blip on their screens becoming two during one of the flying-bomb raids. With Heinkel He 111s identified as the type of aircraft being used, research was undertaken into how this was possible. It would eventually transpire that the bombs were fixed underneath the bombers' starboard wings, which meant their detection at German airfields in aerial photography was – understandably – virtually impossible.

At Venlo airfield in the south-eastern Netherlands province of Limburg, close to the German border, three square buildings similar to those found at the French launch sites were spotted. It was noted that they had appeared since May 1944 and they were seen to be hidden on the ground by tall brushwood fences. Two heavy bombing attacks on the airfield did not affect the square buildings but did put the runways at Venlo out of commission. When the area was later ground-checked, on 16 March 1945, and despite the Germans having demolished the buildings, evidence was found and reported by the Army Photographic Interpretation Section of 8 Corps Headquarters. As the photographic interpreters in the field they were able to confirm that the interpretation carried out at Medmenham had been correct, with construction features consistent with the Crossbow storage sites present. They even managed to discover a number of V1s and jet-propulsion units. Medmenham managed to locate similar Crossbow activity at thirty other airfields, and even spotted a Crossbow road convoy entering Handorf airfield in northern Germany, in photography taken during September 1944.

The first time a V1 was photographed being unloaded from a lorry is recorded on Interpretation Report BS594 of 29 June 1944, which provides interpretation of sortie 106G/1148, flown the previous day by Flight Lieutenant Geoffrey Crackenthorp in a PR Spitfire when on a mission to photograph No-Ball Targets in the Abbeville-Boulogne area. During the sortie he made two runs over a site at Belloy-sur-Somme. In the first run the flying bomb could be seen mounted in a 30-foot-long lorry, with another 22-foot-long lorry in front, both drawn by a 'Mechanical Horse'. By the second run the missile had been taken off the lorry and was being turned around, while the train of three vehicles was 45 feet further down the road.

Despite great improvements to the anti-V1 defences in July, the proof of how deadly just one flying bomb could be was proven on a number of key occasions. On 28 July at 9:41 a.m. south-east London suffered its worst attack when a doodlebug landed on the marketplace in Lewisham. It was a busy Friday when the flying bomb landed, and its characteristic destructive blast effect killed 51 people and seriously injured 124. The Lewisham attack was followed a few days later on 2 August when 44 people were killed after a bomb crashed onto a busy restaurant in the south-east London borough of Bromley.

By 23 August, when the Allied Armies liberated Paris, the British and Canadians were sweeping through the Pas-de-Calais. This resulted in Colonel Wachtel and his launch crews retreating eastward into the Netherlands and abandoning their French firing sites. This meant that London would now be effectively out of range of a major V1 onslaught. The German retreat was rapid, and this meant that the biggest doodlebug threat now lay with air-launched flying bombs. But since the Allies had almost total air superiority this danger was considered to be relatively small and led many to believe that the battle against Hitler's secret weapons was coming to an end.

By the time of the Ministry of Information's 205th Home Intelligence Weekly Report – covering the period from 29 August to 5 September – spirits were reported to be soaring at the amazing Allied advances in Western Europe. As a foretaste of what was to come, though, the Red Army's liberation of the Majdanek concentration camp on the outskirts of Lublin in eastern Poland produced shocking revelations that were met with universal disgust. This managed to further fuel the widespread discussion about the future of Germany, the hatred felt towards the German people and the extent to which the desire for retribution should be tempered with rehabilitation.

The capture of so many V1 launch sites in France and the hope that the remainder would be captured soon, coinciding with the recent lull in their onslaught, as the Germans retreated hastily, was reported to have given British people a great sense of optimism. But there was still some scepticism noted from those who thought that 'Germany still had something up her sleeve', with gas and rocket bombs noted as possibilities. The feeling among most civilians was that there would be an early end to the war, and while October 1944 was the favourite guess for this eagerly anticipated date some expected the announcement of victory any week, day or almost hour. Some people were even reported to be already making plans for celebrations and for what they would do after the war. Duncan Sandys certainly felt confident enough to announce to a crowded press conference on 7 September

1944 that 'Except possibly for a few last shots, the Battle of London is over'. Explaining that over eighty days the Germans had deployed 8,000 flying bombs of which 2,300 had reached London, he singled out the people of Kent, Sussex and Surrey as deserving of praise for their readiness to accept their share of London's dangers. He also thanked the Americans involved in the defence of Britain, who were described as having thrown themselves into the job as if it had been New York or Washington D.C. that they'd been defending. The possibility that more flying bombs could be deployed from aircraft was mentioned, but the fears of many German officers about the effectiveness of such an approach was noted.

In that week's Home Intelligence Weekly Report, the publicity from Duncan Sandys's announcement was recorded to have been received by the public with widespread satisfaction. The people were said to be pleased to hear about just how effective the defences had been and the promptness and honesty of Duncan Sandys's announcement. In London the cessation of attacks was greeted with deep relief, although some scepticism was still noted from people who thought the attacks would recur. That was a similar fear among people in the south-eastern district, and in Folkestone there was a feeling that they should be entitled to a greater share of building labour since they had received bombs destined for London. The shelling of Dover from the German coastal batteries in France generated anxiety among people there who were believed to be using shelters more than at any time since the Battle of Britain. There was also resentment among some Dover folk who felt that the 'premature' announcements by the British about taking French ports had caused the Germans to open fire 'out of bravado'. Many of the people who had evacuated from London were reported to be returning, and this prompted the observation from the Ministry of Intelligence that it should be made compulsory for children to be left in reception areas for the time being. In the south-east two explosions were heard, sparking off rumours that rockets were now being fired by the enemy from the Netherlands or Denmark. Opinion was reported to be divided at the Ministry about whether or not a statement should be made at that point about the threat posed by rockets. A majority of staff favoured an announcement being made to quell rumours and keep evacuees away from London, although a considerable minority were against the move. The breakdown in Russo-Polish relations was causing bewilderment and distress, given the role played by so many Polish airmen during the recent Battle of London. The Russians were said to be regarded with widespread suspicion and there was fear, whatever the rights or wrongs, that the present situation would lead to dissension between the Allies.

The people of Kent were told by the *Kent Messenger* on 15 September that more flying bombs had landed on their county than on London. With a map published in the newspaper showing where every doodle-bug had landed in Kent, public interest was so great that the newspaper quickly sold out. Despite being reprinted the following week, within days orders for 100,000 copies were received, such was the intense interest in a map that would be hung on the walls of houses, offices and pubs as a reminder of the battle. Even after the capture of the Belhamelin Sites in northern France, however, the Germans were determined to continue their attacks with flying bombs on Britain, which duly resumed on the night of 13 September using Heinkel He 111s, which would continue sporadically until January 1945. But the front line was now East Anglia and, after consultation with Fighter Command, the Anti-Aircraft Command redeployed all available defences from the south coast to a strip running from the Thames estuary to Great Yarmouth in Norfolk.

This was another enormous logistical task but by the middle of October 300 static guns, 542 heavy guns, 503 40mm guns and 18 searchlight batteries were achieving a high success rate. During this time only thirteen flying bombs landed in Norfolk, injuring just eleven people and killing nobody. In the event several more, including some air-deployed V1s, would ultimately land on Kent before the county's battle against the flying bomb was over.

Throughout the war, only a small number of conventional bombs were dropped in the Medmenham area. The closest a flying bomb would come was Chalk Pit Lane in nearby Great Marlow on 22 July 1944, when houses were damaged and people injured. The personnel at Medmenham apparently expected more flying bombs to have appeared in their district, given their location on the edge of the target area. On that fateful day in July 1944 the modelmakers heard the distinctive noise of the engine followed by silence as it cut out and are reported to have stood still or to have had their tools or brushes poised motionless as they all waited to hear the loud explosion, hoping that it had landed in open countryside.

As an indication of how difficult the Belhamelin Sites were to spot, particularly if they had never become operational, on 21 September 1944 the Air Ministry compiled – as a result of ground checking – a list of thirty such sites that had never been spotted in aerial photography. Twenty-two of them were on the Cherbourg peninsula and in Normandy south of the Seine, and although one site at Berneuil, in the Somme *département*, had been fired from (despite being very cleverly located evidence of this *could* be seen in aerial photography) the remaining seven sites showed no evidence whatsoever.

In total, 8,564 flying bombs were launched against England, just over 1,000 of which crashed on take-off due to lack of sufficient speed. The ones that did reach the target area killed 6,184 people in London and seriously injured about 18,000. The number of houses damaged was about 750,000. For every V1 launched, approximately one person died and three others were seriously injured. In addition, 2,000 British, American and Canadian airmen lost their lives in the Allied counteroffensive. And a total of 50,000 tons of bombs were dropped against the V-weapon sites. The enemy achieved an average of just under 100 launchings in every 24-hour period, or about one every 15 minutes – day and night.

CHAPTER 11

The V2 Rocket

Since all the interpreters ever saw in the Peenemunde photographs at the beginning of the investigation into secret weapons – once they knew what they were looking for – were rockets lying in the open, it was eminently logical for the original investigation to focus on the threat posed by the long-range rocket rather than the flying bomb. This early analysis led to the Bomber Command attack on Peenemunde in August 1943, followed in the same month by the US Eighth Air Force attack on the Heavy Site at Watten. As 1943 progressed the investigation into the long-range German rocket became Operation Bodyline and in turn Operation Crossbow. At Medmenham the Crossbow team grew from a handful of key people – Norman Falcon, Douglas Kendall, Hamshaw Thomas, André Kenny, Robert Rowell and Neil Simon – to the creation of an entire new Sub-section (B2) and the deployment of other specialist interpreters from across the Medmenham spectrum as required.

In addition to monitoring Peenemunde, efforts were made to discover other sites used for rocket experiments, component manufacture and storage. The rail-served Heavy Sites in northern France from where it was suspected that the rockets would be deployed were also the subject of intense search. Whilst northern France was photographed and rephotographed in the hunt for Bois Carré firing sites and the associated supply network, the remaining Heavy Sites had been located and reported on throughout the remainder of 1943. In the Pas-de-Calais region Mimoyecques was first reported on 25 September, with Siracourt on 5 October, Lottinghem on 2 November and Wizernes on 5 November, while on the Cherbourg peninsula, Martinvast was reported on 22 October and Sottevast on 31 October. The sites were originally reported on by the Army Section in the routine course of their work, but once their significance to Operation Crossbow became apparent BS reports were issued. The major causes for suspicion came from ground intelligence: the fact that a site did not resemble or have a connection with any known military installations, the fact that all sites were started at around the same time, that they were all rail-served, that – with the exception of Watten – they were all aligned towards London or Bristol, and that the scale and urgency of the work was a clear indication of how important their completion was to the enemy.

Since April 1943, the third-phase interpreters in D Section, notably André Kenny, had been responsible for the manufacturing aspects of the secret-weapons investigation. The industry experts were responsible for: monitoring all areas at experimental sites not concerned with the actual launching; identifying the network of factories involved in the manufacture or assembly of Crossbow weapons and their component parts; identifying factories producing chemical propellants used in weapons – notably hydrogen peroxide and liquid oxygen; and studying the electrical supply to sites. Locating Crossbow factories was to prove a particular challenge, as typically no external characteristic distinguished them from any other factory and this meant that ground intelligence, from MI6 and their Pingpong reports, usually proved vital in the first instance. The combination of ground intelligence and aerial photography proved essential throughout Operation Crossbow. It allowed the interpreters to investigate whether a factory was operational and whether there were any telltale signs of Crossbow activity – such as the particular type of railway wagons used for the transportation of components.

With their expertise in electrical engineering, the Industry Section had investigated the power supplies that fed all the Crossbow experimental, manufacturing, supply and firing sites and assessed how electricity was used within the different types of site. At Peenemunde this involved studying the transformer station at nearby Kroslin and while the flying-bomb attacks were in progress an investigation was made in July 1944 into the power and transformer stations and overhead power lines in northern France. From their reports, detailed target information was prepared, identifying specific items of industrial plant that when destroyed would cause the maximum disruption possible.

The Industry Section knew that destroying an entire industrial process was difficult to achieve and that, whilst aerial photographs of bomb-damaged factories might give the impression of total devastation, a lot of perceived damage was often superficial and the study of post-strike photography often revealed how quickly repair work could begin and productivity return. If, however, a critical industrial component – such as an electricity transformer – was destroyed, it could take weeks or even months to repair.

Reports were issued by D Section on approximately sixteen factories suspected to be making or assembling Crossbow components and which were mostly located in Germany but also in France and the Low Countries. From April 1944 an intensive search was made for factories that were suspected of manufacturing liquid oxygen required for the V2 rocket, and this resulted in reports being written on thirteen sites in occupied Belgium. An investigation was also carried out – under the code name SPLEEN – into factories suspected of being involved in the manufacture of the hydrogen peroxide necessary to fire both flying bombs and rockets. It was the discovery of a railway wagon that they nicknamed the 'egg rack' at a factory in Bitterfeld, Saxony-Anhalt, Germany, that provided the investigation with a kick-start. The egg racks measured 28 feet by 10 feet and each contained twelve cylinders. They had previously been spotted at Peenemunde and linked to the flying bombs. The factory was owned by the chemical conglomerate I. G. Farben, and soon more egg racks could be seen at the company's other factories.

Four days after Operation Hydra – the attack by Bomber Command on Peenemunde – on the night of 17/18 August 1943, Adolf Hitler and Albert Speer, Minister of Armaments and War Production, resolved to make greater use of concentration-camp prisoners for the mass production of the secret weapons – and opted to relocate much of the rocket testing to Poland. Despite the risks to the civilian populations below, the chief advantages of moving the German Army's rocket experiments eastwards and inland were the opportunity to test the missiles' impact on dry land and to take the operation beyond the range of British-based photographic reconnaissance and heavy bomber squadrons. Dr Wernher von Braun was anxious to begin testing with live explosives and at a meeting of the 'Long-Range Ballistics Commission' on

9 September 1943 he noted the importance of no further experiments being undertaken at Peenemunde except at night or under the cover of artificial fog – a reference to smokescreens – so that the Allies could not benefit from photographic reconnaissance. Whilst the Army relocated much of the rocket testing from Peenemunde to the Heidelager SS military training area at Blizna in south-eastern Poland, the mass production of rockets would be relocated to an underground factory in the Harz Mountains, the highest mountain range in northern Germany. Heinrich Himmler, Reichsführer of the SS, took the opportunity of ensuring that Major General Hans Friedrich Kammler, who had earlier supervised construction of the Auschwitz concentration-camp crematoria, should be responsible for the construction work.

At the same time, the Luftwaffe Experimental Centre at the airfield – known as Peenemunde-West – was partially relocated to Bruesterort, 30 miles north-west of Königsberg in East Prussia. This would eventually be discovered by the Allies as a result of ground intelligence and was first photographed on a mission flown from Leuchars on 19 February 1944 – sortie L/41 – which revealed the existence of a Bois Carré site located in the modern-day Russian exclave of Kaliningrad, between Poland and Lithuania on the Baltic Sea. Photographs revealed a pair of rectangular buildings, two square buildings, one ski-building and three Bois Carré launching ramps aligned towards the German-occupied Danish island of Bornholm. The Bruesterort experimental site had been developed in record time through the exploitation of thousands of Soviet and Polish prisoners of war. The early manufacture of the flying bomb was beset with production problems and the Reich Aviation Ministry put the East Prussia testing centre under equally intense pressure to complete technical development.

Manufacturing V1 components also involved prisoner labour at the Volkswagen factory near the village of Fallersleben. Near the city of Braunschweig between Hanover and Berlin, north of the Harz Mountains, the Nazis had before the war built a planned town around the village of Fallersleben called *Stadt des Kdf-Wagens bei Fallersleben* – City of the KdF Car at Fallersleben – that was occupied by people working in the neighbouring Volkswagen factory. Kdf is an abbreviation for *Kraft durch Freude* – Strength through Joy – the large state-controlled leisure organisation in Nazi Germany, and Volkswagenwerk GmbH were responsible for manufacturing the

KdF-Wagen as an affordable car for the German people. During the war, the Fallersleben factory became part of the war economy and as well as using slave labour to manufacture component parts for flying bombs, it built a range of military vehicles. When the town became part of the British Occupation Zone after the war it would be renamed Wolfsburg and, thanks largely to Major Ivan Hirst of the British Army, Volkswagen would be reborn as a vehicle manufacturer and the KdF-Wagen would become reincarnated as the Volkswagen Beetle.

Another experimental site, analysed by D Section and discovered through ground intelligence, was near the city of Friedrichshafen on the northern shore of Lake Constance – which provides a natural border between Germany, Austria and Switzerland – at the northern foot of the Alps. Ferdinand von Zeppelin had established his famous dirigible factory there at the end of the nineteenth century, and with other aviation companies attracted there, including Maybach-Motorenbau GmbH, the city produced many of the German aircraft engines used during the First World War. Although the punitive controls of the Versailles Treaty prompted Maybach to manufacture luxury motor cars, under the Nazi regime the aircraft manufacturer Dornier GmbH would rapidly expand. An established centre for the German aviation and automotive industries, a site at the nearby village of Raderach was chosen for a secret test facility by the German Army and by mid-1942 construction of a rocket motor test stand had started close to a liquid-oxygen plant code-named the *Porzellanfabrik* – porcelain factory. When testing was carried out, however, secrecy was immediately compromised as the engine roar and exhaust flames could be heard and seen from neutral Switzerland. This resulted in ground reports, photographic reconnaissance and Interpretation Report DS20 – dated 22 July 1943 – that identified three installations similar to those at Peenemunde. With a full report and plan of the site issued on 5 August 1943, the sighting of 'egg rack' railway wagons in May 1944 similar to those seen at Peenemunde confirmed the continued association. While the flying bombs were still being fired on London and fear of bombardment from the A4/V2 rocket was reaching fever pitch in official circles, Raderach was attacked on 16 August by eighty-nine B-24 Liberator heavy bombers of the Italian-based US 15th Air Force.

In April 1944, when the experimental site at Blizna was photographed for the first time, Peenemunde had been photographed approximately every fortnight for a year and this schedule would continue until April 1945. For Operation Crossbow the interpreters were monitoring a network of experimental sites that extended well beyond the Baltic Sea coast of northern Germany. In addition to Peenemunde and nearby Zinnowitz, the site at Bruesterort on the Baltic coast of East Prussia was being used for the technical development of the flying bomb and Bois Carré-type site, while the 'porcelain factory' at Raderach near Lake Constance was implicated in the manufacture of fuel for secret weapons and rocket-engine testing. The discovery of Blizna as a major experimental facility principally for the V2 rocket – but also, since it featured a Belhamelin-type site, the V1 flying bomb – would be followed by the discovery of launch ramps in Germany, from one in the military training area of Unterluss near the city of Münster in Westphalia to another experimental Belhamelin Site at Altenwalde, near the port of Cuxhaven, where the ramp was aligned on the German island of Heligoland. From photography taken during sortie 106G/4230 on 8 February 1945 the ground report identifying Altenwalde was confirmed, while another ground report the following month about two suspicious-looking square buildings at the nearby Emergency Landing Ground at Neunwalde would similarly be confirmed by photography.

In late 1943 Dr Wernher von Braun concentrated his energies on perfecting the V2 rocket and meeting the deadlines for its deployment. To maintain a link between construction, testing and production, he regularly flew between Peenemunde, Blizna and the underground rocket-production facilities then being developed at Nordhausen in the Harz Mountains. Testing began at Blizna on 25 November 1943 and was monitored throughout by the Polish resistance who would document 139 rocket launches up to 24 June 1944. Much to the frustration of the rocket engineers, around 60 per cent of the rockets that were successfully launched disintegrated two to three kilometres above the intended point of impact. Collecting samples and component parts was vital for the Allies if Scientific Intelligence was to understand the secrets of the rocket, but it would ultimately cost the lives of almost 150 members of the Polish resistance. By the beginning of 1944, when the rocket went into mass production, the tests to perfect the rockets continued. They were then fired northwards from Blizna over inhabited areas towards the Sarnaki region and the river Bug.

The breakthrough in Poland for British Intelligence had come thanks to Armia Krajowa – the leading Polish Resistance movement during the war – who had received reports from local people about suspicious activity at the military training area near the village of Blizna, in the middle of a large forest just north of the main east-to-west railway running from Cracow to Lviv. At the outbreak of war the Poles were in the process of building an ordnance establishment there and certain buildings had already been built to a high standard, complete with blast walls, central heating and a railway siding that provided good access. Despite MI6 having received Polish ground intelligence from the Blizna area since October 1943, when the villages around the testing facility were being evacuated, it would not be until March 1944 that Medmenham learned of its existence. A Polish field agent, with the code name Makray, had covertly monitored the railway line at Blizna and observed a heavily guarded freightcar carrying 'an object which, though covered by a tarpaulin, bore every resemblance to a monstrous torpedo'.

When MI6 passed Medmenham the ground intelligence, in Pingpong reports 659 and 734, the first PR mission over Blizna – sortie 60PR/325 – was flown by Captain Kay in a PR Mosquito, with his navigator Lieutenant Snyman, on 15 April 1944. With Blizna in such a remote part of Poland, the task of photographing it fell on 60 Squadron, the South African Air Force (SAAF) squadron responsible for photographic reconnaissance then based in Italy. The resultant Medmenham Interpretation Report BS370 observed that Blizna had clearly been used for flying bomb experimental work, had a complete Bois Carré-type site, and was clearly still active. The site had a railway line of standard gauge that connected directly with the Polish railway system and six miles south-west of Blizna there was an extremely large hutted camp, the main buildings of which were all arranged in a distinctive horseshoe pattern with some protected by blast walls. When Blizna was photographed for a second time on 5 May 1944 – during sortie 60PR/385 – Interpretation Report BS485 recorded that the Belhamelin ramp had been dismantled and in its place was a line of seven staggered objects of approximately 18 feet by 5 feet, with a crane at one end. Seventeen passenger wagons, plus three flat wagons that carried a cylindrical object that was possibly a tank, similar to those reported at Peenemunde and Raderach, were also observed, while on frame 4054 a large water-filled crater, 2.8 miles north-east of Blizna and measuring 74 feet across and orientated on the line of the now dismantled firing ramp could be seen.

60 Squadron were then part of the Mediterranean Allied Photo-Reconnaissance Wing (MAPRW), which could trace its origins back to the Casablanca Conference held at the Anfa Hotel, Casablanca, between 14 and 24 January 1943. Following the success of Operation Torch – the November 1942 invasion of French North Africa – Prime Minister Churchill and President Franklin D. Roosevelt met to make far-reaching decisions about the Allied forces in the Mediterranean. Also in attendance was General Charles de Gaulle representing the Free French Forces. The conference led to the creation of the North-west African Photo-Reconnaissance Wing in February 1943 in Algiers, an outfit which grew to include British, American, French, Polish, Czechoslovakian and South African personnel and became the MAPRW on 10 December 1943. With the American President's son Colonel Elliot Roosevelt appointed its Commanding Officer, RAF Wing Commander Eric Fuller was appointed his deputy. Fuller had served with the Royal Flying Corps during the First World War and, after developing an in-depth knowledge of the German petrochemical industry, was recruited by Bomber Command to work in their PI unit just before the outbreak of the Second World War. Fuller had worked for a German oil company who asked for his advice on petrochemical cracking techniques during the interwar years, unaware that he was actually feeding information on them to the Air Ministry and was helping the British develop their knowledge and understanding of the oil plants of the industrial Ruhr region.

Based in the village of Iver in south-east Buckinghamshire, the Bomber Command headquarters occupied Richings Lodge, a mansion built in the 1920s by the entrepreneurial Sykes brothers (Eric, Harry and Friend) as the centrepiece of a garden village they were busily planning – only for their company to go into receivership in 1930. The mansion was commandeered by the Air Ministry and code-named North Side. Fuller later recalled that the PI Unit occupied the top floor of the cold and draughty building that was home to a group of colourful characters led by Flight Lieutenant Peter Riddell. The interpreters included: Anthony Fane, an Old Harrovian friend of Riddell and well-known Frazer Nash racing driver who had driven a

BMW in the 1937 Le Mans 24-hour race; Robert Windsor, a pioneer of commercial air survey who had been a pilot for Airwork Limited at Heston Aerodrome and after the war would work for the aerial-survey company Meridian Airmaps; and civilian Cynthia Wood who gained notoriety among the early interpreters after the day she tipped a full bottle of milk onto Norman Bottomley who was then Air Officer Commanding No. 5 Group Bomber Command. Cynthia used to bring milk to the office every day, and was in the habit of putting the bottle on the mantelpiece – much to the annoyance of Peter Riddell, who one afternoon told her off. In a fit of pique, Cynthia tipped the contents of the bottle out of the window without looking first, only for a messenger from the front door to appear and enquire anxiously where the milk had come from that had landed slap on top of Bottomley, who had just got out of his car.

Bottomley knew Cynthia's family and had been instrumental in recruiting her when Riddell was first building the PI unit in the summer of 1938. When she had her WAAF board and was considered 'unsuitable officer material' she was nonetheless employed as a civilian and managed to get her officer's commission after the unit transferred to the new headquarters of Bomber Command at Naphill, near High Wycombe. After their transfer, Richings Lodge at Iver was badly damaged during a night-time bombing raid – on 23/24 February 1941 – which resulted in its being demolished. While they were at RAF High Wycombe, Peter and Cynthia decided to get married, prompting Riddell to ask Eric Fuller to 'find out how one gets married' and make arrangements for the next day. At noon on 18 June 1940, Riddell and Fuller were having a drink together before meeting Cynthia when they suddenly realised that Peter didn't have a wedding ring. He nipped across the road and bought one, and half an hour later the wedding took place in the High Wycombe Registry Office. The parents were told about it later. Within a few days of the wedding Riddell left Bomber Command, having secured a transfer to work with Hemming at the Photographic Development Unit in Wembley.

Meanwhile Fuller was posted, in November 1940, to RAF Oakington, Cambridgeshire, as officer in charge of intelligence at the No. 3 Photographic Reconnaissance Unit. Once the interpreters relocated to Medmenham, in April 1941, Fuller was posted there thanks to Peter Riddell – who in July 1941 went on a 'world tour' courtesy of the RAF. Squadron Leader Riddell was tasked by Charles Medhurst – the Assistant Chief of the Air Staff with responsibility for intelligence – with writing a report on PR and PI activities. The whole expedition took six weeks: Riddell flew via Catalonia to Gibraltar, and on to Malta, Cairo, India, Burma and Singapore. In each case he talked first with the PRU personnel before inspecting the aircraft, working and living quarters.

After a visit to Medmenham by Group Captain Noel Stephen 'Peter' Paynter from RAF Air Intelligence in Cairo it was arranged that Fuller should visit the Middle East in an advisory capacity and report to the Air Ministry on how photographic intelligence could be more fully adopted there.

Paynter would go on to become an Air Commodore and Chief Intelligence Officer at Bomber Command where he established a close working relationship with Arthur 'Bomber' Harris, before resigning his commission while Director of Intelligence at the Air Ministry when the post-war Attlee government refused to involve Harris in the victory celebrations. With this principled decision costing him the princely sum of £100 from his annual pension he soon found a job with MI5 as a senior officer with the Security Service.

Medmenham Commanding Officer Group Captain Peter Stewart is recorded to have visited Fuller after Paynter's visit and to have pointedly asked him who he thought should be sent to the Middle East. When he did not particularly welcome Fuller's reply that Douglas Kendall might be suitable, Stewart immediately asked Fuller himself whether he would like to go. Taking the hint, Fuller soon found himself on a protracted journey that involved getting stuck in Pretoria, South Africa, before his eventual arrival in Cairo on 23 August 1941.

At this stage in the Middle East there were two PI Units with the Army, one at Jerusalem and one at Alexandria, and Fuller's task was to investigate the requirements of a 'Young Medmenham' for that theatre of operations and submit a report. Another PI unit could be found on the Mediterranean island of Malta where one of the great personalities in PR, Adrian 'Warby' Warburton, was then based. One of the interpreters on Malta was Ray Herschel, a Welshman from Swansea who joined the Aircraft Operating Company in June 1939, from which time he worked in the Wembley office on air-survey work. On the militarisation of the unit, Herschel would later recall Squadron

Leader Walter Heath presiding over the Aircraft Operating Company staff as one by one they were sworn to secrecy. When interpretation work relocated to Medmenham, Herschel was still a civilian but in June 1941 he was commissioned and sent to start the first PI section at Gibraltar. Although there had been a small photographic section there for 202 Flying Boat Squadron no interpretation work had been carried out. With films processed in a converted gents' lavatory – with 'Gentlemen' still written on the door – the first interpretation there as they monitored neutral Spain and the North African coast was evidently a primitive affair, undertaken using the light from hurricane lamps.

In October 1941 Herschel was posted to Malta with Howard Montagu Colvin who arrived directly from Medmenham, and together they started the interpretation section on the island. The Malta-based PRU provided the data needed to fill the gaps in information between that secured by the RAF in the Middle East and the North African Photographic Reconnaissance Wing. Colvin had joined the RAF in 1940 on the basis that it 'seemed less bad than the army' but when his records were lost he and a number of other 'unrecordeds' spent their days marching from Garstang to Blackpool for lunch and were then marched back again in the afternoon. On one occasion he was served lunch by a girl he'd known at University College London. Her father happened to be an under-secretary at the Air Ministry and she just had time to tell him that her father was looking for archaeologists to work as photographic interpreters. This revelation was to lead to his promotion from Aircraftsman Second Class to Flight Lieutenant and to his being posted to Malta. In post-war life he became a noted academic, considered by many to be one of the greatest architectural historians, and would be knighted for his scholarship and public service.

From Malta, photographic reconnaissance was routinely carried out throughout the Mediterranean region, over the Italian mainland and the islands of Sicily, Pantelleria and Sardinia and over the French island of Corsica. With the Mediterranean Sea of vital importance to the war in that theatre, the interpretation of shipping was to prove particularly important to the interpreters there.

In 1943 Sqaudron Leader Adrian Warburton commanded 683 Squadron and had a close liaison with the American PR Squadrons when hundreds of sorties were flown to support the Allied invasion of Sicily – Operation Husky – throughout July and August. A detailed mosaic was created from photography of the entire island, and the vast aerial photomap produced allowed detailed pre-invasion planning. Warby had a particularly good working relationship with the Americans and after they arrived in Tunisia he became an adviser to and friend of Elliot Roosevelt. By October 1943, when the Allies had begun their slow advance through mainland Italy, and whilst the Italian government had capitulated, the Germans had strengthened their forces to such an extent that the Allies would not fight their way through to Rome until June 1944. This coincided with the Normandy landings, and with Warburton being given command of a new photographic-reconnaissance wing of four squadrons. The Italian campaign was vital to the strategy behind the planning for Normandy and at La Marsa, Tunisia, a key headquarters of the Allied Air Forces established after the Casablanca Conference, it was decided to establish a British PR wing inside what became the larger Mediterranean Allied Photo-Reconnaissance Wing.

It was planned that 336(PR) Wing would be based for the time being in North Africa and would comprise the Spitfires of 680, 682 and 683 Squadrons and the Mosquitos flown by the South African Air Force 60 Photographic-Reconnaissance Squadron. With Elliot Roosevelt still in overall command of the MAPRW the British wing came into being on 1 November 1943 at La Marsa and with the twenty-five-year-old Warburton appointed as its first commanding officer a close working relationship with Elliot Roosevelt and Wing Commander Warburton continued. One evening in La Marsa, Fuller was having drinks with Warburton and some friends when some of the party decided to drive the thirteen miles into Tunis. Warburton was behind the wheel on the rough road and in what Fuller would later describe as a typical display of the man's character he decided to frighten the people who were driving cars coming the opposite way. Tearing straight towards them on the wrong side of the road, he managed to get away with it a few times before he crashed and found himself in hospital with a cracked pelvis. Warburton's enforced hospitalisation was his first rest from flying for years, and with his heart and soul in aviation he became increasingly dejected.

When Wing Commander Gordon Hughes, one of the pioneer photographic-reconnaissance pilots, was posted to Italy from Britain to command 336(PR) Wing with only two days' notice it was the start of a period he would later remember as an altogether unhappy one. Hughes had been Flight Commander of the Benson-based 540 Squadron, had frequently photographed Peenemunde, and would be awarded a Distinguished Service Order for his bravery photographing Bois Carré sites in the Pas-de-Calais. Despite his promotion from squadron leader the reluctant wing commander felt something of an interloper, and arrived at the new headquarters of the MAPRW at San Severo in the Foggia plain of southern Italy after a miserable flight via Gibraltar and Caserta before finally landing at Bari on the Adriatic Sea. From there he was driven by jeep and was shocked to find everyone carrying a revolver. Having been born into a Quaker family, Hughes considered himself a pacifist, and with an aversion to armaments, life as an unarmed PR pilot equated perfectly with his beliefs. He was to be even more shocked when his driving companion rammed a little Italian civilian car, quipping that they shouldn't have been in the way.

Gordon Hughes had been sent to take the place of Adrian Warburton. On arrival in southern Italy, and reflecting the fact that Warburton was at heart an operations man, it was not surprising that Hughes inherited something of an organisational mess. His unhappiness was compounded by the lack of equipment available to him, particularly when compared to that of the Americans, his having to live in the dirt of mosquito-infested San Severo, his feeling that the South Africans had a tendency to let things slide, and what he considered to be the scheming of Eric Fuller. Added to this was the fact that the commanding officer of the South African Squadron – Major E. U. Brierly – had hoped and expected to get the command, which meant Hughes was not particularly welcome there.

Warburton's accident had managed to both hospitalise him and bring him into even worse disgrace than usual with the RAF authorities. One of the first tasks facing Gordon Hughes on arrival in Italy was to make the case for a court martial against him. But in the event a friendship that Hughes would later consider to be one of the few highlights of his time in Italy developed. Whether Hughes's later claim that he managed to cook up the evidence that exonerated Warburton from blame not only for the motor accident but also for another incident involving Warburton firing a revolver in someone's house was true or not, his admiration for the man was clear. Warburton's approach to life was graphically illustrated when he was supposed to get a medical check after the accident before he went flying again – and so flew to the doctor's to get one. Although this behaviour may appear strange so many years after the war, during a war it would have seemed natural to many people to get what you wanted by brandishing a revolver and to ignore or taunt authority. Which was what Warburton did.

Posted as the RAF Liaison Officer to the 7th Photographic-Reconnaissance Group then based at Mount Farm in Oxfordshire, Warburton managed to persuade his friend Elliot Roosevelt to lend him an American Lockheed P-38 Lightning. On 12 April 1944 he took off from Mount Farm and, flying alongside another P-38 piloted by Captain Carl Chapman, flew to an area 100 miles north of Munich on the understanding that they would then split, photograph their targets and rendezvous again at the same point before flying on to an American airfield on Sardinia. Failing to make the rendezvous, Warburton was not seen again. His disappearance sparked years of speculation about what his fate had been. The mystery was not solved until 2002 when the clues – in the form of personal effects – were discovered in the cockpit of his aircraft, which was buried two metres deep in a field close to the Bavarian village of Egling an der Paar, 34 miles west of Munich. Since one of the Lightning's propellers was riddled with bullet holes it was apparent that Warburton had been shot down, but because he had been flying an American aeroplane the German wartime authorities had assumed he was an 'unknown American Airman' and had buried him in the village cemetery with the grave thus marked.

Aged only twenty-six when he died, Wing Commander Warburton held the Distinguished Service Order and bar, Distinguished Flying Cross and two bars, plus the American Distinguished Flying Cross. Immortalised in books and films, Warburton provided the inspiration for the 1953 film *Malta Story*, starring Alec Guiness as a young photographic-reconnaissance pilot who meets a Maltese girl working in the RAF operations room. The tale of their growing love is set against the backdrop of the heroic defence of Malta by the islanders and the RAF and climaxes with the arrival of the SS *Ohio* and the island being awarded the George Cross. With the island under siege and blockaded by enemy air and naval forces, Operation Pedestal was the British operation to supply the island that did indeed culminate with the arrival of the SS *Ohio*, an American-built tanker with a British crew which supplied the island with desperately needed fuel supplies. After aviation researchers discovered the true identity of the 'unknown American Airman' his body was interred beside those of Commonwealth airmen in the British Military Cemetery at Durnbach, twenty miles south of Munich, and with the Bavarian Alps providing a backdrop, members of the Queen's Colour Squadron of the RAF lowered his coffin into the ground while a piper's last lament played.

The MAPRW evolved into a major provider of target information for raids throughout the Mediterranean region, including for Operation Tidal Wave, the famous American raid by B-24 Liberator heavy bombers on the oil refineries at Ploesti, Romania, on 1 August 1943. As the war developed, and though Medmenham was always the headquarters of Allied PI, the network of Air Force PI units and the interpreters of the Army Photographic Interpretation Section (APIS) spread around the world. Glyn Daniel had transferred in 1942 from Medmenham to create a nucleus of PIs that could support the Far East Campaign, with detachments in Rangoon to support the Burma Campaign, at Colombo and Kandy in Ceylon (Sri Lanka) and at Bombay and Calcutta in India. Whilst in 1943 the North African Central Interpretation Unit was established to focus operations in the Middle East, which had earlier benefited from the activities of the South African branch of the Aircraft Operating Company at which Douglas Kendall had worked in the 1930s.

After being commandeered, the company became the Transvaal Photographic and Air Survey Squadron, with Managing Director Major Charles Robbins MC DFC now holding the rank of Lieutenant Colonel. Robbins had been an artillery observer in the First World War and was involved in the pioneering 1923 aerial survey of the Irrawaddy Delta for the Forest Department of the Government of Burma. For this work, two de Havilland 9 aircraft were fitted with Siddeley-Puma engines and converted into seaplanes because the terrain that they were photographing meant there would have been no chance to make a forced landing – without causing irreparable damage – in an aeroplane that could only touch down on terra firma. From a base established on military land at Monkey Point, three miles from the centre of the colonial capital of Rangoon, tests were carried out with cameras supplied by Major Frederick Victor Laws, a useful reminder of how small the aerial-survey community then was.

The success of Allied photographic reconnaissance would depend greatly on the First World War veterans who had been inspired by the opportunities presented by PR, whether with Aerofilms photographing country houses and factories for private customers in Britain, the Aircraft Operating Company in South America, Africa and the Middle East – or Charles Robbins in the Irrawaddy Delta, where altogether 1,400 square miles of forest were photographed on 3,795 glass plates over the three-month survey.

On 20 May 1944 the Home Army of the Polish Underground acquired the wreckage of a V2. The rocket had been fired from Blizna and had landed in the river Bug near the village of Sarnaki, 80 miles north-west of Warsaw in east-central Poland. It had exploded and the wreckage was lying in shallow water near the bank of the river, with the missile's fins projecting above the waterline. Getting to the scene before the Germans, dozens of the resistance operators pushed the rocket deeper into the soft mud, while cattle were led into the shallows to muddy the water and hide the rocket's outline. German search parties looked for the missing V2 for three days and then abandoned the quest. The Home Army returned to the river Bug on 23 May, dragged the rocket from the water using horses, photographing the process each step of the way, and hauled the valuable wreckage to a barn in the nearby village of Hołowczyce-Kolonia. The rocket was then dismantled by engineers from Warsaw led by Jerzy Chmielewski and its parts were smuggled into the capital where they were examined by Polish scientists.

By 27 June the Poles were able to provide MI6 with a detailed report: the V2's dimensions were 40 feet long by 6 feet in diameter; its fuel was a highly concentrated grade of hydrogen peroxide. In addition, information about the construction of the main jet and about pieces of radio equipment, including a radio transmitter on a wavelength of 7 metres and a receiver of 14 metres, was also passed on. The radio parts had been analysed by Professor Janusz Groszowski of the Warsaw Polytechnic and were ingeniously hidden in the flat of the German commander of the Luftwaffe's airfield at Warsaw, courtesy of his Polish housekeeper. During the day the professor would visit to study the radio equipment and the housekeeper would ensure that they would not be surprised by phoning the German to ask when and what he would like for his tea. By the time the commander arrived home the radio parts would be safely hidden in his own suitcases. In July the Poles offered to supply the wreckage to the British, if they were able to arrange its collection.

On 22 June 1944 the British Air Attaché in Stockholm reported that another missile had crashed in Sweden but this one was different from anything that had previously landed there. An A4 rocket had been launched from Peenemunde on 13 June, had gone off course, exploded and landed near a farm. A Swedish newspaper reported a crater measuring 5 metres wide by 2 to 3 metres deep. The debris was carefully collected by the Swedish military and just as carefully studied by their experts.

In Stockholm the missile managed to provoke substantial diplomatic activity, with the Germans anxious to retrieve the remains of their rocket and the British even more desperate to get their hands on the missile's parts, or at least on a copy of the reports being written by the Swedes. The British offer of mobile radar sets – a type of technology that the Swedes did not at that stage have – was accepted in exchange for the report and a visit from two British experts to inspect the remains. The Swedish report included photographs of a farmer's wife and three soldiers posing next to the remains of the engine casing and detailed photographs of components, including radio apparatus.

The projectile was much larger than a V1, was rocket-propelled, did not have wings, was directionally controlled by vanes and a radio control similar to that used on the Hs 293 glide bomb, and was recorded to be 6 feet in diameter. This information was incorporated into the next report to the Chiefs of Staff on 9 July 1944, alongside ground intelligence from Poland and the statement that when photography of 5 May 1944 over Blizna – taken during sortie 60PR/385 – had been re-examined a rocket had been spotted. The report concluded that, providing firing sites in France were completed, the rockets could be fired on Britain at any time.

Around 12 July, R. V. Jones was asked to review the state of knowledge about the rocket for the Crossbow Committee. Jones regarded such a report as premature. The evidence was incomplete, he said: it could lead to seriously misleading conclusions being drawn, and he chose to title his paper 'Interim Summary Statement of the Broad Lines of our Knowledge and Ignorance'. But, as Jones observed, there was now little doubt that the Germans had developed a technically impressive missile that they called the A4 and its performance was good enough for the bombardment of London. So although the investigation into the rocket might have been overshadowed for many months by the more immediate threat presented by the flying bombs it had fortunately not been forgotten.

R. V. Jones knew that the A4 rocket was around 6 feet in diameter, a little over 40 feet long, probably had four stabilising fins, that the main fuel source was likely to be hydrogen peroxide, and that the radio control system appeared to resemble that found in the Henschel Hs 293 glide bomb, which had been used against Allied shipping. The total weight was unknown but from 80-foot-wide craters at Blizna, Poland, the warhead was estimated by explosive experts at between 3 and 7 tons. The fuse was thought likely to be triggered by barometric pressure or another proximity device since the rocket was seen to have burst in the air, scattering fragments over a two-mile radius – although the British did not then know that this was not an intended feature of the weapon. The stabilisation of the rocket while in flight was thought to have been achieved through the use of a gyroscopic control system, although the Air Intelligence officers who inspected the crashed rockets in Sweden thought this controlled internal movable surfaces that deflected the extrusion of gases. Polish experts who had inspected their A4 considered that the stabilisation was achieved thanks to external rudders. Jones noted that both sources could be correct.

It was noted that radio control was used on these rockets – although it would later transpire that only 20 per cent of V2s actually were radio-controlled – and it was thought that the control system could be divided into two components, the first a transmitter-receiver enabling the controlling ground station to follow the range and direction of the rocket, the second a receiver with an elaborate decoding device to pick up and unravel its coded signals. The basic principle of the equipment was thought to be similar to that of the guiding receiver in the Henschel Hs 293 glide bomb, which controlled that weapon's movements left and right and up and down in accordance with tone modulations transmitted from the control station. With it assumed that the Swedish rocket had come from Peenemunde, the maximum known range of the weapon was estimated at 200 miles. The furthest that a V2 had been recorded by the Polish Resistance as having travelled from Blizna was

185 miles, while most were recorded as having fallen around 150 miles away. They had observed the weapon to be increasingly accurate, and although wild shots were known still to occur, on the limited analysis of the Polish trials it was noted that about half the shots fell within a circle of 20 miles' diameter. With London such a large target, even if the rocket could not yet be accurately aimed at a specific location its potential to devastate the British capital was clear.

It was known that experimental work had continued throughout at the Army Research Establishment at Peenemunde and at the SS training grounds near Blizna. With the Bomber Command attack on Peenemunde prompting the establishment of the experimental facility in rural south-east Poland, the move there had meant that land targets could be hit using flying bombs and rockets and the results examined. While nineteen V2s were recorded as having been fired in Poland between 20 April and 7 May 1944, probably a dozen were fired at Peenemunde in June and continuous trials were carried out in Poland during the same month. One of the gravest gaps in the Allies' knowledge was thought to be lack of evidence of a rocket launching point, and although it was recognised that at least one must exist at Peenemunde the fact that they had never been able to single it out among all the activity meant that Blizna, which had still not been entirely photographed, represented a greater opportunity. Ground intelligence from a Polish eyewitness had described a missile being placed on a short lattice ramp pointing upwards at a forty-five-degree angle, whereupon it rose with a flash and a roar. Despite this report, R. V. Jones considered any further speculation about rocket launching at the time to be idle conjecture and drew attention to a different version of how the V2s were launched from a prisoner-of-war interrogation report, which suggested that the process involved a fairly simple concrete slab. The POW had even pinpointed one such construction that was then within liberated Normandy, and although the first examination on the ground had proved negative it was noted that it would be examined again. This was clearly a sensible decision, and although modern audiences are used to seeing rockets being fired vertically it has to be remembered that the Allies at this time were struggling to understand what was for them a revolutionary technology that would later come to dominate the Cold War.

Little was then known about the manufacture of A4 components beyond the activities at Peenemunde. This presented the same type of challenge that the Crossbow interpreters faced of identifying factories that were of particular interest to them, and highlighted their almost total reliance on ground intelligence to bring them to their attention. A number of sources had identified the Raxwerke near Wiener Neustadt in Austria that was operated by Henschel and Sohn, the German locomotive manufacturer. Having diversified into aircraft manufacture and missile development under the Nazi regime, the company manufactured the Henschel Hs 293 glide bomb that had been closely linked with the A4 long-range rocket and were suspected of manufacturing rocket components at the Raxwerke. Prisoners from the Mauthausen concentration camp worked in the factory until the work was moved after the plant received collateral damage during a raid by the US Ninth Air Force on 13 August 1944 when sixty-one B-24 heavy bombers attacked the nearby Wiener-Neustadt aircraft factory. The intelligence picture was further confused by the fact that many other factories were reported to be making components for secret weapons and rockets and that there were several different types of such devices. The examination of the Swedish A4 rocket revealed evidence of mass production, and intelligence indicated that although missile manufacture had been interrupted in March 1944 to allow resources to be allocated to fighter assembly it was now reported to be going ahead again.

Since the 65th Army Corps was known to be responsible for the deployment of both the V1 and the V2, even though the German Luftwaffe had been responsible for development of the flying bomb, one assessment of the situation was that – unless the Luftwaffe had been heavily marginalised – the rocket was destined to make a bigger and more final contribution to the German war effort. Since there was no negative intelligence about the relationship between Hermann Goering and Hitler, it was considered improbable that the Luftwaffe would be subordinated to the Wehrmacht unless the rocket campaign was closer to the Führer's heart. Since between thirty and forty V2s had been fired experimentally in June, it was considered possible that a similar number could be fired on London in July, provided that the launching sites had been prepared. Given that none of the Heavy Sites would be ready for some time, however, and only Wizerne and Marquise Mimoyecques were considered to provide any indication of being firing points, it raised the possibility that the Germans were continuing to develop the weapon and that the launching sites could in fact be small and well camouflaged. They might therefore have evaded detection within the firing range of Britain despite the large amount of photographic reconnaissance which had been flown over so many months.

One key fact strengthening that possibility was the activity at Blizna, where it was known that the rocket had been fired. So the evidence was likely to be in the aerial photography – it was just that nobody had yet been able to identify the method of firing. It was also clear that the rocket was still undergoing development, but since Blizna did not appear to have major workshop facilities it could be inferred that a Heavy Site facility was not a necessity. This in turn indicated another possibility: that the Heavy Sites were intended for another secret weapon. Either way, developing countermeasures against the rocket launch sites was of paramount importance, particularly since the V2 could deliver a much greater explosive payload than the flying bomb and, once launched, was then technologically impossible to destroy through air-delivered defences. Against the rocket there was no protection.

Complete and immediate photographic aerial reconnaissance of Blizna was seen as crucial, hopefully to identify what the firing point looked like. This in turn could allow the identification of firing points within range of London, and an assessment of the possible scale of attack. Just as the Normandy landings had accelerated the deployment of the flying bomb, which until then had been constantly postponed because of technical difficulties and manufacturing delays, it was clear that the Allied armies who were now advancing through France could have the same speeding-up effect on the V2. The further the Allies advanced, the greater the demand would be for the early deployment of the weapon regardless of its performance. Three salient intelligence problems were then considered to exist: the identification of launching points, the whereabouts of radio-control stations and the nature of the radio-control system itself. It was accepted that identification of the manufacturing centres, the location of the railway communications system and understanding the order of battle were important, but the time available for a solution was short. As the Russians were rapidly advancing on Blizna, and since this would invariably lead to the site being destroyed and abandoned by the Germans, the immediate and complete rephotography of the experimental site was considered vital. As was recovering the radio systems from the Swedish and Polish V2s and identifying the supply organisation for the V1s in northern France.

It was becoming increasingly clear that the effort required to attack the Belhamelin launching points when measured against the results achieved was much too wasteful. There were too many sites that were either too easy to rebuild or that were not only deliberately camouflaged but increasingly naturally camouflaged because of the growth of summer foliage. Since it was now considered impossible to eliminate them faster than they could be rebuilt, it was now recognised as more important to instead attack the bottlenecks in the flying bomb supply system and the production of the missile itself. Locating the network of factories throughout German-occupied territory that were likely to be involved in the manufacture of component parts for the missile was an enormous challenge, but one vulnerable aspect of the supply organisation was seen as being the production of hydrogen peroxide. Since it was considered by Scientific Intelligence to be the chemical that was used for launching flying bombs and propelling the new Messerschmitt Me 163 fighter – and probably the new rocket – it was believed that destroying the hydrogen peroxide industry would cripple these weapons. It was known that the chemical was only produced in sufficient quantity at Peenemunde, and also in Bavaria at the 'porcelain factory' at Raderach, near Friedrichshafen, and at a factory at Höllriegelskreuth, near Munich. But since it was now summer, with its short nights, the limited capacity of Bomber Command to carry out night-time raids meant that a precision daylight attack by the Americans would be necessary.

Despite the best efforts of MI6 agents they struggled to gather as much intelligence about distribution depots in Germany as they had to obtain ground intelligence about the Supply Sites in northern France. Much the most important site for flying bombs had been identified earlier as being at Saint-Leu-d'Esserent, which was thought to have supplied 70 per cent of the V1s launched against Britain. Medmenham was able to provide detailed target information from photographic reconnaissance and despite the depots being deep underground, they were considered promising targets for Bomber Command on the basis that they were within range of the most accurate radio-navigational systems and the heaviest bombs. The network of Supply Sites included locations at Rilly-la-Montagne, south of Rheims, and Nucourt, plus other underground sites along the Oise Valley to the north of Paris, and knowing which points in the network were operational at any one time was essential. This required MI6 to use its secret agents and ground-intelligence sources to find and watch the Supply Sites, to trace the railway communications between them and Germany, to link road communications in France to the flying bomb launch sites, to investigate the production of hydrogen peroxide in Germany, to locate the places in France where the chemicals were transferred from 'egg rack' railway wagons to storage tanks on lorries, and to identify all the headquarters associated with the V1 programme so that their landlines could be cut.

With radio apparatus from the Swedish rocket on loan to them – it would later prove to be from a test vehicle for the radio control of an experimental surface-to-air German rocket code-named *Wasserfall* – it meant that for a short period the British were under the false impression that the V2s were radio-directed and that radio-jamming countermeasures could therefore be used against them. The continuous cooperation between ADI (Science) and G Section at Medmenham ensured a knowledge of the positions and types of German radar equipment from Skagen at the northern tip of Denmark to Bayonne at the extreme south-western part of France near the Spanish frontier. One of the great successes of their collaboration helped ensure the success of the Normandy landings, when on D-Day the Allies successfully deceived the Germans about their intentions. It had been recognised that German radar would need to be either jammed, deceived or physically destroyed, but that whatever approach was taken – just as was the case with the Allied bombing campaign of the northern French railways at the time – it was imperative that their action did not draw particular attention to Normandy.

It was decided to undertake a trial attack on one of the largest types of German radar equipment known by the British as Chimney because of its physical appearance, and *Wasserman 3* by the Germans. The installation chosen was a Chimney near Ostend in Belgium, on the basis that the Belgian espionage network was so good and their agents were able to observe the effects of the raid by rocket-firing Typhoons. Two attacks were made, which were enthusiastically observed by the Belgian agents who confirmed that the turning mechanism had been damaged and later that the station had to be dismantled for repair. This raid had proved to be a successful means of destroying the German radar network, which included the Jagdschloss – named after a German ruler's hunting lodge – the Coastwatcher, Hoarding, Würzburg and Giant Würzburg – after the German city of the same name – Freya, after the Norse goddess, and Chimney systems.

For several weeks before the Allied invasion of France, rocket-firing Typhoons attacked German radar installations along the French and Belgian coasts from the Netherlands border to Brittany. The key question for the Allies was: would their deception plans work? They were concerned that German U-boats, E-boats, radar and the Luftwaffe might spot the invasion fleet and be waiting for them. To best ensure this did not happen, the night before the Allies landed on the beaches of Normandy, Mosquito squadrons patrolled the French coast ready to shoot down any German aircraft, and as a double safety measure were kitted out with equipment that could jam the frequencies used by German night fighters. That same night, Operation Glimmer, a deception plan that involved Stirling bombers dropping Window – the metal-foil chaff – over the English Channel towards Boulogne in the Pas-de-Calais was activated. While mines were dropped around the headland of Cap d'Antifer in Upper Normandy, 617 Squadron were dropping Window from their Lancaster bombers in the same area. Operation Taxable required the Dambusters Squadron, commanded by Group Captain Leonard Cheshire, to fly in rectangular patterns, with the centre point of each pattern seen to be moving on radar at a rate of 8 knots. With Window dropped throughout, the intention was to give the impression that an area 16 miles long by 14 wide was actually the invasion fleet heading for Fécamp on the Normandy coast.

The planning for radar operations involved a number of key people from the Scientific Intelligence Unit. From the technical side it involved Squadron Leader Derek Garrard, from the tactical intelligence side James Alan Birtwistle – who had been an Intelligence Officer at Fighter Command during the Battle of Britain – and Rupert Gascoigne-Cecil. An outstanding sportsman, Gascoigne-Cecil graduated from Balliol College, Oxford in 1939 with a degree in biochemistry and, having been a member of the University Air Squadron, on the outbreak of war he was called up. Joining 61 Squadron, who were equipped with pre-war Hampden bombers, he carried out many attacks against Germany, and after two tours of duty as a bomber pilot, this wing commander with two DFCs joined the small team responsible for Scientific Intelligence and was given the challenge of neutralising German coastal radar prior to the Normandy landings.

An unconventional pilot, Gascoigne-Cecil would frequently fly himself and R. V. Jones to airfields where they would give briefings to the personnel involved in operations. On one such occasion, while the straight-talking Gascoigne-Cecil was flying south-westwards from Hendon a Spitfire from Northholt decided to surprise them with a dummy attack. Gascoigne-Cecil's combat experience automatically kicked in and, out-thinking the Spitfire pilot, he quickly banked the aircraft vertically – much to the shock of R. V. Jones who later remembered nothing but the sight of wild rhododendrons in bloom on the west Surrey hills – and the sound of Gascoigne-Cecil shouting at the top of his voice, insulting his attacker's performance. With several other aerial incidents that day Rupert was to educate Jones in the two rules of flying. Firstly, ensure that your radio does not work because then you cannot be ordered around by anyone on the ground and you will have absolute discretion over your actions. Secondly, you can get away with anything once.

As the Allied armies advanced through France information began coming to light from German prisoners of war, including from one who had been involved in selecting sites for the storage and firing of V2 rockets. A launch site near the Château du Molay to the west of Bayeux, Normandy, was discovered through such prisoner interrogation and personnel from the Scientific Intelligence Unit were sent to survey it. The young Welsh physicist David Arthen Jones, who had earlier joined R. V. Jones to work on flying bomb matters, inspected the newly liberated site in Normandy. A fast bowler for Glamorgan County Cricket Club and all-round sportsman, he had been recruited to help establish through the analysis of German radar plots the accuracy of the flying bombs that were being test-fired at Peenemunde, and revealed in December 1943 significant improvements to the weapon's accuracy. Together with his colleague, the physicist Squadron Leader David Nutting, he drew a sketch of the site, which featured a simple tree-flanked stretch of road into which concrete platforms had been set and on either side of which parallel loop roads had been built among some trees.

On seeing the sketch R. V. Jones recalled having seen a similar installation at Peenemunde and developed the hypothesis that, far from Heavy Sites being required to launch the rocket, the launching point was actually remarkably simple. The road layout at Château du Molay had been carefully constructed with curves that allowed the transporters – known as Meillerwagens – that carried the rocket to negotiate corners. As the Allies continued their advance, papers came to light providing intelligence about the location of V2 storage and firing sites in Normandy and gave more force to the threat presented by the rocket in the summer of 1944.

While the interpreters would soon find themselves involved in surveying installations on the ground – an activity they called 'ground truthing' – soon after the Normandy landings a directive from General Dwight David Eisenhower at the Supreme Headquarters Allied Expeditionary Force led to the creation of T-Force. This followed the success of 30 Assault Unit, brainchild of James Bond author Ian Fleming at Naval Intelligence. Controlled through the Combined Operations Headquarters, it was tasked with moving ahead of advancing Allied forces, and undertaking covert infiltrations into enemy-held territory – by land, sea or air, to capture intelligence. Thirty Assault Unit was disparagingly labelled 'a bunch of Limey pirates' by General George Patton and was initially deployed on the Dieppe Raid of August 1942. It had taken part in the Allied invasion of North Africa – Operation Torch – before being actively involved with the Normandy landings ahead of the Allied advance through Europe.

T-Force units were attached to three army groups on the Western Front and were ordered to identify, secure, guard and exploit valuable and special information and sources thereof, including documents, equipment and persons of value to the Allied armies. T-Force elements often accompanied combat units, or arrived soon after targets were seized, to prevent any looting or sabotage and ensured that key equipment, documentation and personnel did not disappear. The Allies were determined to discover the Nazis' military secrets – particularly those of the secret-weapons programme – and knew that scientific and technical knowledge was vital to economic recovery after the war.

They also wanted to ensure that such information must not fall into the hands of the Soviet Union.

After having first been photographed on 15 April 1944 when ramps for firing projectiles, looking similar to a Belhamelin-type site, were discovered there, Blizna was again photographed on 5 May 1944, during sortie 60PR/385. In the subsequent interpretation report, written at Medmenham – after the delivery of the results of the sortie from the MAPRW at San Severo in southern Italy – it was noted that the firing ramp had been dismantled, and special tank-carrying railway wagons spotted earlier at Peenemunde and the 'porcelain factory' in Raderach, plus a large crater 2.8 miles north-east of Blizna, could instead be observed. In June 1944 R. V. Jones was living with his wife and family in the south-west London suburb of Richmond, and they felt the impact from flying bombs directly when the blast effect from a V1 blew in the house windows of some nearby family friends. The now homeless friends sought refuge with the Jones family, but with conditions so cramped that they were forced to sleep with their heads under tables, R. V. Jones had limited time and opportunity in the evenings when he would normally have liked to run through new developments in his head.

When his wife and children left for the safety of Cornwall in early July he was able to begin going once more through the rocket evidence in the evenings, evidence which by then was accumulating rapidly. Convinced of the existence of the rocket, Jones contacted the chief doubter, Lord Cherwell, with an ultimatum to the effect that if he continued to deny the reality of the V2, Jones had all the evidence to prove its actuality. Not altogether happy about the situation in which his old tutor found himself, Jones was thankful when Cherwell agreed to his suggestion that he, Jones, should gradually release facts over the next week or so to provide him with the chance to change his position.

Unlike the Medmenham interpreters – with the sole exception of Douglas Kendall – Jones had access to Ultra messages which revealed that the Germans were sending *Geräte* – apparatus – between Peenemunde and Blizna. In the hope of proving a connection between the two experimental sites, Jones would later record in *Most Secret War* that one evening in early July 1944 he analysed his copies of photographs taken by 60 Squadron during sortie 60PR/385 over Blizna on 5 May 1944. Despite the Medmenham interpreters having already issued Interpretation

Report BS485 on the sortie and with knowledge of the evidence from the Château du Molay, he undertook his own photographic interpretation. Looking carefully at frame 3240 – and its associated stereo pair – he identified a rocket, 90 millimetres down from the top of the frame and 26 millimetres in from the right-hand edge. With this dramatic intelligence breakthrough, and fearful that it would be lost if he were killed that night by a flying bomb, Jones telephoned his colleague, the physicist Dr Charles Frank, at his home in Golders Green. Jones gave Franks the exact sortie and frame references, and told him where on the photograph the rocket was. The Chiefs of Staff and Lord Cherwell were quickly informed and this managed to provoke a reaction from Douglas Kendall at Medmenham.

On 15 July Douglas Kendall – in his role as the Technical Control Officer – wrote a letter about 'Crossbow Ground Reports' for the 106 Photographic Reconnaissance Group headquarters at RAF Benson, noting that information had been supplied by ADI (Science) directly to the Chiefs of Staff about Blizna, based on ground intelligence and the interpretation of aerial photography, and that Dr R. V. Jones had claimed to have discovered an A4 rocket on a photograph.

This move by Jones troubled Douglas Kendall for several reasons. Firstly, Medmenham at the time of analysing the photographs had never been warned or briefed that ground intelligence indicated that rocket experiments were taking place at Blizna, which meant the location had never been investigated from that perspective. Secondly, Medmenham had never been asked to provide an opinion on the photographed object that had then been submitted to the Chiefs of Staff as being an A4 rocket. In his later publication, Jones made much of the fact that the interpreters had failed to spot the rocket despite, in his view, the photographs having been exhaustively searched at Medmenham.

At the time this point was refuted by Kendall who politely submitted that in future Medmenham should be fully briefed to ensure that the maximum amount of information was extracted, given the enormous amount of photography being analysed for Operation Crossbow, and that Medmenham should be consulted on issues of photographic interpretation. But he noted his appreciation that their views would not necessarily have to be accepted or would necessarily even be correct. In his letter he noted that the object in question at Blizna was then being examined, that a report would be duly issued and wryly observed that although the object in question *could* be a rocket, it was by no means certain, and there was a quite reasonable possibility that it was merely an excavator.

On 17 July 1944 Peter Riddell forwarded Kendall's communiqué with a cover letter to the Director of Intelligence responsible for Operations at the Air Ministry. Referring to the sending of amateur efforts at interpretation to the Chiefs of Staff as 'undesirable', because it was the surest method of discrediting Air Intelligence in general and photographic interpretation in particular, Riddell suggested that the matter should be taken up in case it led to any misunderstanding. The impact of the interpreters not being fully aware of important information about Operation Crossbow was duly noted and the impact that similarly wild statements had made in the past observed. This highlighted a constant problem suffered by the interpreters during the war, that of not having access to all the available intelligence that existed. This certainly managed to allow R. V. Jones both at the time and later to discredit the interpreters' skill while glossing over the fact that it is very easy to look but not always so easy to see, particularly if you do not know what you are looking for – which was particularly the case in Operation Crossbow given that the interpreters were looking for a revolutionary new technology. Blizna was a useful case in point: if the interpreters had been better informed about what they were meant to be looking for, the diagnosis of the site could have been carried out more quickly – although, in the fog of war, too much information could have achieved the opposite result.

On the very same day that Peter Riddell wrote to the Air Ministry supporting Kendall's observations, R. V. Jones spent his evening revisiting the photographs from sortie 60PR/385 that covered Blizna. He later mentioned the technical belief at the time that the rocket would need to be fired from some kind of gun, in order to ensure its stability in flight. It was assumed that a large tower at Peenemunde had been built for this purpose. Looking for this structure in the Blizna photographs Jones claimed to have been puzzled by the fact that there was not one there but in the process of his examination he happened to spot a concrete platform 35 feet square. This, combined with the evidence from Château du Molay, led him to propose that perhaps the A4 rocket was fired vertically rather than at an angle, from a relatively simple launch pad, and would be stabilised by gyroscopes and deflectable jet rudders. Suggesting that this could explain the hitherto inexplicable 40-foot columns that had been spotted at Peenemunde he once again telephoned his colleague Dr Charles Frank in the evening. With the telephone lines out of action, however, this time he wrote a note in case a flying bomb got him during the night. He later recalled the note as stating that whoever found it should take it at once to Broadway Buildings, 54 Broadway, SW1, where it should be given to Dr Charles Frank. The note somewhat cryptically instructed Frank that on frame 3240 of sortie PR/385 he would find a square concrete platform in the middle of a clearing at Blizna. Keeping all the intelligence to yourself meant you ran the risk of taking it to your grave.

Conveniently for R. V. Jones, the following day a Crossbow Committee meeting was scheduled for ten o'clock that evening, 18 July 1944. It would be chaired by the Prime Minister and required him to provide an account of the general intelligence situation. That evening the Prime Minister was reported to be in a testy mood, fuelled by increased public criticism of the government's handling of the attacks from flying bombs. The revelation that Ultra messages quoting Geräte figures indicating that more than 1,000 A4 missiles had by then been made only managed to fuel Churchill's anger further and prompted accusations that the intelligence community had been caught napping. In the case of the Scientific Intelligence Unit the most important recent concern had been to concentrate on offensive priorities – the destruction of the German radar systems, a campaign that successfully allowed the Allies to cross the English Channel safely on D-Day – rather than defensive ones.

Turning to the question of evidence, Jones records telling the Prime Minister about the photographs of Poland that he had interpreted personally several days earlier, the discovery of how many rockets had been constructed and his discovery of how the missiles were fired. Asked whether Jones had told Lord Cherwell about this most recent development, the Prime Minister was shocked to discover that he had not, given that Jones had only discovered it the previous evening and had been stuck in committee meetings all day. By the time of the next Crossbow Committee meeting on 25 July, the Prime Minister had clearly had time to reflect on the situation and before formal proceedings began he asked R. V. Jones how many meetings he had attended that day. With the reply that it had been five, and this meeting was his sixth, he was immediately given Prime Ministerial approval to excuse himself from any meeting that he did not consider worthwhile.

The following day Douglas Kendall issued a report on his investigation into the measurement of supposed rockets at Peenemunde and Blizna, prompted by R. V. Jones's revelation that he had discovered a rocket at Blizna. Kendall noted that very careful remeasurement had been made of the objects at Peenemunde and Blizna and that a very definite similarity in lengths, both of the fins and the overall length of the rockets, could be observed.

With the rocket connection between Blizna and Peenemunde confirmed by the PIs, attention had also turned at Medmenham to the other revelation made in the report to the Chiefs of Staff. This resulted in a report dated 4 August 1944 about photographic evidence from Peenemunde and Blizna about how the rocket was fired. This revealed that another sortie had been flown by 60 Squadron over Blizna – sortie 60PR/506 – on 19 June 1944, which photographed rockets in the open, with two in the process of being camouflaged.

This was followed with another discovery on sortie USEC R/70 of 27 July 1944 which photographed three circular objects with dark central markings that had appeared since the last cover. They measured approximately 15 feet in diameter. Ground scarring also indicated that the area had been active. When evidence from earlier sorties was consulted it was apparent that similar objects of the same diameter could be seen and, in view of their similarity to the objects seen on later sorties, the possibility of their being firing bases shaped like a lemon-squeezer was noted. In a later supplement to that report, a drawing was prepared based on the careful study of photographs of Blizna, showing the possible shape of a lemon-squeezer, the term coined for the mobile rocket-firing bases.

By the time of the photography of 27 July 1944 all the lemon-squeezers had been removed and all evidence pointed to the fact that the site had been abandoned. That sortie was one of several flown over Blizna by the United States Eastern Command, an American squadron based in Russia during the war. At the Crossbow Committee meeting on 25 July 1944 Churchill was able to report that he had received a civil reply from Marshal Joseph Stalin giving permission for an Anglo-American mission to visit Blizna once the Red Army had overrun the area.

The importance of beginning the investigation at the earliest opportunity after the Germans had left the area meant that, after permission was received from the Soviet government, the mission left England on 29 July before the area had even been captured. Although Blizna fell to the Red Army on 6 August, continued fighting in a region south of it meant that the area was not cleared for entry until 23 August. Permission was finally granted and on the evening of 3 September, when the front line was only five miles away, the mission was able to begin work. It was led by Colonel Terence Sanders, the former Olympic rower who had visited Medmenham in November 1943 to explain the method then believed necessary to launch the A4 rocket. (This was when he had revealed that a projectile launched on ramps was a more likely prospect in the short term.) Also on the mission was Flight Lieutenant Wilkinson, who had inspected the A4 rocket that had landed in Sweden, and an assortment of British, American and Russian technical representatives and translators.

The preliminary report written in Moscow and dated 22 September 1944 recorded that from captured documents it was apparent that the area had been developed as an immense *Heidelager* for SS troops. The original Polish buildings had been modified by the Germans who had also constructed four horseshoe-shaped barracks and many other buildings to provide accommodation for 12,000 soldiers in peacetime and 16,000 during wartime. It was assumed that the SS troops at the *Heidelager* had no connection with the personnel involved with the rocket experiments other than for administrative matters. Around 400 personnel from the Wehrmacht, the regular army, were recorded as being billeted at Blizna, with one or two of the senior officers having their own houses.

Most operations at Blizna were carried out in an area of about one square mile in the middle of a forest, which was surrounded by a wire fence, about six or seven miles from the SS barracks. The central part of the experimental area consisted of a large open space, and buildings were mainly constructed on the fringes of the forest. The village of Blizna had stood at the southern end of the site until it was completely demolished by the Germans when they took over the area and developed a broad-gauge railway line. The two ramps pointed towards two Polish towns and there was some evidence that flying bomb trials had been carried out during a brief period in April. The principal trials were conducted during the period 14 to 18 April, when there was some evidence that Hitler and Mussolini attended the tests. From sortie 60PR/385 it was known that the ramps had been dismantled before 5 May, shortly after the time when the weapon was put into operational use. The Belhamelin Site was studied and what had appeared on aerial photography as dog-leg blast walls were in fact not walls at all but a series of stout poles about 25 feet high which supported camouflage.

A detailed plan was drawn of the site from the information gathered on the survey which showed that the rockets had been fired from a corner of the forest. The method of launching was made clear from an inspection of the firing sites and through the cross-examination of local inhabitants, including one who had worked as a bricklayer on the site. The launch platform normally used was an 18-foot-square concrete platform onto which a steel base-plate was placed, the shape of which was unknown. When the rocket was brought to the platform on a special lorry (the Meillerwagen), which had hand-operated winches, the missile was raised into the vertical position and lowered onto the base-plate.

The fuelling of the rocket was carried out at the firing site, once the projectile was in the vertical position, and it was normal for the convoy to consist of one rocket on a special lorry, one base-plate on a special trailer that was towed behind a lorry, two 'very cold' lorries which carried liquid oxygen, and about five other vehicles. It was noted that these vehicles were all capable of travelling on rough tracks, which indicated the great mobility of the weapon, which could be launched from a relatively simple concrete platform.

The arrangements for transporting components by train were recorded. In the case of the rocket it was transported in one piece on two railway wagons, sheeted up and camouflaged. The liquid oxygen was carried in what most observers described as the 'cold wagons'. These were specially designed and heavily camouflaged railway trucks that were usually distinguished by the intense coldness of their outer surfaces. Flying bombs were recorded as being transported on single flat wagons, with the missile's engine already mounted in its fuselage but with the wings detached. Transportation of projectiles was noted to have been principally carried out under the cover of darkness.

The fact that every scrap of machinery and all the fixed installations at Blizna had been so methodically removed by the Germans indicated that plans had been well in place for the orderly evacuation of the site. There was limited evidence of anything having been blown up, but although the buildings had been left standing all the contents had been cleared. About ten craters, within a radius of four to five miles of Blizna, were located and investigated. They varied in size, dependent on whether the projectile concerned had landed in the mostly sandy soil or in a boggy area, with the largest crater being 100 feet in diameter and 30 feet deep and some of the smaller ones only 30 feet in diameter and 10 feet deep. From these craters around one and a half tonnes of material was collected, which surprised the members of the mission because the Germans were usually meticulous about collecting parts. The most interesting discoveries were part of a tail fin that incorporated the mounting for a wireless aerial, a fuel tank, a burner unit and numerous smaller components. The report indicated that it had now been proved beyond all doubt that the rocket fuel was liquid oxygen, which 'was very good to drink' according to an interview with one captured German soldier.

On leaving Blizna, arrangements were made to visit the villages of Sawin and Sarnaki, 100 miles and 150 miles respectively from the experimental site, to view the impact of flying bombs, which according to local eyewitnesses had arrived during a brief period beginning 7 April 1944. In Sawin five craters were visited and although three had definitely been caused by flying bombs and one was indeterminate, one very large crater – measuring 70 feet in diameter and 25 feet deep – had been caused by a rocket. In the Sarnaki area evidence was gathered that it had been a main target for rockets launched during May and June. The area in which the craters were found was on both sides of the river Bug and extended eight miles wide by eight miles deep and included three villages. Local eyewitnesses reported that most of the rockets in this area had burst in the air, high up, but the warheads had generally landed intact and then had blown up. In a few cases whole rockets had landed intact and then blown up on the ground, creating large craters about 70 feet in diameter by 25 feet deep. In the conclusion to the mission's preliminary report, written in

Moscow, the hospitality shown to the Anglo-American team by the Red Army was duly noted.

It was in the Sarnaki region that the Armia Krajowa, the Polish Home Army, had managed to hide a rocket by sliding it into the river Bug, and this was where – on the night of 25/26 July 1943 – an RAF Dakota was headed, to a landing strip near the village of Zaborów twelve miles north-west of the town of Tarnów. The strip was used by the Luftwaffe during the day but was being used secretly by the partisans that night. The purpose of the operation, which had been given the code name Wildhorn III by the Special Operations Executive, was to smuggle out the rocket wreckage that had been concealed in the river Bug and also to bring out a number of members of the Armia Krajowa.

Complicating factors were the fact that 400 German soldiers were known to be camped only one mile from the landing point, that a considerable number of other German soldiers were known to be on the move in the vicinity, and that the Russians were fast advancing. The Dakota was piloted by New Zealander Flight Lieutenant Guy Culliford, who was later awarded the Virtuti Militari, Poland's highest military decoration for heroism and courage in the face of the enemy. Alongside him was his navigator, Flying Officer Szrajer of the Polish Air Force, and nineteen suitcases for the rocket wreckage, accompanied by Captain Kazimimierz Bilski, Second Lieutenant Leszek Starzynski, Major Boguslaw Wolniak and the courier Lieutenant Jan Nowak. Culliford was part of No. 267 Squadron RAF, which operated principally around the Mediterranean theatre, providing general transport services, casualty-evacuation flights and drop missions to support partisans.

Guided by the torches of the partisans and with an S-Phone – the ultra-high-frequency duplex radio system that had been developed during the war for use by Special Operations Executive agents to communicate with friendly aircraft and to coordinate landings – the Dakota landed successfully, was unloaded and reloaded 'with incredible rapidity' and in less than five minutes was ready for take-off. But as Culliford opened the throttles, the machine stood stationary as the brakes jammed and the wheels sank into the soft ground. After cutting the connections supplying hydraulic fluid to the brake drums they tried again but again failed to get airborne. When a spade was produced, the partisans frantically dug out each wheel, but with the engines once again started Culliford tried once more and failed as the machine slewed slightly to starboard and stopped. Given the absolute importance of exfiltrating the V2 parts, they dug for thirty minutes and on the fourth attempt Culliford managed to get the aircraft moving. With only the port landing light on, he opened the throttles and after ploughing along over soft ground made it into the air. Having lost hydraulic fluid he was unable to raise the undercarriage and, travelling hazardously slowly, they poured water from their emergency rations into the mechanism and managed to manually pump the undercarriage back up. After an hour's delay on the ground, there was no choice but to set a course through an area known to be infested with enemy night fighters.

On board, alongside the V2 wreckage, were drawings, diagrams and over eighty photographs in the custody of Lieutenant Jerzy Chmielewski – code name Raphael – who had earlier supervised the dismantling of the rocket. Alongside him were another four Poles, including Tomasz Arciszewski, a Polish socialist politician who would become the Prime Minister of the Polish government-in-exile in London from 1944 to 1947; Józef Retinger, a Polish political adviser who earlier that year had been parachuted into Poland to talk with leading political figures and who would become the first Secretary General of the European Movement International between 1948 and 1950, an organisation that would help lead to the foundation of the European Union when Duncan Sandys was its President; Second Lieutenant Tadeusz Chciuk (code name Celt) who had trained in Scotland as a junior officer in the artillery before going on commando missions by parachute into occupied Poland; and a fourth man, Czeslaw Micinski. Early in the morning as the sun was rising, they made a successful if somewhat hair-raising landing at Brindisi in southern Italy – the Dakota's lack of brakes required them to land into the wind on a runway still under construction.

After landing, Arciszewski, Retinger and Chciuk flew on to Rabat in Morocco to meet the then Prime Minister of the Polish government-in-exile, Stanislaw Mikolajczyk, who was then en route from London to Moscow. With the Russians advancing through their country rather than the western Allies, the Poles feared that Stalin would impose Communism on Poland and take possession of their eastern territories, which had a population made up of Poles, Ukrainians and Belarusians. Despite Winston Churchill's attempts to broker talks between Mikolajczyk and Stalin, efforts would break down on several occasions, particularly over the matter of the Katyn Massacre. The mass execution of an estimated 20,000 Poles in 1940 on the orders of the Soviet Politburo, including its leader Joseph Stalin, and the later discovery of their graves in the Katyn Forest in 1943 was used as political capital by the Germans at the time and led to the end of diplomatic relations between Moscow and the London-based Polish government-in-exile. Meanwhile Lieutenant Jerzy Chmielewski and the fourth man, Czeslaw Micinski, arrived in London on 28 July 1944 with their precious cargo. On arrival at Hendon airfield, R. V. Jones recorded that Jerzy Chmielewski – who could not speak a word of English – refused to let anyone near the V2 wreckage until he had authority from one of only two Polish officers he knew in Britain. One was General Bór – the code name used by General Tadeusz Komorowski during the war – a commander in the Armia Krajowa, who on his arrival managed to mollify Jerzy Chmielewski who had been sitting on a sack containing V2 wreckage, with his knife drawn when anyone made a move towards him.

Although R. V. Jones was frequently considered a divisive character by many of the Medmenham interpreters, the camaraderie enjoyed between the Wireless Section and ADI (Science), who by Christmas 1944 were based in Monck Street, Whitehall, was clearly shown in a memorandum from G Section, dated as 'The Night Before Christmas' and referenced 'MDM/S.BIN/OPS.UMPTEENTH'. It continued thus:

A CHRISTMAS CAROL – 1944

(Words by 'G' Section – Music to taste)

As Christmas comes but once a year
Now do we all express good cheer
To those for whom our craft we ply
The lords of science, A.D.I.

First comes Doctor R. V. Jones
Dispensing 'gen' in friendly tomes
Joad-like, never at a loss
Sitting in his old 'Jagdschloss'

Dr. Smith, whose trips we covet
Liberating (how we'd love it!)
Würzburgs, Chimneys, Freyas, Giants
And other things well known to science

The acumen of Dr. Frank
For revelations we must thank
Of secrets of those toys of hate
Vengeance weapons 1 to 8

Then there is Professor Wright
Whom we hear sits up all night
Performing mental acrobatics
Upon inversion mathematics

Wing Commander Gascoigne-Cecil
With whose radar floats we wrestle
In a controversial storm
Whether disc or cruciform

Squadron Leader Whistle Burt
(Hope he thinks us not impert)
Keeps us always on the hop
Won't these pinpoints ever stop?

Squadron Leader Gerrad's name
To us on 'D' day became
A talisman – we'd but to ask it
For 'shots' of any broken 'Basket'

Another year of good liaising
Introducing 'gen' that's quite amazing

With special cover and games to please us
Like 'Oranges and Lemons' . . . !!

We wish you well both one and all
We hope to hear the 'demob' call
The downfall of the Hun to see
A happy Christmas – Section 'G'!!

CHAPTER 12
From V1 to V2?

The German Army originally planned to launch A4 rockets from bombproof bunkers at Watten and Wizernes in the Pas-de-Calais and Sottevast and Brécourt on the Cherbourg peninsula, where they would also be stored and serviced prior to launch. It was originally intended to provide bombproof accommodation for trains that would deliver A4 rockets and supplies, a liquid-oxygen factory and facilities to launch the rockets. With the discovery of the Watten facility in aerial photography, followed by a ground report the following month indicating German long-range rocket development at the same pinpoint, more photographic reconnaissance was laid on, leading in mid-July to the first BS Report on the site being issued by Medmenham.

From this point Watten was photographed weekly and by 29 July considerable progress was reported from the most recent photography, which resulted in a detailed plan being prepared. The last report before the site was first attacked indicated that shuttering was in place ready for the concrete to be poured for an upper floor that, once constructed, would have created a bunker safe from conventional bombs. This resulted in the site being attacked, ten days after the Bomber Command attack on Peenemunde, by the US Eighth Air Force on 27 August 1943, when 185 Flying Fortresses severely damaged it. The raid had been timed on the advice of Sir Malcolm McAlpine, of the civil engineering company Sir Robert McAlpine Limited, to take place at the most critical stage of construction, when the concrete had been poured but had not hardened. The damage-assessment photographs were subsequently examined by McAlpine, and showed a tangled mess of concrete and steel reinforcing girders which solidified within days into a rigid mass. He assessed the setback to the Germans as being around three months, and said that in effect it would be easier for them to begin construction again.

Soon after the raid it was noticed that anti-aircraft batteries had appeared on the site, and despite work being at a standstill after a month the air defences continued to be reinforced. In October 1943 it was decided to reduce photographic reconnaissance to once a month and by November new activity could be seen. By the end of that month it appeared that the Germans had been listening to McAlpine since the site was being cleared and by the end of the year work was progressing on the construction of new and stronger buildings on the southern part of the foundations for the earlier building.

The principal contractor at Watten was the German construction company Philip Holzmann AG. They worked alongside the Organization Todt who had earlier constructed large parts of the Atlantic Wall, Reichsautobahn, parts of the new Reich Chancellery in Berlin and various Nazi rally grounds. When Albert Speer – Reich Minister of Armaments and War Production – was interrogated after the war, he confirmed the purpose of each individual Heavy Site. According to him, both the first and second Watten buildings in the Pas-de-Calais were intended for the manufacture of liquid oxygen, the production of which had always been a bottleneck in the V2 programme. Watten was intended to be a bombproof source of supply in the centre of the projection area that could supplement other supplies where appropriate.

After Crossbow sites were captured by ground forces, and following his earlier mission to Blizna, Colonel Terence Sanders led what became known as the Sanders Mission, which carried out detailed surveys of key Crossbow sites. Scale drawings were made and photographs were taken before many were then deliberately blown up by the Allies to limit future access to them. Information was gathered from the interrogation of key German personnel, including Speer, and from those employed by construction companies that had worked on the Heavy Sites. Employees of Philip Holzmann AG were able to confirm that the attack of August 1943, coming just after the concrete had been poured, had been timed to perfection by the Allies: if the attack had been delayed by a week the building would, in their view, have been completely bombproof. When the time came to rebuild Watten, they conceived an ingenious plan to build the roof of the new building on the ground, which was then raised into position once the concrete had set. The objective was to create a bombproof roof at the earliest stage, but it did require the creation of an enormous hydraulic jacking system that was capable of lifting sections of the roof that weighed as much as 3,000 tons.

The German engineers had learnt from the Watten experience and built the remainder of the bunkers according to the *Verbunkerung* design, which involved construction in six simple stages. First, large parallel trenches were dug into which concrete was poured to create the side walls of the bunker. As a relatively narrow and long structure – in Second World War bombing terms – its destruction by aerial bombardment before, during or after the concrete had been laid was difficult. Even if it had been hit, repairing the damage would have been relatively easy. With the side walls in place, another deep trench was dug beside the inside walls of the bunker, with the spoil heaped in the middle to form a solid earth core. The side walls of the bunker were then completed by pouring more concrete into this new trench, which was keyed into the existing wall with a substantial recess. With a massive concrete roof slab then laid, keyed into a recess on each side, the earth core was excavated to reveal a cavernous bombproof main chamber. With these structures not only more difficult to bomb effectively than the original Watten building had been during construction, the fact that they were also three-quarters underground made them less vulnerable to bombing when completed. With other parts of the *Verbunkerung* bunkers also system built, if parts were bombed and destroyed during construction they could be easily replaced. This highly effective bunker design would prove a great challenge to the Allies and would require an equally ingenious countermeasure.

The challenge of destroying *Verbunkerung* bunkers was of particular interest to the interpreters responsible for damage assessment. With the number, range and navigational accuracy of bombers increasing dramatically throughout the war, the number of interpreters involved in damage assessment also grew. From there only being nine in 1939, at their peak in 1944 there were fifty-four. Just as Bernard Babington Smith found himself relocated from Bomber Command to Medmenham – where he continued to develop interpretation techniques and extracted intelligence from night photography – the challenge for K Section was to develop techniques that could accurately assess and quickly report damage.

Since the main objective in planning air operations against the enemy was to ensure maximum damage for the minimum cost, the work of the section was essential. They reported on the success of strategic and tactical bombing; monitored the enemy's clearance, repair and reconstruction of damaged areas; plotted bombs to check navigational accuracy and the effectiveness of bombing techniques; undertook detailed analysis of industrial damage and checked the accuracy of damage reported by ground intelligence sources; while a sub-section was responsible for the preparation of special reports – including many on *Verbunkerung* bunkers – and other top-secret installations.

With round-the-clock bombing at the height of the war – the American heavy bombers flying in daylight and Bomber Command at night – K Section was operational twenty-four hours a day. Members of the section on the British side were predominantly commissioned officers from the RAF and WAAF and on the American side were from the United States Army Air Force and Women's Army Corps. The section was subdivided into four watches of between six and eight people who were responsible for all work of an immediate nature, while a specialist group worked on Detailed Reports, research and special projects. The shift pattern was on a twelve-hours-on-twenty-four-hours-off basis, and meant that one shift was always on duty at night while two were on duty during the day. All other members of the section worked a normal day shift. Although the Americans were responsible for their own photographic-reconnaissance squadrons and the Joint Photographic Reconnaissance Committee at RAF Benson – chaired by Peter Riddell – ensured efficiency and no duplication of effort between the Allies, the interpretation of the photography created was carried out through an almost unique, and on the whole friendly, cooperation between the Allies. For this reason many of the sections at the Allied Central Interpretation Unit, including damage assessment, were headed by Americans.

In civilian life the Columbia University-educated Major Geoffrey Platt was a prominent New York architect who worked alongside his brother in their prestigious New York architectural practice. Married to the granddaughter of an American Ambassador to the United Kingdom – the young woman was described by *Time* magazine on their marriage as a Manhattan poetess and socialite – Geoffrey Platt was responsible for designing houses for the wealthy of upstate New York and also designed a number of important commercial buildings, perhaps most notably the radical Steuben Building, which appeared phoenix-like on a Depression-era Fifth Avenue, New York, in 1937. The critically acclaimed retail shop for Steuben's glassware replaced a traditional New York brownstone with a building featuring 3,800 glass blocks that allowed light to flow into the building during the day and out onto the street at night.

Before the war, Geoffrey Platt had been a keen rower and had competed at Henley-on-Thames where he discovered that the exclusive Phyllis Court country club was out of bounds to common fry. Douglas Kendall records Platt as being in awe of the place and utterly astonished on arriving in wartime England with a group of fellow officers to find that he was not only billeted there but, as the most senior officer, was in charge. In addition to being a home to the model-makers, Phyllis Court also served as a billet during the war. Douglas Kendall recorded that all the country-club furniture had been replaced with bunk beds, with the exception of three stags' heads which gazed down from the walls. It was not long before they had been relabelled, and three new species of antelope appeared: Ibex Plattibus named after Major Geoffrey Platt, Ibex Kendalis after Wing Commander Douglas Kendall and Ibex Fullobeerus after Wing Commander Eric Fuller.

When Geoffrey Platt joined the United States Army Air Force he was shown a list of jobs considered appropriate for qualified architects. Since he already had some knowledge of photographic interpretation he knew it was what he wanted to do. Straight after PI training at Harrisburg, Pennsylvania, he travelled to England and worked as an interpreter, where his architectural training would prove useful to Operation Crossbow. When the purpose of different features at various Bois Carré sites was being considered, the wide variety of experts at Medmenham who might be able to help were consulted. This meant that deductions could be made through a process of elimination. Douglas Kendall consulted Geoffrey Platt about the significance of deep trenches at the sites. With information about temperature conditions in northern France and the depth of the trenches he advised – correctly, it would later transpire – that they were for water pipes that had to be sunk deep enough to avoid freezing. Platt would prove useful to Kendall during the investigation into the Bois Carré sites when garbled messages were reaching General Ira Eaker, Commander of the US Eighth Air Force. To help clarify matters, Geoffrey Platt was fully briefed by Kendall about Operation Crossbow and given the task of briefing the American general, who reportedly listened and reacted with two words: 'Target Them'.

At Medmenham, Platt served as a second-phase interpreter in Z Section, where he was responsible for developing techniques for interpreting strike photographs that were taken by the cameras housed in bombers, during the release of bombs and their immediate impact on the ground. Strike photographs were routinely compared with pre-strike and post-strike aerial photographs to provide a complete assessment of the damage caused. At Medmenham the damage-assessment interpreters in K Section had been commanded by RAF Flight Lieutenant Andrew Lyall – and when in May 1944 Geoffrey Platt transferred to the headquarters of the US Eighth Air Force as senior PI Officer, all K Section was transferred from Medmenham with him. Anxious that this decision might upset his British colleagues, Andrew Lyall later recalled his initial conversation with a contrite Geoffrey Platt: 'Andrew, I am so very embarrassed about all this. I know nothing of what you and your lot do, but I'll leave you alone and do all I can if you need anything' – and so began a harmonious special relationship.

Awarded the OBE for his wartime service, Geoffrey Platt returned to New York where he continued his architectural career, which included a project for the American Battle Monuments Commission. The Suresnes American Cemetery on the western outskirts of Paris – within view of the Eiffel Tower – had been built to commemorate the American dead of the First World War. After the Second World War, the Federal Government Agency decided that the cemetery should commemorate the dead of both world wars and commissioned the addition of memorial rooms and ground-floor loggias, designed by the Platt brothers, to adorn the memorial chapel. The masterpiece of First World War commemorative architecture had originally been designed by their father, the eminent New York architect Charles Platt, and had only been completed in the 1930s as another world war loomed. After having witnessed so much destruction in Europe it was perhaps fitting that in his native city Geoffrey Platt was the driving force behind the creation of the New York City Landmarks Preservation Commission – the city government agency responsible for the designation and protection of landmarks and historic buildings.

Shortly before D-Day – at the peak of the bombing operations – K Section transferred to the headquarters of the US Eighth Air Force – code-named Pinetree – in the requisitioned Wycombe Abbey Girls' School at High Wycombe, Buckinghamshire. Within the space of a few hours on 5 May 1944 the team moved, and found on their arrival various reminders of its previous occupants – including, if rumours are to be believed, a notice in one of the dormitories next to a bell – *Ring if mistress required* – that became the source of much ribaldry among the Americans.

The report most frequently issued by the Damage Assessment Section was the Immediate Interpretation Report. This gave a comparison of pre-raid and post-raid photographs and provided a general statement on the distribution and severity of damage followed by brief details on the condition of the more significant elements of the target. The report was typically accompanied by a photograph showing the principal points of damage. Referred to as the K Reports, there was a lengthy distribution list for them. They included the Command responsible for the attack, and a variety of air force intelligence, research, planning, targeting and public relations sections.

An indication of the immense scale of the Second World War was the fact that when both Bomber Command and the US Eighth Air Force attacked a target, upwards of five hundred K Reports would be distributed by the section. At the same time a teleprinter signal was sent immediately to the same distribution list, to speed the delivery of information. For targets of strategic importance a Detailed Report would be issued, dependent on other work pressure, within twenty-four to forty-eight hours of the Immediate Report. This required a detailed examination of the target area, and provided in-depth assessments of damage to urban areas, as well as to individual targets considered as priorities by the Ministry of Economic Warfare, and to those of industrial, communication, military or civilian significance which were summarised and assigned one of the five damage categories depending on the severity of destruction. To assess damage accurately, large-scale photographs were required, and ideally post-strike photography was obtained within a few hours of a raid in order that the immediate disorganisation caused by a raid could be assessed before the enemy had a chance to begin any repairs.

Detailed Reports were accompanied by a colour-photo-lithographed Damage Plot produced on a Rotaprint machine. For German towns these were at a scale of 1/25,000 and were produced in two colours – red and blue – with damage from the attack under review shown in red and that from previous attacks in blue, providing a visual indication of the extent of damage. In Detailed Reports on precision targets Damage Illustrations were reproduced at 1/6,000 scale on the same basis, but with reconstruction indicated with a hatched blue line. As the photographs relentlessly passed through K Section, the Damage Assessment interpreters became accustomed to witnessing the mass destruction of Continental Europe by the Allied attacks from the relative serenity of the Home Counties. High-explosive bombs were shown in the aerial photography to effectively demolish buildings but they displaced so much material that they left areas littered with debris that deposited a characteristic layer of white dust from building rubble on the ruins before post-strike photographic reconnaissance was taken. Damage from fire left a somewhat cleaner legacy, but burning depended largely on the material from which buildings were constructed and from their contents. While collapsed roofs gave bombed towns and cities a honeycomb-like appearance on vertical aerial photographs, the shadows of windowless ruins cast distinctive eerie long shadows on the street.

K Section were witness to the bomber offensive in a way that not even the bomber crews themselves were. From the highly detailed post-strike photographs they witnessed the human impact on the civilian population, and would often see piles of furniture and other personal belongings from the buildings appear quickly on the street in front of their homes as people began salvaging what they could after a raid.

In the early years of the war, damage was largely assessed by house counting, but the tremendous increase in attacks necessitated another means and led to the invention of the Damometer. Just as Squadron Leader Claude Wavell in the Wireless Section had been the brains behind the Altazimeter, which reduced lengthy mathematical calculations to only a few minutes, in K Section Alick Heron invented, designed and constructed the Damometer as a means to assess large areas of damage in built-up areas. The implausible-looking contraption – something of a Heath Robinson affair – accommodated a stereo pair and while looking through a fixed stereoscope the operator could move anywhere around the three-dimensional image by using controls. By moving a pointer in millimetres, distance and therefore areas of damage or no damage could be calculated. By methodically working through a German town, block by block, the percentage and actual amount of damage caused could be relatively quickly calculated, saving hours of soul-destroying work. This resulted in what became known as the Statistical Analysis.

In June 1944 the Joint Chiefs of Staff agreed to an American proposal for a comprehensive analysis of the bomber offensive in the European Theatre. The United States Strategic Bombing Survey (USSBS) had been authorised by President Roosevelt who hoped that lessons from bombing strategy in the European war could help win the war in the Pacific. By August 1945 when the atomic bombs were dropped on Hiroshima and Nagasaki – missions that would be covered in detail by a later USSBS study of the Pacific Theatre – investigations were nearing completion in Europe.

In late 1944 the Americans had created twelve specialist teams that investigated the impact from bombing in the subject areas of: Aircraft; Geographical Areas; Civilian Defence; Equipment; Military Analysis; Morale; Munitions; Oil; Overall Economic Effects; Physical Damage; Transportation; and Utilities. Headquartered in London, the survey involved the collective effort of 1,150 officers, enlisted men and civilians. While the war was ongoing it was considered essential for survey teams to follow closely behind the front line to ensure that vital records were not irretrievably lost in the fog of war. This inevitably meant that survey personnel suffered several casualties, including four who were killed in the course of their work. Key personnel recruited to the USSBS included the President of Prudential Insurance, Franklin D'Olier, who was appointed chairman, Paul Nitze – a Wall Street financier who would become an important policy chief during the Cold War – and rising stars of academia including the economist John Kenneth Galbraith. The survey inspected hundreds of key German factories, towns and cities and amassed volumes of statistical and documentary material. Information contained in aerial photography

was considered vital, alongside information gleaned from the interrogation of surviving political and military leaders. With two hundred
detailed reports written, the overall report on the European War
chronicles Germany as being scoured for its war records, which
were sometimes but rarely found where they ought to have been.
Sometimes they were discovered in safety-deposit boxes, often in
private houses, in barns and caves, once in a hen house and on two
occasions hidden in coffins.

The report is unequivocal in its assessment of the contribution of air
power to the war. It records the Allies as having dropped 2.7 million
tons of bombs, and as having flown more than 1.4 million bomber and
2.8 million fighter sorties in the European War. Whereas in the First
World War air power had been in its infancy, in the European Theatre
of the Second World War air power had reached adolescence and a new
relationship now existed between air power and strategy. Some 3.6
million homes in Germany – approximately 20 per cent of the total –
were recorded as having been destroyed, with 300,000 civilians killed,
780,000 wounded and 7.5 million homeless. When these figures are
compared with the damage inflicted by the Germans on Britain, the
vast superiority of the Allied Air Forces and the relative insignificance
of the German secret-weapon programme becomes clear. With an
entire USSBS report dedicated to the V weapons and Operation
Crossbow, however, the importance of this revolutionary technology
was accepted and its potential to dominate the Cold War was made
clear. When the European report was released in November 1945, *Time*
magazine ran a feature article, 'Awesome and Frightful', describing the
report as the definitive reference source of man's inhumanity to man
in the pre-atomic age. The great lesson identified by the survey was
considered to be that the best way of surviving a war was to prevent
it from happening in the first place, but maintained that now the
combination of the atomic bomb with remote-controlled projectiles of
ocean-spanning range stood as a possibility which was awesome and
frightful to contemplate.

While the USSBS utilised an organisation of more than a thousand
personnel, in December 1944 Prime Minister Churchill refused a
request from the Air Ministry for anything similarly grand. Although
the United States government was actively using the survey to help
formulate the development of its military doctrine and superpower

role in the post-war world, for as long as the war in Europe continued Churchill was only prepared to sanction a team of twenty to thirty people for the British equivalent of the task. Norman Bottomley – then Assistant Chief of the Air Staff – chose nevertheless to circumvent what he considered a draconian limitation on numbers and secured specialist personnel already working on many of the issues at RE8, in Princes Risborough, and specialists from other government departments. With the net result being only seventy to eighty personnel from the RAF and a hundred other experts, the British efforts fell far short of those of the Americans. Professor Solly Zuckerman, the architect of much of the bombing policy during the war, was recruited as principal author of the British report – *The Strategic Air War Against Germany 1939–1945* – by the British Bombing Survey Unit.

Unlike the rapidly completed USSBS, the British survey was an altogether smaller and slower production that was finally completed after the war when Britain assumed a diminished role in the new world order. The report provided a review of bombing policies and directives, the scale and distribution of the bomber effort, the effects of the bomber offensive and an assessment of its contribution to the winning of the war. It revealed the relatively small scale of raids, the low destructive power of bombs and the frequent navigational inaccuracy of aircraft early in the war which made damage assessment a laborious task, and went on to show how the scale and devastating force of the bombing offensive increased significantly during 1942 and 1943 and exponentially for the remainder of the war. This made detailed cataloguing of bomb damage neither necessary nor practicable and saw the replacement of house counting and the identification of individual streets and public buildings by the area categorisation of damage caused by carpet bombing.

In addition to the K Reports, which formed the principal workload of the Damage Assessment section, special reports on damage known as K(S) Reports were written on request, and when K Section moved from Medmenham to Pinetree, Flight Lieutenant William 'Bill' Seaby was asked to take over responsibility for them. After studying art under his father, the Professor of Fine Art at Reading University, Bill had a short-lived career, working for B. A. Seaby Limited, the antiquarian coin dealer on Regent Street in London's West End. But, deciding the life of a coin dealer was not for him, he went to work as an archaeologist for the Birmingham Museum and Art Gallery. Post-war, the antiquarian scholar who had an interest in horology and numismatics relocated from the Midlands to Belfast and worked as the enthusiastic Director of the Ulster Museum. After having joined the RAF in 1940 and working initially as a lowly Radio Transmitter Operator, he became another archaeologist with an enquiring mind who would find himself relocated to Medmenham with an officer's commission. The special jobs in K Section often involved a particularly high level of secrecy and included writing K(S) Reports on specific targets, reconstruction, bomb performance, cumulative damage, the status of particular buildings, bomb-fall distribution and the impact of German bombing on targets in Britain. Operation Crossbow targets, particularly the Heavy Sites in northern France, received special attention, given how difficult such heavily reinforced sites were to destroy.

After the first attack on Watten in August 1943, which had successfully destroyed the first bunker during its construction, the site was attacked again on 2 February 1944 before the roof had been constructed. This attack caused only light damage to the railway line and a few of the buildings at the site, and by 13 March it was reported by the interpreters that the roof had miraculously appeared. Quite how this had been possible was only apparent later, when the tidying-up of the site could be observed. This resulted in unsuccessful attacks on 19 and 20 March, followed by a successful raid on 26 March when three direct hits on installations were achieved. From this time on, Watten was attacked more or less continuously and thanks to damage-assessment photography, activity on the ground could be observed throughout. The key problem was that although the 1,000- and 2,000-pound bombs being used could effectively destroy the infrastructure around the main building they were ineffective against the reinforced concrete structure of the building itself.

This resulted in the 12,000-pound Tallboy bomb being deployed on 20 June 1944. The brainchild of Barnes Wallis, inventor of the bouncing bomb used by 617 Squadron for their attack on the German dams, this heavy bomb aimed to destroy targets through an earthquake effect. It worked on the principle that a suitably designed high-power bomb could penetrate the ground near a target and explode underground, sending shock waves through the ground like an earthquake, which hopefully would weaken the target building's foundations and undermine the structure. The task facing the bomb aimers was to ensure that their bombs were near misses, since hitting the main building would probably have meant that their bombs would just have bounced off, thus limiting their impact. From damage-assessment photography after the first raid, a near miss was thought to have damaged the southern wall of the structure. The site was subsequently attacked with Tallboy bombs on 6 July 1944 and when the site was photographed on 25 July a hit was seen to have caused considerable damage to the main building, and it was assumed that the site was out of commission.

In addition to the data they had from the USSBS, Sanders Mission and British Bombing Survey a number of the interpreters had the opportunity to ground-truth sites that they had previously observed on aerial photography. This included Bill Seaby who joined two missions to France and one to Germany to study at first-hand the impact from the Barnes Wallis bombs. In addition to the bouncing bomb used on the Dambusters raid – code-named Upkeep – 617 Squadron also deployed the 12,000-pound Tallboy and 22,000-pound Grand Slam bombs against a range of targets. With each of these bombs so tremendously powerful, each of their impact craters was carefully plotted by K Section and they were subsequently inspected on the ground.

In France the Heavy Sites and heavily reinforced U-boat pens were attacked after the Tallboy was first deployed against a railway tunnel after the Normandy landings. On the night of 8/9 June 1944 the railway tunnel at Saumur in the Touraine region – 125 miles south of the fighting in the Normandy battlefields – was hastily attacked after it was suspected that a German Panzer unit was about to head northwards by train through it. That night, 483 aircraft – 286 Lancaster bombers, 169 Halifaxes and 28 Mosquitos – attacked French railways to prevent German reinforcements reaching Normandy. Twenty-five Lancaster bombers of 617 Squadron dropped their Tallboys accurately onto the target. The bombs exploded successfully under the ground, sending shock waves through the soil that resulted in the roof of the tunnel collapsing and blocking it. The resultant damage-assessment photographs – taken during sortie 1066/765 on 9 June 1944 – showed the immense impact of the attack which required thousands of cubic metres of spoil to be excavated from the tunnel. On the later mission to Germany, Seaby – who by then had been nicknamed 'Crater Chaser' by his PI colleagues – found himself alongside an Air Vice Marshal, two Group Captains, a Wing Commander and Barnes Wallis himself, inspecting first-hand the damage to the Mohne and Sorpe dams inflicted by Upkeep bombs.

The Tallboy bombs were used by 617 Squadron against railway tunnels, Crossbow sites, U-boat and E-boat installations, German-controlled ports, canals, railway bridges and dams, and were even used to sink the *Tirpitz* battleship, the *Lützow* battlecruiser and in attacks on the Berghof, Adolf Hitler's home at Obersalzberg in the Bavarian Alps near Berchtesgaden in southern Germany. In Germany, Barnes Wallis inspected their dramatic impact on the Dortmund-Ems Canal and the railway viaduct at Bielefeld, which had also been attacked with the Grand Slam bomb. Crossing the river Were, the viaduct, which carried the main railway line from Berlin to the Ruhr, was a strategically important target that had hitherto managed to escape destruction. It stood as a reminder of how difficult it was during the Second World War to destroy long narrow structures – such as *Verbunkerung* bunkers – however accurately the bombs used against them could be dropped. The first of the forty-two 22,000-pound Grand Slam bombs that were to be dropped in the war, along with Tallboy bombs, managed to destroy more than 100 yards of the viaduct through an earthquake effect on 14 March 1945. A low-level oblique photograph taken post-strike reveals easily one of the most dramatic examples of bomb damage during the war, showing enormous circular bomb craters and the destruction of arches, with railway lines dramatically dangling in mid-air. Barnes Wallis was suitably impressed after viewing the bridge on the ground and is recorded to have extended his thanks to K Section for their accurate assessment of the damage recorded in their interpretation reports.

Following the attack on Watten with Tallboy bombs the site would also be attacked by the Americans during Operation Aphrodite, when unmanned bombers – drones – with automatic pilots were used against Crossbow targets and U-boat pens. The operation involved taking old aircraft that were stripped of all non-essential components and loaded with 25,000 pounds of TNT and the powerful nitroglycerine explosive Torpex. After development at the Air Proving Ground at Eglin Field in Florida, operations in England began on 23 June 1944 from remote RAF Fersfield – with its conveniently long runway – in rural Norfolk. The mother ship was a Lockheed Ventura bomber and the drones were usually ageing Boeing B-17 Flying Fortress bombers. Getting them airborne involved a two-man crew flying an aircraft in the conventional way. Once airborne it would be joined by the mother ship and a Lockheed P-38 Lightning, which would fly alongside in case, as happened on a number of occasions, the drone went out of control and there was no alternative but to shoot it down.

With remote-control equipment installed, along with two television cameras that provided views of the ground and of the instrumentation in the cockpit, once the drone was airborne the mother ship would control it. With equipment tests complete – and the canopy removed to aid their descent – the two crew members would parachute over the English countryside while the mother ship controlled the drone on the way to its target. The first drone attacks were on Heavy Sites at Mimoyecques, Siracourt, Wizernes and Watten on 4 August, but with one crashing in England, the second destroyed by anti-aircraft fire, the third undershooting the target and the fourth overshooting, it was an inauspicious start to the operation.

Watten would be attacked again on 6 August 1944 when two more drones were directed onto their target and exploded. This was followed on 12 August with a mission to attack the Heavy Site at Mimoyecques, when Lieutenants Wilford Willy and Joseph Patrick Kennedy were to pilot the drone initially before bailing out. Flying immediately behind them was an American PR Mosquito filming the drone in flight and hoping to photograph its immediate impact on its target. While flying over Suffolk the drone exploded prematurely and instantly vaporised Wilford and Joseph, the elder brother of future US President John F. Kennedy, who had been expected to be the family's political standard bearer. Wreckage landed near the village of Blythburgh and the Mosquito rose a few hundred feet in the air from the force of the blast, which also scattered debris in its path, forcing it to land at the nearest airfield. This succession of disasters made the British authorities understandably doubtful about the effectiveness of this revolutionary technology.

The Watten bunker was later used to test the Disney bomb, a rocket-assisted bunker-buster bomb which had been developed by the Royal Navy to penetrate hardened concrete structures such as submarine pens. It was devised by Royal Navy Captain Edward Terrell, a Liberal politician who had been a successful barrister before the war – he had a natural flair for invention and by 1940 had registered patents for objects ranging from pens and ink bottles to peeling knives. On joining the Royal Navy Volunteer Reserve, he would find himself at the Directorate of Miscellaneous Weapons Development where, among other things, he worked on the Disney bomb. Unlike Barnes Wallis's earthquake bombs, the Disney bomb was designed to directly penetrate a heavily reinforced target courtesy of its construction from a thick steel shell and its impact speed of almost 1,000 miles per hour. Successfully tested against Watten, the Disney bomb saw limited use with the United States Army Air Force during the war, having arrived on the scene too late to be used in the Allied bombing offensive against Germany.

Watten was the first bunker to be constructed for the storage, servicing and launch of V2 rockets and it would be followed by another in the Pas-de-Calais at nearby Wizernes and two on the Cherbourg peninsula at Sottevast and Brécourt. In a quarry near the town of Wizernes, ten miles to the south-west of Watten, the Organisation Todt built a bunker code-named *Schotterwek Nord-West* – Gravel Pit North-West – that was constructed using a version of the *Verbunkerung* system. Like Watten it was intended to provide a secure bombproof environment for the storage, servicing and launching of V2 rockets. The site was of a sufficient scale that a liquid-oxygen plant and a hospital were planned. New railway-construction work was first reported there in November 1943 when new accommodation huts were also spotted. Medmenham issued a full report on the site, accompanied by a plan, on 5 December 1943. The next day entrances in the cliff face were seen and by the end of the year excavation work for the construction of a vast dome and more railway lines was sighted. Conventional bombing of the site began in February 1944 and continued, but throughout these attacks construction of the one-million-ton concrete dome continued unaffected.

Even the first attack using Tallboy bombs on 24 June 1944 failed to have much impact. On 7 July 1944, Detailed Interpretation Report BS 643 was issued, providing a history of the site accompanied by plans and a suggested perspective of the cliff face since no oblique photographs of the site existed. To learn more about the cliff face, some of the most spectacular low-level oblique photography of the entire war was carried out during dicing sortie 106G/1347 on 10 July 1944 by Flight Lieutenant P. J. King in a PR Mosquito of 540 Squadron, alongside his navigator Flight Sergeant Bowlen – who took photographs using a nose-mounted forward-facing oblique camera. Using this information to confirm the perspective view of the cliff, on 17 July 1944 sixteen Avro Lancaster bombers of 617 Squadron followed Group Captain Leonard Cheshire – who was flying a target-marking Mosquito – and attacked Wizernes again with Tallboy bombs. The bomb aimers ensured that their bombs were near misses, so that their weapons didn't just bounce off the massive dome, and on the following day it was reported that two, possibly three Tallboys had fallen in the quarry west of the dome, with the resultant explosion bringing down much of the adjoining cliff face. It would later transpire that the foundations of the site had been so weakened that the Germans were forced to abandon the location. This success was confirmed in large-scale photographs taken on 19 July 1944 and by the fact that no further work was undertaken at Wizernes before the site was overrun at the end of August 1944.

With the interpreters mindful that the Heavy Sites were rail-served, a particularly close watch had been kept on the railway system in northern France during Operation Crossbow and this led to the identification of Sottevast on 31 October 1943. At that stage the site on the Cherbourg peninsula consisted of a railway spur and excavations aligned at right angles to Bristol. The site was attacked on a number of occasions, and with work progressing slowly, it appeared to have finally been abandoned in April 1944. The other bunker site on the Cherbourg peninsula was called Martinvast by the interpreters – despite actually being located near the village of Couville – and was first reported on 22 October 1943. Like Sottevast it was aligned at right angles to Bristol. After an initial attack during November 1943, work was seen to continue, as it did despite attacks throughout January to April 1944. Unlike Sottevast, however, it appeared never to have been completely abandoned until it was overrun by the invading Allied armies at the end of June. The reason turned out to be that the place was being converted into a bombproof site for the launch of flying bombs against Bristol, missiles which in the event were never deployed before the location was captured.

In the Pas-de-Calais there were two further such sites at Siracourt and Lottinghem – discovered in October and November 1943 respectively courtesy of telltale railway construction. On 24 February 1944 the bombing of Lottinghem began and although it damaged the site it also damaged the neighbouring hutted camp – a common feature near a Heavy Site – that accommodated the many prisoners of war who undertook much of the construction work. Bombing continued until the end of April by which time the site was completely abandoned, shortly in advance of the Normandy landings.

At Siracourt the first bombing had taken place in January 1943, when the walls of the *Verbunkerung* bunker were complete but only a small portion of the reinforced concrete roof had been completed. Despite the bombing raids, the effectiveness of the *Verbunkerung* design was demonstrated by the fact that construction continued unaffected, with the main part of the roof completed by mid-April. This prompted attacks on the bunker with the Barnes Wallis-designed Tallboys in late June and through July and August as the Allies struggled to advance through France. In September 1944, by when the Heavy Sites associated with the deployment of the A4 long-range rocket had been destroyed and overrun by the advancing Allied armies, many people would be fooled into thinking that the threat from the German rocket was over.

CHAPTER 13
Big Ben

'This new form of devilry continued until our armies
captured The Hague . . .'

Lord Ismay, Memoir

The original investigation by Medmenham into the enemy's secret
weapons focused on their rockets, and since that was what they saw
lying on the ground at Peenemunde it seemed eminently logical that
it should. It was only later when a flying bomb was discovered and
identified as a more immediate threat that the focus of day-to-day
activity on Operation Crossbow changed. Despite this, Medmenham
never neglected research into rockets. After all, the continued effort
being made by the Germans to build the elaborate Heavy Sites in
northern France, in spite of the sustained attacks on them, was
indication enough of the importance placed on the weapon and its
means of deployment.

But since the interpreters had been advised from the outset that
the revolutionary weapon would be dependent on rail-served, heavily
reinforced bunkers, it is not altogether surprising that this clouded
their thoughts on how the weapon would be deployed. For reasons
that Douglas Kendall put down to the interpreters involved not having
sufficiently open minds, and their being handicapped by the scientists'
belief that the rocket would be guided during launch by rails, they
remained ignorant for a long time about the simple means by which
the rockets were to be launched and the particular difficulties they
would face in developing countermeasures. In the event it would
turn out that much of the evidence about how the rockets would be

deployed had been in their hands since André Kenny first investigated Peenemunde. It just took a long time to assess the significance of various different features recorded in the photographs.

At Peenemunde the large crane-like structures, surrounded by enormous blast walls constructed from earth, were readily apparent in aerial photographs. More discreet was an area of reclaimed beach some distance off, accessed by a track and with a fan-shaped piece of tarmac laid on it. With the benefit of hindsight it is easy to criticise the interpreters for never fully questioning why the Germans had gone to the effort of building this. The fact that the interpreters even diligently recorded a column, rocket-sized in height, on the tarmac only helped to fuel later criticism, most notably from R. V. Jones. After all, had the full significance of this area been appreciated it would have been immediately apparent that the rocket did not need any special firing ramp, nor any fixed handling gear or the supposed guide rails. But although this would undoubtedly have saved a great deal of trouble during the Allies' subsequent investigations in northern France, and might have saved a certain amount of bombing effort, it would have been unlikely to have changed the German deployment of the weapon. By the beginning of September 1944 the Germans were ready to deploy the rocket, rebranded by now as the *Vergultungswaffe 2*, roughly fourteen months after it had first been spotted at Peenemunde. Only one month before this, after having had the evidence for well over a year, it was realised that the whole firing mechanism for the rocket was mobile and that it could be fired from any relatively small flat area rather than just from the Heavy Sites. This revelation meant that Medmenham faced a challenge way beyond anything they had faced when searching for the Bois Carré and Belhamelin sites.

By early September 1944, by when Paris had been liberated, and as the Allies continued their seemingly inexorable advance through France, for many the end of the Second World War was in sight. On 23 August 1944 the Führer had feared this development and issued his orders to the Commanding General of Greater Paris – General Dietrich von Cholitz – on the fate of the French capital. The defence of Paris was seen as being politically and militarily vital since the loss of the city would lead to the loss of the entire coastal plain north of the Seine and would limit the Germans' capacity to launch rockets in their long-distance war against Britain. Throughout history the loss of Paris had inevitably brought with it the fall of France and so, in defending the city, the most energetic tactics were to be employed. The position of the Führer – 'Paris must not fall into the hands of the enemy, or if it does, he must find there nothing but a field of ruins' – was unequivocal. In the event, the City of Light was *not* burning when freed from Nazi rule on 25 August by the 2nd Armoured Division of the Free French Army and American Fourth Division. But with flying bombs and long-range rockets at their disposal, Paris was not yet safe from German attack, in a development of the Medmenham operation that would come to be known as Continental Crossbow.

A fortnight after the liberation of Paris, the 206th Home Intelligence Weekly Report recorded a further rise in the spirits of the British people for the sixth week in a row, with the boost in morale linked to the thrilling news of fighting on German soil, the relaxation of the blackout, and a reduction in Home Guard, Fire Guard and Civil Defence duties. Particular satisfaction was recorded at the capture of flying bomb sites, rapid Allied advances on the Western Front and the liberation of places synonymous in the British psyche with years of stalemate in the First World War. One perhaps unexpected consequence of the newsreels, press stories and photographs was a widespread impression that the French, and to a lesser extent the Belgians, were a great deal better off than was generally supposed. The sight of newly liberated Parisians looking well dressed and fed was further compounded by details in soldiers' letters home, generating resentment among many. But however widespread this feeling might have been, an overriding spirit of optimism and a sense that victory would come in weeks, or even days, was recorded by the network of contacts who had been secretly listening to Britain throughout the war. While the month of October was considered a favourite guess by many, a more cautious minority expected that victory would come by Christmas at the latest.

That week, Duncan Sandys's announcement to a crowded press conference on 7 September 1944 that 'except possibly for a few last shots, the Battle of London is over' was enthusiastically reported.

Although Sandys warned that the enemy might attempt to launch V1 missiles from aircraft, the widespread relief and satisfaction of people was recorded in the Home Intelligence Report. The efficiency and courage of those manning the defences and the resourcefulness of the planning that had ensured so many bombs were destroyed were sources of tremendously improved morale.

On Thursday, 7 September 1944 the SS arrived in the Wassenaar district of The Hague and forced local civilians to evacuate their properties within two hours. The third largest city in the Netherlands in the south-west of the country was the seat of the Dutch government and royal family, and had been extensively modified by the Nazis to accommodate the Atlantic Wall. Now the aggressive actions of the Germans would make the city a target for the Allies and would bring further destruction, and death, to Dutch civilians.

On the morning of Friday, 8 September 1944, six trucks and launch vehicles arrived and began preparing for the launch of two V2 rockets. Just after 6:30 p.m. the first V2 rockets to be fired against Britain were targeted on the Fire Station in Southwark Bridge Road, London. At 6:40 p.m. the first rocket landed in Staveley Road, Chiswick – nearly eight miles off target – where it killed three people and seriously injured another ten. Sixteen seconds later the second rocket fell harmlessly to Earth near Epping, Essex, eighteen miles from the aiming point.

Within an hour of firing, the suburb of The Hague was deserted and there was little evidence that any rockets had ever been launched. Despite constructing the elaborate Heavy Sites in northern France for the launch of their V2s the Germans had been unable to develop and use the technology quickly enough and equally had failed to build bunkers capable of withstanding the earthquake effect of the Barnes Wallis bombs. But this first launch against London was nonetheless a salutary lesson that, despite all the elaborate arrangements, the rockets could be launched by mobile units – and their mobile nature meant that Britain was still wide open to attacks.

Although the rocket attack was clearly expected in official circles, and was not altogether a surprise to a cynical civilian population, its timing one day after Duncan Sandys had so publicly announced that the Battle of London was over was certainly unfortunate. The sounds of the two explosions in the south-east of England prompted widespread speculation and rumour there, with most people presciently believing they were due to rockets launched from the Netherlands or Denmark. Although the opinion of people in the south-east appears to have been divided about whether or not the government should make a statement, and though a majority appear to have favoured one in order to quell rumours and keep evacuees from returning home, a minority was noted to have been against it since the attacks from flying bombs had got worse after Mr Churchill's statement.

Despite a complete news blackout, for those who witnessed a white vapour trail hanging vertically in the air and who heard a loud double thunderclap when the rockets broke the sound barrier and exploded on the ground the conspiracy of silence during the weeks that followed concerning this and the following attacks must have made a mockery of the official euphemism that a succession of gas mains had blown.

The credibility of Duncan Sandys was to be tested further when, within days, Doodlebug Alley moved eastwards to the county of Essex. Although he had warned about the possibility of flying bombs being air-launched, his dramatic press conference was taken by many evacuees as the signal to return home. On their return many discovered wrecked or roofless homes and at dawn on Saturday, 16 September there was another attack by nine bombs of which five were shot down, two crashed in open country, one got through to Woolwich and the ninth crashed in Barking, east London, killing thirteen people and injuring a hundred or more. Although there was understandable anxiety and disappointment that the attacks had not ceased entirely, the general reaction recorded was still one of relief that large-scale attacks had stopped and the general expectation was that no more than spasmodic attacks and stray bombs would follow.

But the air-launching system developed by the Germans involved using specially modified Heinkel He 111 medium bombers and represented a dramatic new threat. It greatly extended the effective range of the V1 and when considered alongside the V2 rocket meant that the Germans could deploy their secret weapons beyond London and the south-east. In retrospect Sandys's statement was all the more surprising, given the British knowledge that air-launched flying bombs had already been targeted against the country. A short-lived campaign from airfields in France, which came to an abrupt end when the Allies broke out of Normandy, had been waged against targets including Southampton. But then Dutch airfields at Venlo and Gilze-Rijen from which flying-bomb operations were mounted against London were selected. In order to conceal the Heinkels as much as possible from British radar installations, flights were made at 300 metres over the Continent and at 100 metres over the sea before the German aircraft climbed to a height of 500 or 600 metres. On levelling out, their airspeed was increased to 170 mph before the flying bomb's ramjet was started and the cruise-style missile was released.

On 8 September 1944 the first long-range rocket fired in active operations had fallen on the south-eastern outskirts of Paris, followed some hours later by the rocket that fell on Staveley Road, Chiswick. From this date the V2 offensive against the United Kingdom continued and, after a brief lull, rockets would continue to be fired against targets on the Continent. The possibility of Continental Crossbow attacks was foreseen by the Supreme Headquarters of the Allied Expeditionary Force, who faced the challenge of coordinating air-defence measures in the North-west European Theatre of Operations.

On 17 September, the twenty-fifth rocket landed on Adelaide Avenue, Lewisham, and killed five members of one family. The worst attack to date destroyed eleven houses, seriously injured twenty-nine people and killed fourteen, on the same day when British airborne forces began landing on their dropping zones at Arnhem. The Allied operation against overwhelming odds to capture the bridges across the Rhine was destined to fail. But because Operation Market Garden threatened at first to cut off the rocket units, the SS forces manning them withdrew hastily. Although this would give London a brief respite, moving the V2s from The Hague to a densely wooded area near the hamlet of Rijs in the Gaasterland district of south-west Friesland now brought East Anglia within range. After attacks on the United Kingdom began again on 25 September, one of the more dramatic days of the V2 campaign was Tuesday, 3 October when six rockets were fired at regular intervals against Norfolk, which, given its largely rural nature, helped ensure that no populated areas were hit nor people killed. During the East Anglian offensive the limited capacity of the weapon to be effective at causing death and destruction anywhere except in densely populated areas was illustrated when almost thirty rockets caused no fatal casualties and only fifty-one people were injured, mainly by flying glass. By the time the V2 teams abandoned Rijs Wood, the strategically vital Belgian port of Antwerp had fallen into Allied hands and was now destined to become a target.

With the V2 attacks continuing unabated, on Wednesday, 10 November 1944 the Prime Minister finally admitted publicly that

Britain was under attack from long-range rockets. The British were now resigned to another winter of war, the onset of which, combined with the usual bad weather, had a somewhat depressing effect. Although people continued to be pleased at the liberation of Antwerp, the Scheldt estuary and the island of Walcheren – from where rockets had been fired – there was a general war-weariness and widespread lack of interest in the fighting. Against this background the Prime Minister's statement was recorded to have been generally welcomed, both in affected areas and elsewhere, because it allayed the anxiety and uncertainty associated with official silence. Before the statement there had been fantastic rumours in circulation – that ninety-six rockets had fallen on London in the last week of October alone, for example – when in reality the figure was less than a hundred, in total, for the whole country. Although many felt that the statement was well overdue and merely confirmed what was already known or suspected, for others who had not heard the rumours the news came as a great shock. In northern regions there was discussion and speculation about whether the enemy had succeeded in extending the range of their secret weapons, while others questioned the wisdom of standing down the Civil Defence services. For the people of London and the south-east the chief topic of conversation was still, understandably, the rocket attacks, which left many people worried, nervy and apprehensive at the possibility that they might become more frequent. Although the lack of warning of any imminent attack was considered to be trying, some people even showed a preference for rockets over flying bombs on the basis they were less disturbing.

Though many had thought in the aftermath of the Normandy landings that the fighting would be over by Christmas 1944, an announcement by the Prime Minister that the war might not end before Easter meant that any lingering hopes of an early finish to hostilities disappeared. Whilst the German counteroffensive in the Ardennes to recapture the strategically important port of Antwerp had pushed the Allies into retreat. On the Home Front the occasional doodlebug was still hitting the south-east, in particular Essex, and the V2 rocket attacks continued. On top of all this, the harshness of the European winter of 1944/1945 contributed to the sense of war-weariness.

While many in southern England suffered, and many in the German-occupied part of the Netherlands actually starved to death in what became known as the *Hongerwinter*, those in northern Britain could have been forgiven for feeling relatively immune. This was certainly the perceived public mood recorded by the Ministry of Information, with northerners characterised as feeling concern for those in the south, particularly when they had relatives living there. But being beyond the perceived range of the weapons gave them a sense of safety. This was all to change on Christmas Eve 1944 when the Luftwaffe outflanked the UK's east coast defences and created panic in a part of the country hitherto immune from the doodlebug attacks.

In this one single onslaught, more than thirty flying bombs were targeted on northern England, particularly the Manchester area, with the worst incident occurring in Oldham. At 5:50 a.m. a doodlebug exploded, destroying terraced houses and killing thirty-two people, including three small children who had been evacuated by their parents from Streatham to the perceived safety of Lancashire, after being made repeatedly homeless in south London. The Manchester raid startled people who assumed that flying bombs could not reach the North, was the last major flying bomb attack – and was notable for being used to deliver Christmas letters from British prisoners of war. From late August 1944, some of the V1s had been equipped with a device enabling them to discharge a payload of propaganda leaflets. On the Christmas Eve raid the idea was taken an ingenious stage further.

Selected British prisoners of war were given the opportunity to write an extra letter home, in addition to their normal allowance that was sent through the Red Cross. This particular load of mail was delivered by the aptly titled *V1 POW Post*. With added assurances, supposedly from the prison camp commandants, that the prisoners were being well looked after, the letters carried requests that their finders should send them on to the addressees. More a masterful piece of propaganda than a serious attempt to plot where the flying bombs dropped, it was largely foiled because so many of the letters when dropped were still bundled together, while those that were released successfully failed to carry far in the cold frosty December morning.

After the Crossbow sites in northern France were overrun, the Germans were determined to use flying bombs and rockets as a means of attack not only against Britain but also against the advancing Allies in Europe. When the air-launched V1 attacks on the United Kingdom were resumed on the night of 13 September, ground reports had already been received that flying-bomb launching sites were being constructed in Germany and the Netherlands. In mid-October, Medmenham was even asked to keep watch on certain parts of Denmark and photographic interpreters attached to the Second Tactical Air Force, now operational in Europe, were asked to watch for V1 and V2 activity.

The importance of Antwerp as a supply base for the Allied Expeditionary Force, and the potential for other Continental targets to emerge, made it imperative to extend the remit of Operation Crossbow. The threat of Continental Crossbow led the Allied Expeditionary Force to create V Section at its Supreme Headquarters and after summoning Douglas Kendall and Lieutenant Colonel William O'Conner – Kendall's American counterpart in the Technical Control Office – three interpreters from the Medmenham Crossbow team were sent to France, Belgium and Holland to train the first- and second-phase interpreters attached to the tactical photographic-reconnaissance squadrons. Three more were sent to Supreme Headquarters and briefing materials were sent to the various training units. This included the School of Military Intelligence at Matlock, Derbyshire, who continued training the Army interpreters who were now operational in the field.

However important the discovery of Continental Crossbow sites was to the Allied Expeditionary Force, the interpreters who were operational in the field struggled to find them. Although V Section might have supplied them with information about the suspected firing areas, tactical photographic-reconnaissance squadrons had to cover such vast geographical areas and were assigned so many more urgent tasks that even just getting the areas photographed proved a challenge. This required the Crossbow team at Medmenham to keep a complete copy of all good photography covering the suspected firing areas taken by the British and American photographic-reconnaissance squadrons based in England. The challenge for the interpreters was as great as ever, for despite the large number of flying bomb attacks on Belgium only a relatively small number of firing sites directed against that country were ever built, particularly when compared with the large number of Bois Carré and Belhamelin sites built in northern France that were targeted on Britain. In Germany, sixteen were discovered in the south of the Rhineland, while in the Netherlands ultimately fourteen were found in the east of the country near Almelo and Deventer, and six in the west near Rotterdam and The Hague. These discoveries were collectively made by the interpreters attached to the tactical air forces and newly established V Section and by those in the Medmenham Crossbow team.

While Antwerp was the key target and Liege the secondary one, reports of a renewed attack on London in mid-February 1945, a month after the last of the air-launched flying-bomb attacks, created a flurry of activity at Medmenham. For a short time the hunt for launch sites intensified and Crossbow photographic reconnaissance and interpretation became a top priority again. The difficulty was that the Germans had learnt their lessons well from the bombing of the Bois Carré and Belhamelin sites and those built in Germany and the Netherlands were becoming progressively more difficult to spot. The vast majority of sites were again hidden in woods, but special care was now taken to ensure that the piston and cradle which carried the flying bomb along the ramp fell either on rough scrubland or into another wood on the target side of the ramp. This ensured that the telltale signature marks, which were obvious on the cultivated French farmland, were harder to detect. The sites near Rotterdam and Delft were built on entirely new principles.

Two were hidden in the enormous oil refinery at Pernis, surrounded by the port of Rotterdam, with the result that the long slender ramps and shadows merged into the background of tanks, overhead pipelines and chimneys. The whole effect was marred, however, by placing the two non-magnetic Square Buildings, which had to remain isolated, out of alignment with their surroundings. Another of the sites was put in the grounds of a sugar factory at Puttershoek, where the ramp was placed beside a large factory building. Here again the Square Building was all too evident, while the pistons and cradles fell on open ploughed land leaving telltale scars that were made still more obvious by the wheel marks created by the vehicles sent out to retrieve them.

Of the ramps that fired projectiles onto London, one was cleverly hidden between two buildings in the grounds of the Lever Brothers soap factory to the west of Rotterdam, but easily the most crafty was hidden on the premises of a glue factory to the south of Delft. But although the ramp here was constructed between two overhead pipelines orientated on London, and the Square Building was neatly

t Vignacourt, in the Somme *département*, a Belhamelin type site is hidden in the village. Whilst the
unch ramp can be seen in the trees, top right, a servicing building has been craftily hidden under
n existing barn.

Above: The modification of Heinkel He 111
medium bombers, to accommodate a flying-
bomb under their port wing, dramatically
increased the effective range of the weapon, as
people in Lancashire – hitherto immune from the
attacks – experienced on Christmas Eve 1944.

Left: WAAF Flight Officer Nora Littlejohn, a
leading member of the Crossbow team, points
at launch rails hidden within an orchard in
Vignacourt.

Soon after Britain came under attack from flying-bombs in June 1944, one landed intact and was immediately sent to Peter Endsleigh Castle. A technical artist with the Air Ministry, his cutaway drawing of the Fieseler Fi 103 flying-bomb would be methodically studied by those planning countermeasures against this revolutionary new weapon.

The silhouettes of Supermarine Spitfires manoeuvring alongside flying-bombs in an effort to 'wing tip' the projectile off-course. The bravery of fighter pilots during Diver missions would go down in aviation folklore.

Above: The cruciform silhouette of a flying bomb, juxtaposed with the dome of Methodist Central Hall and the spire of St. Martin-in-the-Fields, during its silent descent onto the Victoria district of central London. The 'deafening silence' – before the inevitable explosion – had dire consequences for the morale of Londoners after the morale-boosting Normandy landings.

Right: A low altitude oblique photograph showing one of the concrete-reinforced main entrances to the St. Leu D'Esserent mushroom caves, used by the Germans to store flying-bombs.

Duncan Sandys M.P. announces to a packed press conference on 7 September 1944 'that except possibly for a few last shots, the Battle of London is over'. Sitting, left to right, are Air Vice-Marshal William Gell (Balloon Command), Lieutenant-General Sir Frederick Pile (Anti-Aircraft Command) and Brendan Bracken, the Minister of Information.

The experimental site at Blizna, photographed during sortie 60PR/385 on 5 May 1944 by 60 Squadron, South African Air Force. Looking carefully at this frame – and with the knowledge that both flying bombs and V2 rockets were being tested at the site in Poland – Dr R. V. Jones identified a rocket ahead of the interpreters.

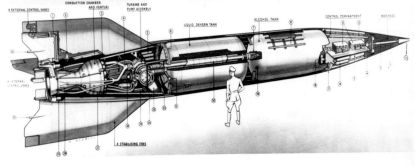

1 CHAIN DRIVE TO EXTERNAL CONTROL VALVE
2 ELECTRIC MOTOR
3 BURNER CUPS
4 ALCOHOL SUPPLY FROM PUMP
5 AIR BOTTLES
6 REAR JOINT RING AND STRONG POINT FOR TRANSPORT
7 SERVO-OPERATED ALCOHOL OUTLET VALVE

11 NOSE PROBABLY FITTED WITH NOSE SWITCH, OR
   OTHER DEVICE FOR OPERATING WARHEAD FUZE
12 CONDUIT CARRYING WIRES TO NOSE OF WARHEAD
13 CENTRAL EXPLODER TUBE
14 ELECTRIC FUZE FOR WARHEAD
15 PLYWOOD FRAME
16 NITROGEN BOTTLES

21 OXYGEN FILLING POINT
22 CONCERTINA CONNECTIONS
23 HYDROGEN PEROXIDE TANK
24 TUBULAR FRAME HOLDING TURBINE AND PUMP
   ASSEMBLY
25 PERMANGANATE TANK (GAS GENERATOR UNIT BEHIND
   THIS TANK)

United States Army cutaway drawing of the V2 long-range rocket, showing engine fuel cells, guidance units and warhead.

STORAGE & LAUNCHING OF A.4. ROCKET PROJECTILE.
(BASED ON AVAILABLE INFORMATION)

Left: As the Allies advanced through Normandy a launch site near the Château du Molay, to the west of Bayeux, was discovered through POW interrogation and soon received the attention of R. V. Jones and the Scientific Intelligence Unit.

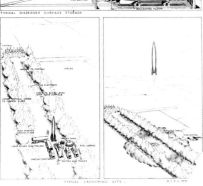

Above: Low altitude oblique photograph of the bunker at Wizernes - built according to the ingenious Verbunkerung design principle - with its immense concrete dome.

METHOD OF CONSTRUCTION OF "VERBUNKERUNG"

BASED ON SITES OF THE SIRACOURT TYPE.

METHOD OF CONSTRUCTION OF "THE VERBUNKERUNG."

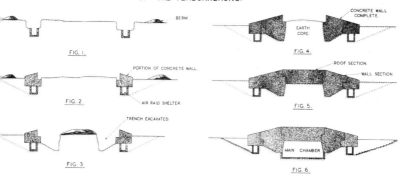

Geoffrey Stone served with the Army
Photographic Interpretation Section (APIS) Unit
attached to HQ 11th Armoured Division in the
European Campaign.

Serving with an APIS field unit typically involve
living in a tent alongside the office truck.
Geoffrey Stone would remember Captain Derek
van den Bogaerde as something of a dilettante,
who would charm his way through any situatio
As Dirk Bogarde the film actor, he would achie
matinée idol status in the 1950s.

The Crossbow Team: In most operations during the war photographic reconnaissance played a greate
or lesser part, but in no other major operation was that part so complete or all embracing as in the
battle against Hitlers's secret weapons.

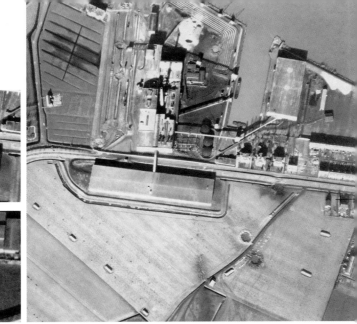

A V1 site hidden amongst the sugar mill at Puttershoek in South Holland. Whilst the ramp was cleverly tucked in beside industrial buildings, once operational the tell-tale 'signatures' of ground-scarring and roof damage soon revealed its existence.

Left: A V2 long-range rocket before take-off at the British testing area near Cuxhaven, on the shore of the North Sea, 2 October 1945. The British-led Operation Backfire was part of the scramble to acquire and understand German rocket technology.

Above: Completed Fieseler Fi 103 flying bombs at the Mittelwerk, following its capture by Allied troops. By the time the American Army captured Nordhausen, more than 20,000 slave labourers from the neighbouring Mittelbau-Dora concentration camp had been worked to death in the vast underground factory.

Found amongst the American-captured records was GX photography of Britain. A Heinkel He 111 bomber flying over east London on 7 September 1940, with the West India Docks and the Isle of Dogs below.

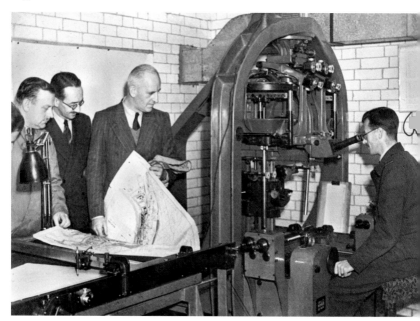

To celebrate the triumphs of photographic intelligence during the war, after VJ Day journalists were invited to visit Medmenham. In this photograph Flight Lieutenant Harry Williams provides a demonstration of the Wild Autograph A5.

placed among existing structures, when the site became operational, damage was soon caused to the roofs of sheds across the line of fire, and there were the piston marks on the ground as well. The accidental detonation of a flying bomb on the ramp completed the history of the site.

The Germans had certainly progressed a long way in the techniques of site concealment since the days of early 1944 but, even if sites were not discovered before launching began, the defects of the firing system soon revealed their presence. The V1 firings finally ceased on 28 March 1945, four days after the Allies crossed the river Rhine in force, but investigations into the flying bomb continued until the surrender of German forces in The Hague area.

When the first V2s were fired against Britain, on 8 September 1944, the knowledge about the rocket from photographic intelligence had advanced little from the information contained in sortie N/860 flown over Peenemunde on 23 June 1943. This showed rockets in vertical and horizontal positions and the Meillerwagen used for transporting them. On 7 August 1944 Medmenham was made aware that the Meillerwagen was also associated with the actual firing, but that a concrete platform was still necessary for launch. Knowledge about how V2 rockets were launched had accumulated slowly, particularly after the Normandy invasion when the Allies began overrunning known Crossbow sites in the Pas-de-Calais and on the Cherbourg peninsula. After this time a flood of information was supplied by the Combined Services Detailed Interrogation Centre – the interrogators of detainees, defectors and prisoners of war who were known or suspected of having worked for Nazi Germany – who reported to the War Office. In one of their reports, dated 20 June 1944, about the interrogation of a German POW who claimed to have strong anti-Nazi sentiments, information was volunteered in the hope that it might shorten the war and bring about the destruction of the Nazi regime. The data supplied by the POW, who had been an architecture student before being called up, was considered reliable and was duly circulated around the War Office, Admiralty and Air Staff.

In December 1943 the POW had been transferred to the Sonderstab Berger – Special Staff BERGER – a unit under the control of Lieutenant Colonel Berger who was responsible for the selection, surveying and construction of sites for rocket installations and supply dumps for rocket projectiles. Around this time the construction of sites had begun in the Operation Overlord area and was being rushed ahead so that the installations would be ready by June or July 1944. While the raids on the Heavy Sites were ongoing, secrecy around these new rocket sites was high, with admission only possible by means of a pass, and security reportedly so tight that arrangements for the safe keeping of plans and documents were routinely checked by the *Geheime Feldpolizei*, the secret military police of the Wehrmacht. When the POW developed a friendship with a local French girl – and was spotted with her in a local café – he was accosted by the *Feldpolizei* and ordered to break off the relationship. The POW was of the clear opinion that these sites in the Overlord area that were constructed in great secrecy by Wehrmacht personnel, with security in the hands of the *Geheime Feldpolizei* and probably also the Gestapo, were from where the attack on the United Kingdom would have been launched.

Information was supplied by the POW about the network of rocket installations, each of which contained three concrete firing platforms, one behind the other at intervals of 40–50 metres. As an aspiring architect the POW could recall constructional details from memory and supplied drawings of firing platforms that were always built along a stretch of road. Construction involved removing the surface and excavating to a variety of depths – possibly an austerity measure to limit the amount of building materials used – before concrete was poured and then specially toned to match the colouring of the road. When finished, the firing platform – known as an *Abschussplatte* – was completely level with the original surface of the road and scarcely distinguishable from it. Since December 1943, work on the supply dumps for rocket projectiles had also, allegedly, been ongoing and included a quarry that had been requisitioned by the Wehrmacht and into which twenty to thirty tunnels had been bored and a light railway installed. The plan was for the projectiles to be wheeled out into an assembly shed where the ballistic cap would be attached and propellant loaded. Then, by means of a crane, the complete projectile would be hoisted onto a Meillerwagen and towed to the firing site. To launch the projectile it would be placed on metal tripods, with the main force of the engine's thrust taken by the thickest and most heavily reinforced part of the concrete platform. Tests had apparently even been carried out with a 25-metre-long tree trunk to ensure that the rocket could navigate local roads.

In June 1944 British Intelligence faced the dilemma that, if the statement from the POW was true, then the Heavy Sites were clearly not necessary for the launch of the rockets. Even if the Heavy Sites were overrun the V2 would still be a massive threat not only to British civilians but also to Allied forces on the ground. These developments were followed in July by the dramatic meetings of the Joint Intelligence Committee at which R. V. Jones was lambasted by Duncan Sandys, with support from Herbert Morrison, for not sharing Air Intelligence data about the rocket. Operation Crossbow reached another crisis on 30 July when the Air Ministry Director of Intelligence responsible for research – RAF Group Captain Jack Easton – was also made responsible for Crossbow Air Intelligence. This was a direct snub to R. V. Jones that resulted promptly in his resignation letter, which he quickly thought better of on reflection, given how much this problem was now a race against time. The timely purchase – by trading – of V2 wreckage from the Swedish government (and courtesy of the Polish underground) gave Scientific Intelligence vital information, although the interpreters faced the common Operation Crossbow challenge of not being completely in the intelligence loop.

Although suspicion persisted, the theory that a concrete platform was necessary for firing would only finally be overturned when the first known launching positions were overrun on Walcheren Island in November 1944. This revealed that any reasonably level piece of ground could be used, that all the launch equipment was mobile, and that no blast marks of any description were visible. It became clear that even in open country no launching site could be confirmed from photographic evidence alone, unless the rocket was actually photographed during the fuelling process or in the vertical position before taking off. In wooded areas even this chance was small. This meant that even though it might be convenient for the Germans to use the same site regularly they could move quickly to another flat section of ground if they felt they had been observed. This all meant that the only positive detection of firing points in the early stages of preparation, in October and November 1944, were at Rijs in the Netherlands where two fresh clearings were spotted in an area where radar plotting had revealed suspicious activity and at Metelen,

Germany, where a pilot had seen a rocket emerge. In both these cases, however, the photographic evidence was of no value and the sites would certainly never have been found without some other evidence to direct attention to the spot, in these cases verification from radar plotting and visual observation.

It was not until 29 December 1944 that rockets were seen on photographs in the operational area. By this stage the enemy had been driven out of most territories within range of Britain and that had good enough communications, by road and rail, to permit supply. The Germans were compelled to use the Haags Bos, a wooded park in the centre of The Hague, for forward rocket storage. On that particular day at least thirteen V2 rockets could be seen on their Meillerwagens, dispersed under the trees. Unfortunately for the Germans it was winter and the leaves had now fallen so the striped appearance of the projectiles with their dazzle-painting camouflage only succeeded in drawing attention to them. The Haags Bos also became the main firing area for the V2 rockets and, in an attempt to make the area reasonably untenable, a request was made to Fighter Command who had hitherto only been bombing targets on the evidence of radar and ground intelligence. Attacking Spitfires were frequently accompanied by Mustangs of an Army Liaison Squadron for photographic-reconnaissance purposes in order that evidence about the launch sites could be compiled at Medmenham to assist countermeasures. One of the pilots from No. 602 Squadron involved in Operation Big Ben, who dive-bomb attacked V2 launch sites in Spitfires, was Flight Lieutenant Raymond Baxter. A man of action who at the age of fourteen had flown with Alan Cobham's Flying Circus, he would later become one of television's best-known outside broadcasters and the first presenter of the popular BBC science programme *Tomorrow's World*. Baxter later recalled the pilots' reliance on aerial photography to locate V2 launch sites, and the difficulties they had throughout the operation because of the bad winter weather. Ideally a stream of four aeroplanes, flying very close together, would drop in turn on the target. Starting at around 8,000 feet, they would dive at between 70 to 75 degrees, dropping the bombs at 3,000 feet before pulling out.

In late February the Germans temporarily abandoned the Haags Bos, which caused a few days' relief from the rockets for their targets, only for the attacks to resume from elsewhere. After all their grand attempts at subterranean bombproof storage in France, a lack of time and the high water table in the Netherlands meant underground storage was impracticable and the Germans were forced to keep their rockets in the open. The standard method adopted was to disperse the rockets on Meillerwagens at various points throughout woods. Once these points were discovered they were attacked by Fighter Command and only a small part of the bomber force, given that the bombers' main responsibility at this stage of the war was to support the Army. Since it was not possible to divert large numbers of bombers, it had to be accepted for the time being that some rockets would still make it through. The first operational rocket-launch site positively identified was found on aerial photographs taken during sortie 106G/4538 on 26 February 1945. It was situated on a rough roadway running through a sparse wood south-east of the racecourse at Duindigt in the north-eastern suburbs of The Hague. The area had been photographed as part of the effort to discover any new V1 launch sites in the west of the country, only for a rocket to be spotted in the vertical launch position and actually in the process of being fuelled.

Interpretation Report BS1039 was written by interpreter George Reynolds, who had found the Duindigt site five minutes before going off duty on 27 March 1945. The Yorkshireman later recalled phoning Douglas Kendall at home to let him know about the discovery and to inform him that a stereo pair (frames 3279 and 3280) had recorded the preparations for launch perfectly in 3D. Copies of the photographs were swiftly taken to W Section who drew a sketch plan which recorded: a rocket in a vertical position, the launch platform, Meillerwagen, a probable tanker and fuelling vehicles, and a tractor. While working in B2 Section, Reynolds would recall how frequently the team were relocated around RAF Medmenham, depending on the relative priorities of tasks at any one stage, and how for a time he shared a little room with WAAF Officer Pamela Bulmer in one of the towers at Danesfield House. On the day when the first V2 rockets landed on Britain, he remembered the section being in a ground-floor room at the west end of the house – and how Douglas Kendall burst into the room and said, 'It's happened!'

The relative positions of the attendant vehicles in Reynolds's report matched very closely plans that had been captured after the Normandy landings and the concreted areas found at Château du Molay, west of Bayeux, Normandy, which had been inspected by the Scientific Intelligence Unit. While an unloaded Meillerwagen and a large amount of wheeled-track activity in the area indicated it had been used for launching purposes previously, very large craters in the surrounding square mile testified eloquently to the danger of aborted launchings. The search for new storage areas showed that a small area at Ravelijn, a short distance north of the racecourse at Duindigt, contained three rockets on Meillerwagens – seen to be dispersed under tree cover close to a house known to be used for hospital purposes. The bombing of Duindigt and Ravelijn by Fighter Command, on information received from Medmenham, quickly prompted the Germans to abandon the sites and once again the rate of fire against London eased. The unremitting search of aerial photographs would continue until the attacks stopped, not only those against the United Kingdom but also the parallel attacks that were being made with far greater intensity against Antwerp.

With the SS hiding in the wooded Haagse Bos in the centre of The Hague, the challenge that Fighter Command faced in pinpointing their attacks resulted in an alternative bombing strategy being adopted that would result in tragedy. For the people living in the adjacent neighbourhood of the Bezuidenhout, Saturday, 3 March 1945 was another day of starvation and cold during the Dutch *Hongerwinter*. Since the Normandy landings, conditions had grown steadily worse and although the Allies had liberated many in the south of the country their liberation efforts had ground to a halt with the failure of Operation Market Garden.

When a national railway strike followed, encouraged by the Dutch government-in-exile who hoped that it would help the Allied campaign, the retaliation from the German administration was an embargo on food transports to the western Netherlands. Although partially lifted in early November 1944, by then the unusually harsh winter had set in and canals were frozen solid and impassable. Living in the main western battlefield in which a retreating German army was hell-bent on destroying locks to flood agricultural land and crippling communications systems to cause maximum disruption, the Dutch people suffered. For many in The Hague this meant spending their days on foraging expeditions into the countryside, looking for wood to keep themselves warm and for food to keep themselves alive.

The Dutch famine would ultimately result in the British mounting Operation Manna and the Americans Operation Chowhound in early May 1945, when thousands of tons of food was dropped onto still-unliberated parts of the western Netherlands.

On that Saturday morning in early March, by when some Dutch people had resorted to eating flower bulbs for nourishment, the RAF took part in their first and only raid on the Crossbow sites in the Netherlands, using the bombers of the Second Tactical Air Force. They had played a key role in the counteroffensive against Crossbow targets in France after developing their effectiveness at bombing small targets in occupied territory. What the squadrons called Ramrod operations – bombing operations designed primarily to destroy a specific target – had seen No. 2 Group RAF fly 4,710 sorties against mostly flying bomb sites from November 1943 until May 1944. But the sites' camouflaged nature had meant that radio navigation would prove vital to locating and destroying so many of them.

The Second Tactical Air Force had already visited The Hague in spectacular fashion on 11 April 1944 when Mosquitos of 613 Squadron attacked the Gestapo headquarters at low level. The Dutch underground had been growing bigger and bolder every day and so, unsurprisingly, the number of stolen or forged identity cards, food cards, work certificates and passes of all kinds reached a level prompting the Nazis to recall them. Fearful that this process would help identify members of the Dutch resistance, 500-pound delayed-action high-explosive bombs and incendiaries were hurled against the five-storey building where the checking was taking place. The particular challenge for the RAF pilots was its proximity to the Peace Palace, with its large tower that dominated the city skyline. Its occupants were the members of the Permanent Court of Arbitration that had been established during the first Hague Peace Conference. Success was dramatically recorded in the strike photographs, confirmed by the post-strike photographic reconnaissance which showed a smouldering wreck, and by the fact that no Dutch civilians outside the building were reportedly killed and there was only one smashed window at the Peace Palace.

On the morning of Saturday, 3 March 1945 three squadrons of Mitchell medium bombers and No. 342 Squadron in their Boston light bombers were on another Ramrod mission, this time against rocket storage. The bombers had been ordered to use Gee-H radio navigation, only for them to change to visual target-identity recognition over an area that was obscured by cloud cover from their altitude of 10,000 feet. Spitfires from 603 Squadron were also over the city at the time and knew that clouds were sitting between 3,000 and 8,000 feet and that the bombs were being dropped onto a residential area. With the crystals in their radios set to another channel they could do nothing as the bombs dropped onto Bezuidenhout.

Soon after, at 09:45, Flying Officer Maksymowicz of No. 303 Polish Fighter Squadron led an attack on the Haagse Bos by four Spitfires and witnessed black smoke rising to 8,000 feet from buildings south-east of the location and as far as Rotterdam. As had been the case throughout the war, combat films were made by fighter pilots during the Big Ben operations, including Raymond Baxter and Maksymowicz. Unlike the precision raids by the fighter pilots, who made every effort to avoid civilian casualties, the bomber attack devastated the entire neighbourhood, with 535 Dutch civilians killed and over 3,200 homes and other buildings destroyed.

On hearing the news Winston Churchill wrote a personal minute, whose circulation included the Secretary of State for Air and the Chief of the Air Staff, condemning the feeble efforts to neutralise the rockets and the extraordinarily bad aiming that had led to so much slaughter. The Prime Minister demanded reports and explanations about why, when there had been so many accounts of pinpoint bombing of Gestapo targets in the Netherlands, and Air Intelligence had such good information about where the rockets were stored, the railway lines supplying the Haagse Bos were not instead being attacked with precision and regularity. In his view, all that this raid had achieved was to scatter bombs around this unfortunate city without the slightest effect on the V2s but with much impact on innocent human lives and the sentiments of a friendly people. Not surprisingly, the reaction of the Netherlands government-in-exile was a formal protest for the manner in which the attack had been planned and executed, along with an appeal that no repetition would take place.

Although no further bomber attacks were planned, Fighter Command continued their campaign while official investigations into the incident began. Charles Portal, Chief of the Air Staff, made his feelings clear to Archibald Sinclair, Secretary of State for Air, that despite 'one gross error' the campaign should continue. By 18 March the investigations had revealed that an Intelligence Officer in the Second Tactical Air Force had incorrectly transposed one set of coordinates and placed the aiming point in the centre of the Bezuidenhout rather than on the Haagse Bos. Added to this, an entry in the Second Tactical Air Force Duty Operations Officer's log mistakenly recorded – based on information received from the Air Ministry – that houses within 500 yards of the target area had been cleared of inhabitants. While this would add to the potential for disaster, by this stage the problems associated with attacking Big Ben sites were overtaken by events. The advancing Allied armies would soon put the weapons beyond range of the United Kingdom.

From sites east of Bonn, Germany, and Hellendoorn and Enschede in the eastern Netherlands, projectiles were spotted on Meillerwagens in transit through woods on two occasions. In both cases camouflaged railway stock was visible at a neighbouring railhead that was most probably used for offloading directly onto Meillerwagens for transport to launching sites. The study of communications systems would prove an effective means of tracking down where weapons manufacturing occurred. This involved Major Moody and his band of rail, road and waterway specialists in F Section studying the communications systems that fed back from the firing areas into Germany itself. While this approach would prove successful with the transportation of the V2 rocket it was never applied to the flying bomb, since the Allies had been able to control its forward storage through bombing attacks and, unlike the rocket, the V1 was considered comparatively simple to manufacture and would therefore have been constructed by hundreds of small dispersed factories. After flying bombs began dropping on the United Kingdom, a few of them without detonating, this understanding was confirmed. But although the Allies were not in a good position to identify and attack the network of factories involved in the manufacture of V1 parts, they were very successful in identifying where the V2 rockets were manufactured, and how and where they were transported.

With aerial photography confirming the assumption that rockets would have to be delivered to forward storage sites by railway because of their great size and weight, all railway lines feeding back from the firing areas in the Netherlands were closely studied. F Section had evolved techniques for studying all the different marshalling yards, loading points and sidings in order to determine what traffic was moving – both military and civilian – and by studying such factors as traffic congestion could effectively determine the effects that bombing attacks were having. From studies of Peenemunde and later Blizna, the railway vehicles used to carry the liquid oxygen used in the V2 had been identified.

Between 1 August 1944 and 2 April 1945, a flat car around 60 feet long and loaded with a tank around 27 feet long was seen on seventeen different occasions in Germany and the Netherlands. On all occasions when the items were spotted in the Netherlands they were tracked back to Niedersachswerfen in central Germany. At around the same time a special unit was discovered that had been designed for transportation of the rocket and which was spotted at twenty-nine different places between Niedersachswerfen and the Netherlands. This led to the conclusion that an important part of the rocket-manufacturing process had been discovered. A number of industrial plants were investigated in Germany – based on information received from ground-intelligence sources and prisoner-of-war interrogations – where there was suspicion that complete flying bombs, rockets or components for them were being manufactured.

Medmenham issued their first report on the factory at Niedersachswerfen on 31 August 1944 after a request was made to locate and report on a factory suspected of secret-weapon activity. A ground-intelligence source reported that it was in an old limestone quarry in the area and that German political prisoners and foreigners were working on the production of rockets or flying bombs. Aerial photography of the area was taken in September and October 1944 and revealed two large underground complexes for storage and manufacture.

At Medmenham a calculation was made of the floor area likely to be used by considering the shape of the hill in relation to the entrances and by studying the geology, since tunnelling normally takes place on a single geological level. The amount of electrical power being fed into the factory, which could be worked out from the size of the transformers, was a further indicator that this was an immense underground factory. The sole feature which gave credence to the assertion that it was connected with the manufacture of long-range rockets turned out to be the presence of unusual rolling stock. While it was known that the rocket could easily have been transported in sections on normal railway wagons and constructed before launch, it was considered more likely that it would have been transported when fully assembled. Since the overall length would have been 45 feet, 10 inches long, it was understandably felt that such an object would have been particularly obvious and easy to spot on aerial photography, particularly given how many rockets would be travelling by rail. Equally, since the load would be relatively fragile, purpose-built or modified wagons would be required and could be spotted by the rail experts.

Outside the tunnels at Niedersachswerfen an unusual configuration of wagons was spotted on the sidings, where three wagons had been linked together to effectively create an oversize one, the like of which had never been seen before. Although the wagons were fully sheeted to prevent their loads being seen, a theory was developed, based on the peculiar shadows thrown by the load, that two projectiles were being transported per unit of three wagons with their noses overlapping on the centre wagon. Confirmation of this was quickly forthcoming from a prisoner-of-war interrogation and this type of triple-flat wagon was added to the photographic-recognition material.

On the supply side Medmenham carried out an intensive search for the chemical plants supplying the liquid oxygen. Rather than the Germans building new ones, it was discovered that thirteen existing chemical plants in Germany and Belgium were responsible, all of which were placed on the target list. Because a considerable amount of hydrogen peroxide was required for driving the fuel pumps on the rocket, and since there would have been limited requirement for the chemical before the war, a special investigation was carried out. Sources of production were again located and target information compiled, but once more other bombing priorities ensured that these too would be spared.

In the short time available, all the main manufacturing points, the likely sources of fuel and the communication systems used to move the weapons from the manufacturing area to the front, had been identified. This provided the Allies with three methods of retaliation in addition to attacking the operational areas. Of these, communications were arguably the hardest to disrupt in the long term, although they could be temporarily put out of action through the bombing of marshalling yards and bridges. The Germans had become remarkably quick at repairing bombed bridges and getting railway lines back into operation and often the most that could be expected from such disruption was a respite of a day or two.

The destruction of industrial plant had been proved to be more worthwhile because it could stop all production for several months at a time. At this relatively late stage in the war, however, the bombing effort was being directed towards the enemy's synthetic oil plants and refineries and so, although bombing the liquid oxygen and hydrogen peroxide plants remained a potential form of retaliation, the higher priority of other targets ensured that they were never attacked. In the closing months of the war, the Germans' daily output of fuel was infinitesimal to the point where Panzer units were brought to a halt and the Luftwaffe's fighter aircraft could not take off without special permission unless Allied bombers appeared in force. For this reason single aircraft, such as the photographic-reconnaissance aircraft, were left more or less alone.

As the German rocket campaign continued, the sightings of liquid oxygen and triple-flat units began to form a pattern which indicated Niedersachswerfen as the central if not the only assembly plant for the long-range rocket. Almost all the recorded pinpoints of V2 railway traffic proved to be along the two main railway systems from the underground factory, which supplied the northern and southern groups of rocket launching sites. To prevent their discovery by photographic reconnaissance in the Netherlands, the Germans adhered rigidly to the practice of 'stabling' the rockets in covered rail yards during the day and only moving units at night.

The mounting instability of the enemy position in March 1945 caused a general relaxation and in quick succession rockets covered by camouflage netting were spotted in railway stations at Leiden, The Hague, Rotterdam and Amsterdam. During March the advance of 21 Army Group meant that time was running out fast for the rocket troops who would fire the last rocket at the United Kingdom – on 27 March 1945 – before retreating. On that spring afternoon twenty-three people from Kynaston Road and Court Road, in the suburban town of Orpington on the south-eastern edge of London's urban sprawl, were seriously injured and one woman was killed. Thirty-four-year-old Mrs Ivy Millichamp was in her kitchen when the rocket fell and, catching the full force of the blast, was the last person in Britain to be killed by enemy action.

Although the V2 rocket offensive might have been over this was not to prove the case for the flying bomb. The third and final phase of the V1 offensive began on 3 March 1945 when doodlebugs were fired from ramps at Ypenburg, Vlaardingen and a site near Delft. Though the first two sites would be discovered by Medmenham and were attacked by Fighter Command, the Delft site was never discovered and was responsible for the last flying bombs targeted against the United Kingdom. In the early hours of 29 March 1945 a V1 exploded in open countryside near the village of Iwade, close to the town of Sittingbourne in Kent, and would be the last doodlebug to land on the United Kingdom. Another sign of the rocket campaign would come the following month, on 10 April 1945, when liquid oxygen wagons were photographed alongside uncamouflaged and unsheeted triple-flat wagons on railway sidings at the German port of Wilhelmshaven. The photographs showed rockets assembled and loaded in position as had been deduced from the first examples seen at Niedersachswerfen.

In early November 1944 the Medmenham Technical Control Office took the decision to create a new team in the Army Section, known as B6, that was responsible for the study of underground factories, in particular those connected with the manufacture of secret weapons. Although the Allied Central Interpretation Unit was sufficiently large and important to have an RAF Group Captain in overall command, the success of Allied Photographic Intelligence in the Second World War was very much the result of its operational direction. At Medmenham this was primarily the responsibility of Wing Commander Douglas Kendall, working alongside his American counterpart Lieutenant Colonel William O'Conner, with representatives of the armies and navies of the various Allied forces who best ensured that conflicting priorities were minimised. A man then only in his early-thirties, Kendall had pioneered the development of photographic interpretation since Wembley, and had fought to ensure Medmenham remained an inter-Allied intelligence unit, much against the opposition of Elliot Roosevelt at one stage. Kendall would not only play a critical role in Operation Crossbow but as the only Ultra handler at Medmenham he was in the unique position of understanding the bigger picture. Although the relationship between Kendall and MI6, most notably

R. V. Jones in their Scientific Intelligence branch, was frequently tested to breaking point, Kendall was able to play a guiding role throughout the Second World War.

In the middle of 1943 a decline in the quality of the intelligence on German fighter production was recorded. What made assessment of German air strength difficult was the incomplete nature of the intelligence as well as the conflicting interpretations made of it. The most useful source of information at the time was undoubtedly photographic intelligence, but even when the interpreters' evaluations of the actual impact on a target's productive capability and the structural damage inflicted on it were correct, there remained a tendency to overestimate the level of damage. Ultra could play only a limited role, since German aircraft-production information was rarely passed through the military channels from which Bletchley Park garnered most of its intercepts. For Colonel Richard Hughes, Chief Targeting Officer for the US Eighth Air Force, the problems were compounded by the fact that British Air Ministry Intelligence could not agree among themselves how to interpret the data that came to them. Hardly the archetypal American officer, Hughes had been born a British subject, in Salt Lake City, but had been raised in England and after schooling at Wellington School, Somerset, attended Sandhurst and served as a British Army officer. After service in the First World War, he served with the Indian Army before meeting an attractive woman on a world cruise, marrying her and moving to St Louis, Missouri. After losing his life savings investing in aircraft, the naturalised American citizen worked for Air Force Intelligence in Washington D.C. where, at the beginning of the Second World War, target intelligence simply did not exist.

After Pearl Harbor, when they had been so spectacularly caught off guard, the Americans spent an enormous amount of money developing a bombing force. They knew that a large air force was a prerequisite, but had thought little about how it would be deployed against targets throughout the world. This meant that in the short term they were dependent on RAF Target Folders, but were soon prompted to question whether the target information was good enough and whether they should adopt a different approach. Central to much of this early thinking was Lieutenant Colonel Tommy Hitchcock, an assistant Air Attaché at the US Embassy in London. The seeds were often sown on Sunday afternoons when General Carl Andrew 'Tooey'

Spaatz – Commander of the Eighth Air Force – would entertain at his house in the affluent London suburb of Wimbledon.

Hitchcock was a millionaire polo player from Long Island who joined the Lafayette Flying Corps – the American volunteer pilots who flew for the French during the First World War – and was shot down and captured by the Germans, only to escape by jumping from a moving train and walking to freedom in Switzerland. After the war he returned to study at Harvard, married Margaret Mellon, daughter of the Gulf Oil founder, and as the friend of fellow polo player Robert Lehman became a partner in the Lehman Brothers Wall Street investment firm. While in London he was instrumental in championing the development of the P-51 Mustang long-range single-seat fighter – which would accompany the Eighth Air Force bombers on their daylight raids over Europe – as fighter bombers and tactical reconnaissance aircraft. When commander of the first P-51 long-range fighter group he would lose his life while test flying near Salisbury, Wiltshire.

To galvanise the Americans' strategic thinking the Enemy Objectives Unit was established and financed by the secretive Office of Strategic Services – which in time would be superseded by the Central Intelligence Agency – and operated with relative impunity from the American Embassy in London. The differences between the British and Americans were to prove significant, with the American advocacy of precision over area bombing and the targeting of oil plants over railways a source of constant antagonism between the main protagonists on both sides. The tension was heightened further by the personalities involved, notably Professor Solly Zuckerman whose academic reputation had been built on the study of the social life of primates, and Colonel Hughes who was known to have the habit of never using a briefcase. The consequence was that he would walk around with papers bulging out of his inside coat pockets – altogether safer than a briefcase, in his view – which meant that tobacco would burst spectacularly onto meeting tables alongside the paperwork. The Casablanca Conference of January 1943 might have proven to be a moment of power shift with the Americans when Churchill accepted the American preference for precision bombing, but division was destined to continue about what the targets should be. In the view of Hughes, it was essential that the European air war should be fought in conjunction with the ground war, and any target system had to be seen in that light. In the Americans' view oil was so essential to the German armies in the field that it was unquestionably the optimum target system, with the 'wall around the orchard' being the aircraft industry.

B6 Sub-section was to prove important to Operation Crossbow. It was formed to study subterranean industrial complexes after an interpreter involved in locating underground flying bomb storage dumps, who worked in the Crossbow team, was given permission to study underground factories suspected of being involved in the manufacture of V weapons. Although the Industry Section had been interpreting surface factories involved in the manufacture of V weapons, nobody had been analysing the underground factories that were known to exist. The Aircraft Section, under Constance Babington Smith, had issued a report in September 1944 about suspected subsurface aircraft factories. When the Crossbow interpreters began cooperating with the Americans the result was a joint conference with Captain Walt Whitman Rostow at which a decision was taken to make a test case of the value of photographic interpretation in the targeting of underground industry.

A former Rhodes scholar at Balliol College, Oxford, Rostow was a lifelong anti-communist born in New York to fervently socialist Russian-Jewish immigrants. An economic historian and powerful Cold War adviser to Presidents Kennedy and Johnson, he would play a decisive role in decisions about American involvement in Vietnam. He volunteered for army service during the Second World War when he served with the Office of Strategic Services. The predecessor of the Central Intelligence Agency had by then developed a particular interest in the work of Medmenham, notably the Aircraft Section, which helped bolster the relationship with the American Enemy Objectives Unit.

The importance of obtaining up-to-date photographic intelligence, and of Medmenham and the PIs being kept in the intelligence loop, was impressed upon the Americans after an attack on the Focke-Wulf aircraft factory at Bremen on 17 April 1943. Although Medmenham had been keeping a close watch on the factory, they were not informed about either the raid or the fact that the Focke-Wulf Fw 190 fighter aircraft was newly off the production line. In the weeks that followed the raid which successfully destroyed the factory – at the cost of sixteen

American bombers – only minimal amounts of repair work were observed by the Aircraft Section at Medmenham. With this throwing doubt on whether the factory had been operational immediately before the costly attack, this helped focus the American consensus about bombing priorities and sent Rostow and Captain Charles Kindleberger to the Air Ministry to investigate.

Kindleberger was in London on a mission from the Office of Strategic Services to bring aircraft analysis to the notice of the Enemy Objectives Unit, something that would in time prove critical to Operation Crossbow. At Air Intelligence 2(a) they found a system of recording intelligence that they considered quite fantastically inadequate and inaccurate, consisting as it did of a card index on which references to reports from a variety of sources, with an invented code of references, along with garbled and inaccurately transcribed entries had been written. With the information equally bad for the Bremen raid, it was arranged for Rostow to join Air Intelligence where he rapidly started a new system while the intelligence officer responsible was on leave, only for him to show concern on his return when nobody consulted him and his card index any more.

On investigation it would transpire that at the time of the raid the factory was inactive and production had ceased a short while before. The influence of the Americans was to prove vital in ensuring that Medmenham was kept better informed and led to the Allies establishing the Jockey Committee in June 1943. The purpose of the committee was to monitor through intelligence analyses the German aircraft industry and Luftwaffe operations and to recommend appropriate targets. With this happening in advance of the Crossbow countermeasures it would improve the intelligence background of the operation and meant that with such powerful allies the Medmenham Aircraft Section operators were able to discover that certain main centres of aircraft development had never been photographed. This resulted in fifty factory airfields where prototypes were likely to appear being added to the photographic-reconnaissance programme.

But while intelligence which identified key targets for destruction began to rapidly accumulate, factors including the lack of long-range fighter support for daylight bombers flying deep into enemy territory, and bad weather, meant that it was not until February 1944 that a week of attacks would cause immense damage to the German aircraft industry. A consequence of the attacks was the subsequent dispersal of the factories, many of which were underground, so that the Germans' coveted jet aircraft could be developed and manufactured in safety. Operation ARGUMENT had been postponed repeatedly since November 1943 because of bad weather. Its objective was a series of aerial assaults on German fighter production, from ball-bearing manufacture to engine and airframe assembly, airfields and aircraft-storage areas. It was both an attempt to relieve pressure on the Combined Bomber Offensive and to cripple the Luftwaffe in advance of the Normandy landings. For the six days from 20 to 25 February Allied bombers attacked in what became known as Big Week, when the intensity of the attack pushed aircrews to the limit.

At the joint conference with Captain Walt Rostow in October 1944 the decision was made to make a test case of the value of photographic intelligence in the targeting of underground industry. An underground factory near the village of Langenstein, in modern-day Saxony-Anhalt, Germany, was selected for the test. While the model-makers prepared a detailed model at 1:2,000 scale, W Section created a plan with revised contours courtesy of the Wild Machine, and a Medmenham geologist studied the area on geological maps and aerial photographs before producing section drawings showing the thickness, disposition and quality of the rocks that were being tunnelled through. From these a vulnerability plan was drawn, which showed that much of the factory was protected by less than fifty feet of sandstone.

The Aircraft Section provided an analysis of an adjacent airfield, while local obstructions and anti-aircraft batteries were identified by the Cover Search interpreters. By the time the report was issued on 18 October 1944, forty underground factories had been discovered on aerial photography and ground-intelligence sources indicated the large-scale dispersal of German war industry. This justified the creation of the new sub-section, which specialised in the interpret-ation, cartography and geological understanding required to target underground factories.

By December 1944 the growth in number of underground factories meant that the section consisted of seven interpreters from the Army, RAF, WAAF and USAAF, and two cartographers from the Royal Engineers. The key questions that the section had to answer were:

- Is the project intended to house a factory?
- If so, estimate its size and importance, and if possible its function.
- What is its degree of readiness?
- Estimate its vulnerability to penetration bombing.
- Provide briefing material for low-level attack.

In order to give the best possible chance of answering the first three questions, it was found essential to hold copies of all available ground intelligence, plus an associated card index and situation map in the section. For the same reason it was often necessary for interpreters to be present at prisoner interrogations.

Human intelligence was supplied by a variety of sources during the war, including the Combined Services Detailed Interrogation Centre at Trent Park, Cockfosters in north London, where captured German generals had been secretly recorded discussing secret weapons. Throughout the war RAF Officer Denys Felkin ran a branch of the Air Ministry that ultimately became the Assistant Directorate ADI (K) – which saw him rise to the rank of Group Captain – and which specialised in the interrogation of German aircrews. Under Felkin's jurisdiction every legal method of obtaining information beyond name, rank and number was tried, including hidden microphones.

In a document date-stamped 7 November 1944 by the Central Registry at Medmenham there is the statement of a prisoner of war described by his interrogator as a 'somewhat muddy-minded Pole' who worked at Peenemunde from November 1943 until March 1944 and at the Mittelwerk, in Niedersachswerfen, from March 1944 until July 1944. Pages of technical information were supplied by the Pole about the testing procedures at Peenemunde before he was transferred to Niedersachswerfen in which he explained the physical characteristics of the tunnels, the range of different weapons being manufactured there, details about factories that supplied component parts, and the fact that most of the workers were concentration-camp prisoners. In an appendix to the document, which contains a sketch of the factory, the interrogator makes a *nota bene* scorning the Pole for being a 'very poor subject as far as photographic interpretation is concerned, being quite unobservant and completely incapable of visualising how those things which he saw on the ground appear in an aerial photograph'.

*

The largest underground factory monitored by Medmenham was constructed near the central German town of Nordhausen, on the southern edge of the Harz Mountains in modern-day Thuringia. At the suggestion of the industrial giant I.G. Farben in the 1930s, the Nazi Germany-owned company Wirtschaftliche Forschungsgesellschaft Gmbh – referred to as 'Wifo' – was established and began creating a centralised fuel and chemical depot deep inside the Kohnstein mountain just north-west of the town. Tunnelling began in 1936 and by July 1943 forty-five huge galleries for the depot had been completed, at which point the Führer declared it was now required for the secret-weapon programme.

The revenge weapons were now considered more urgent than all other armament programmes, and this was a decision only reinforced by a personal meeting between Generalmajor Dornberger and Dr Wernher von Braun who visited the Führer on 7 July 1943 at the *Wolfsschanze* – Wolf's Lair – complex. Built near the East Prussian town of Rastenburg for the 1941 German invasion of the Soviet Union – Operation Barbarossa – it was one of several heavily fortified Führer headquarters scattered throughout Europe. With all the industrial resources of the Reich now available for rocket production, however, the Reichsführer of the SS, Heinrich Himmler, persuaded Hitler that he, Himmler, should be responsible for security against any espionage and sabotage directed against the project. With the decision about Nordhausen quickly followed on 17/18 August by the Bomber Command attack on Peenemunde, Himmler capitalised on his opportunity to play a greater role in the rocket programme. His offer was to supply a workforce from concentration-camp inmates that could rapidly get the underground factory ready and begin production work. Since the prisoners had no contact with the outside world, secrecy was practically guaranteed.

Hitler approved, but within days stipulated that the original V2 factories should continue to produce rockets while the underground factory was being established: the target was for 900 rockets a month overground and 900 a month underground. In early September he changed his mind and ordered that all 1,800 missiles should be manufactured at Nordhausen.

In September 1943 the nondescriptly named Mittelwerk Gmbh

– Central Works Limited – was established to run the secret underground factory. That same month a contract was agreed with Wifo ensuring that their personnel, who had the technical knowledge and experience, would continue to oversee expansion of the underground facility. The difference was that the orders would now come from Mittelwerk Gmbh and the majority of the workforce would be concentration-camp inmates. Although an enormous logistical exercise, every effort was made to conceal the existence of the factory, with a large number of code names and false addresses. The company's mail was even routed through a post office box sixty miles away. With the whole area code-named Mittelraum, the term *Mittelbau* was given to construction projects and *Dora* to the main camp housing the slave labourers. Mittelbau-Dora and Nordhausen were destined to become synonymous with human wretchedness. They were also the centre of a complex of underground factories where the Nazis' best-known secret weapons – the V1 flying bomb, V2 rocket and jet engines for the Messerschmitt Me 262 and Arado Ar 234 – were built.

Although the number of operational Messerschmitt Me 262 aircraft was small, they were growing in number and proved a particular menace to the photographic-reconnaissance pilots who had been used to flying the fastest aircraft in the sky. This was particularly the case in southern Germany where there was an odds-on chance, if flying a PR Mosquito, that you would be attacked. This prompted the redeployment of the new Spitfire Mark XIX onto flights over the trouble spot, since it stood a better chance of coping and gave the crews more confidence. The prospect of the Me 262 problem spreading to north and north-east Germany, and its potential to impact on photographic intelligence, only highlighted the importance of the work being undertaken in the underground factory.

The inmates worked simultaneously on the tunnels and on the construction of their own camp, which was initially a sub-camp of Buchenwald. But in 1944 its scale saw it become an independent camp, and the last of the large concentration camps created by Nazi Germany. By the end of 1943 10,000 prisoners had been transferred from Buchenwald – mostly Russians, Poles and Frenchmen – but with the Dora camp then still under construction over 4,000 of them lived a subterranean existence in the tunnels. With such bad accommodation, heavy labour and maltreatment, the death rate among the inmates would prove horrendous to the point where far more people would die working in the tunnels than were ever killed directly by the secret weapons – something which in March 1944 necessitated the construction of a crematorium on the site. By this stage the Nordhausen factory was operational and had supplied the SS-controlled testing site at Blizna with rockets that proved largely unsuccessful during tests.

Since the V2 rocket consisted of 20,000 components it is not altogether surprising that the rocket scientists struggled to perfect this revolutionary new technology. While directives from Peenemunde concerning technical improvements and a visit from Dr Wernher von Braun followed, it was also recognised that many possibilities for sabotage by the inmates existed. Anti-sabotage measures were increased, with the extensive use of informers and planted spies. With negligence considered equal to sabotage, punishment by incarceration or death was commonplace.

From only fifty rockets manufactured in January 1944 – when their electrical systems had to be installed in the Demag vehicle works at Falkensee, near Berlin, which required the camouflaging and trans-portation of the missiles by rail – by the month of May 1944 it is recorded that 437 rockets were manufactured at Mittelwerk where the electrical components were then also installed. When the V1 offensive began against Britain the following month, in the aftermath of the Normandy landings, the fact that the flying bomb was being actively deployed reinforced an earlier decision that V2 production should be drastically cut in favour of V1 and jet-engine production. As a con-sequence, V2 production fell to 132 in June as V1 production was partially relocated from the Volkswagen factory at Fallersleben, and Junkers jet-engine production now occupied space in the safety of the Nordhausen underground factory. Despite the V2 factory having to progressively make room for the manufacturers of other secret weap-ons, as they perfected the manufacturing process towards the end of 1944 they reached a steady production of 600 missiles a month and by the end of the war had manufactured 5,940 rockets at Nordhausen.

As 1944 progressed, increasing numbers of factories from across the Reich relocated to central Germany, as the region around the Harz Mountains rapidly became the unlikely home to the industrial hope of Nazi Germany. By the end of the war the Mittelwerk company employed some 8,000 civilians and 25,000 slave labourers in the

Nordhausen area. In 1945 this included more equipment and personnel from Peenemunde, who travelled by train, car, truck and river barge, before settling ten miles south-west of Nordhausen.

Fearful of the advancing Soviet Army and the well-known Soviet cruelty to prisoners of war, von Braun and his planning staff opted for surrender to the Americans. While on an official trip in March 1945, von Braun suffered a complicated fracture of his left arm after his driver fell asleep at the wheel, with the result being an elaborate plaster cast.

The following month, as the Allied forces advanced deeper into Germany, the rocket scientists were moved by train to the town of Oberammergau in the Bavarian Alps where they were closely guarded by the SS who had orders to execute them if there was any danger of them falling into enemy hands. But von Braun was able to persuade an SS major that dispersal of the group into nearby villages would mean that they were less of a target for enemy attacks. After surrendering to the Americans he would be held briefly at the British-American detention centre at Kransberg Castle to the north of Frankfurt – as part of Operation DUSTBIN – where senior German scientists and industrialists were interrogated. As an important catch for the Americans, von Braun and many other rocket scientists of Nazi Germany were quickly recruited and would find themselves relocated to America as part of Operation Paperclip.

By the time Nordhausen was captured by the Americans in April 1945, more than 20,000 slave labourers had been worked to death in the tunnels. In the days before the Yalta Conference decreed that Nordhausen would fall within the Russian zone of occupation, British and American intelligence units, scientists and politicians converged on the intact factory. The most active team is reported to have been led by Major Robert Staver, chief of the Jet Propulsion Section of the US Army Ordnance Corps in London.

Before the tunnels were surrendered to the Soviets on 1 July 1945 the factory was thoroughly explored and investigated. Although all the assembled rockets, blueprints and senior scientists were evacuated, since the production line and enough component parts and junior personnel remained, the Russians were able to recommence production for a period. The Soviets used the Dora camp to intern captured Nazis and, later, Sudeten Germans from Czechoslovakia before it was broken up. In the summer of 1948 they attempted unsuccessfully to destroy the tunnel system with demolition explosives but not before they had systematically dismantled the factory which, along with its remaining German personnel, was transported to the Soviet Union before the 1949 establishment of the German Democratic Republic.

The overriding objective of Operation Paperclip, masterminded by the Office of Strategic Services, had been to deny German scientific knowledge and expertise to the USSR. Many Peenemunde rocket scientists would work for the Soviet Union and the Americans during the Cold War and although von Braun would become central to the development of rocket technology by the US Army and NASA – culminating in man landing on the Moon in 1969 – he understandably never escaped his past. Notwithstanding his many achievements, von Braun's past as a member of the SS, the exploitation of slave labour to build the rockets, and the death and destruction caused through the missiles' deployment were always indelible stains on his reputation. A 1960 biographical film – *I Aim at the Stars* – told the life story of the man who was by then an American citizen and head of the George C. Marshall Space Flight Center. But the unwillingness of people to change their minds about him was amply shown by posters for *I Aim at the Stars* commonly being graffitied with 'But Sometimes I Hit London'.

CHAPTER 14

Open Skies

'The wind of change is blowing . . .' Harold Macmillan, 3 February 1960

At the close of the war in Europe, British and American forces uncovered a chaotic mass of German photographic intelligence, consisting of prints, negatives, diapositives, microfilms, anaglyphs, vectographs, spheroids, maps and postcards. And though the story of its capture and exploitation was to prove dramatic, fascinating and complex, its role in galvanising the Anglo-American Air Intelligence partnership at the beginning of the Cold War was to prove positively crucial. And so, while wartime victory was being celebrated and the majority of those involved in Operation Crossbow were demobilised and returned to their civilian lives, photographic intelligence was not ignored as it had been after the First World War.

Signals Intelligence from Bletchley Park confirmed that the Luftwaffe had extensively deployed photographic-reconnaissance aircraft over eastern Europe throughout the war. Since this photography was likely to be the only photographic intelligence of their key post-war adversary available to the Western powers, there began a frantic Anglo-American race against the Russians in the closing days of the war to locate it. Unsurprisingly, the Red Army was equally determined that no aerial photography of Mother Russia or the Eastern Bloc countries would fall into the hands of the western Allies – particularly any covering their most sensitive industrial and military complexes – and that if it did they would take steps to recover it.

Towards the end of the conflict, when every effort was being made to capture the spoils of war, Army interpreter Captain Humphrey

Spender worked on a report about German photographic intelligence. After working in the Operation Crossbow Section at Medmenham, he had been seconded – alongside Ursula Powys-Lybbe – to the secretive Theatre Intelligence Section in Somerset House, Strand, where selected interpreters provided detailed intelligence on the Normandy coastline from aerial photography to assist the planning of Operation Overlord.

After the successful Allied invasion and during the long-drawn-out advances through Europe, information had been gathered from the interrogation of prisoners of war involved with German photographic intelligence as well as from captured documents and ground checks in newly liberated areas. Only months after Spender had been watching the coastline of Normandy in 1944, he discovered that during Operation Sea Lion – the abortive German plan to invade Britain in 1940 – the Germans had been closely watching the whole English south coast and the Thames estuary in much the same way, and that defence maps and large-scale models were made of towns including Portsmouth. As a detailed picture of the German approach to photographic intelligence began to emerge, a number of key differences became apparent.

With a more decentralised approach to their work, and the dispersal of the principal film and print libraries throughout German-controlled Europe, it became apparent that there was effectively no equivalent to Medmenham in the Third Reich. And when stereo photography was created, the German decision to utilise technically elaborate stereoscopes – which ultimately proved clumsier to use when compared with the relatively basic but highly effective pocket stereoscopes used by Allied interpreters – meant the German practice was to interpret from single negatives and prints using a basic magnifier. This disregard of 3D information and relative lack of organisation, combined with the decision that interpretation should generally be undertaken by Other Ranks – compared with the subject specialists who were deliberately recruited and given officer commissions at Medmenham – meant that Allied photographic intelligence had a competitive advantage throughout the war. The prophetic observation by the German General Werner von Fritsch in 1938 that the military organisation with the best photographic intelligence would win the next war had been proven correct.

When it became apparent that Germany had lost the war, photography considered to have no ongoing intelligence value in their network of libraries – such as that taken during the North African Campaign – was 'segregated' and burnt. From later prisoner interrogations, it was confirmed that the surviving photographs were then mostly transported southwards to the Nazi stronghold of Bavaria in southern Germany. In one case, photography from a unit based in the north German village of Schönwalde am Bungsberg, in modern-day Shleswig-Holstein, was evacuated and buried in the Bavarian city of Kaufbeuren, only for it to be dug up and loaded onto lorries for hiding in a coal mine in the closing stages of the war. When it was realised that defeat was inevitable, the junior officer in charge was ordered to destroy the films in a nearby quarry rather than let them fall into the hands of the Allies. But given the perilous state of Germany at the time, no petrol was available for the journey and the officer was forced to sell his watch in order that the burning could be carried out.

Although the decentralised organisation of German photographic interpretation meant that smaller aerial-photography libraries like this were scattered throughout German-controlled Europe, prisoner interrogation identified the biggest prize as being the Central Air Film Library – *Zentrales Archiv Für Fliegerfilm* – located forty miles north of Berlin.

In the early part of 1945 Douglas Kendall was appointed chairman of the committee responsible for investigating German photographic intelligence, which aimed to discover as much information as possible about their organisation, equipment and records. Recently awarded the OBE for his work throughout the war, Kendall and other Medmenham interpreters attended some of the apparently gentlemanly POW interrogations and ensured that all the right questions were being asked by the Air Ministry Assistant Directorate ADI (K), who were responsible for the interrogation of enemy airmen. Throughout the war Group Captain Denys Felkin had been responsible for this branch of intelligence, which at the end of hostilities worked jointly with US Air Interrogation and created a mass of intelligence reports that provided a detailed chronicle of the wartime experiences of adversary aircrews. At Medmenham preparations were made for Squadron Leaders David Linton and Ramsay Matthews to fly from RAF Benson at 08:30 on 8 May 1945 on a thirty-day mission to inspect liberated photographic-intelligence installations in Germany. Working alongside American Air Force colleagues, and with permission to carry cameras, they were instructed to make meticulous notes of everything they saw, and to make every effort to acquire maximum information.

Priority was given to likely places, including the manufacturer of optical systems Carl Zeiss AG in Jena and the main school of photography and interpretation in Hildesheim, from where equipment and records would be captured. The frantic race to capture such material also resulted in a reconnaissance sortie being flown by the Americans of 13th Reconnaissance Squadron – sortie US7GR/149/B on 18 May 1945 – which it was hoped would establish the existence of the Central Air Film Library. From a sky-blue P-38 Lightning flying 19,000 feet above the Brandenburg hamlet of Dannenwalde, the countryside, forestry, munitions bunkers and houses were all perfectly photographed and were precisely described in the second-phase interpretation report at Medmenham, made from rush-developed prints. But having been given the hopeless task of determining whether enemy photographic negative storage and interpretation facilities existed there, the interpreter completed the report with the merest hint of sarcasm, writing that there was 'no obvious way a negative store can be identified from the air'.

Interrogation of a German officer revealed that 'wild confusion' had reigned in German photographic intelligence in the early part of the war, with the Blitzkrieg mentality encouraging a belief that as long as the German Army was advancing on the ground there was little use for aerial photography once tactical information had been extracted. This meant that whilst in theory all the aerial films were accompanied by a trace and form detailing all the critical information about the squadron making the sortie – the flying height, date, time, observer's name, map sheets, cameras used and so on – the reality was that this information frequently never existed. When the Germans were forced into retreat, bringing them back into territory they had already photographed thoroughly, it was recognised that requests for prints and mosaics could be provided from existing stocks. To allow this to happen a cataloguing expert was appointed and the films were moved northwards from Berlin to Dannenwalde where they were stored in a disused munitions bunker.

Geographically catalogued according to a gridded key-index map of the world, the process would ultimately be completed using an enormous team of women – akin to the WAAF at Medmenham – by the summer of 1944. With the Red Army forcing the Germans into retreat, that same summer the Dannenwalde bunkers were again required for the storage of munitions and the film library was loaded onto ten 250-ton barges moored in a lake at nearby Himmelpfort, where cataloguing and photographic work continued. It was soon considered prudent to relocate the barges and so, alongside three more from Berlin loaded with valuable records, the flotilla headed southwards along the German waterways. With the Americans also advancing on them, a last-ditch attempt was made to reach the Czechoslovakian river port of Aussig from where it was planned to transport the 90,000 rolls of film to a coal mine at Marienstein, thirty miles south of Munich. While travelling on the river Saale the barges were attacked from the air, and with five barges on fire the officer in charge made the decision to destroy them all.

Just as the Aircraft Operating Comany and its Aerofilms subsidiary came to play a critical role in British photographic intelligence, it proved equally the case with German aerial-survey companies who were militarised on the outbreak of war. Hansa Luftbild GmbH had been pioneers in photogrammetric mapping and were renamed Sonderluftbildabteilung – shortened to 'Sobia' – but were allowed to continue operating from their headquarters next to Tempelhof Airport in Berlin.

With an expanding Third Reich, the decision was taken that Sobia would produce rectified mosaics, as they had in peacetime, from which mapping of the totalitarian empire could be created. And although British Air Intelligence might have had a passing interest in surveys of Denmark, Holland, Belgium and France the fact that prisoner interrogations indicated Albania, Dalmatia, Poland and large parts of Russia had also been scrutinised meant that securing the company's film library in Berlin was vital. Since the Sobia building in Tempelhof, like Paduoc House in Wembley, had been hit during an air raid but not totally destroyed, there remained hope that 10,000 rolls of film had survived. But when the film vault was ultimately discovered the use of oxyacetylene equipment to open it resulted in an explosion and the unfortunate destruction of vital intelligence. From the surviving cataloguing the indications were that the company had a vast archive of aerial photography covering places throughout Germany and the rest of Europe.

Although the German system of photographic interpretation might not have been as advanced as that of the Allies, the capture of German aerial photography proved to be an intelligence coup. Due both to their decentralised approach and their attempts to hide it, photography was discovered in different parts of Germany and though the Red Army was suspected of capturing some, they would prove equally as modest as the Allies about whether or not they had. In May 1945 the first and largest collection was discovered, inside a farmer's barn, hidden underneath straw, near the spa town of Bad Reichenhall in the German Bavarian Alps. Various official records indicate that an American team moved it only hours before the Russians, who knew of its existence and were also looking for it, arrived on the scene. The farmer had reported it to American soldiers, fearful that it might get him into trouble, which in due course resulted in the American interpreter Harvey Brown – based at 'Pinetree' in Wycombe Abbey – setting off 'like a bat out of hell'. He flew immediately to Stuttgart and travelled onwards by road to the foothills of the Bavarian Alps. Arriving at the barn, he recorded the discovery as feeling like a Robert Louis Stevenson story of pirate's treasure as he burrowed under the hay to look into more and more crates stuffed with methodically packed prints.

Despite a view among some of the Americans that the find should be kept secret from the British, the overriding feeling was that the close working relationship should continue. So Harvey Brown immediately contacted Douglas Kendall who immediately flew out in the nose compartment of a PR Mosquito. By the time Kendall arrived, the wooden crates – containing a collection that would have legendary status among post-war photographic interpreters – had been shifted and were being used as improvised tables by American soldiers in a mess they had set up nearby, located near Berchtesgaden and the Nazis' Obersalzberg mountain retreat. This was where famous visitors had been courted at the Führer's Berghof – including the 1938 appeasement visit by Prime Minister Chamberlain – and where images of Hitler with senior Nazis had been immortalised in the colour cine-films of Eva Braun.

The Allies knew that the Obersalzberg, situated high above the market town of Berchtesgaden, had a subterranean network of bunkers and was a last-stand fortress. For this reason it had been attacked by RAF Lancaster bombers, including 617 Squadron who had played such a pivotal role in the destruction of the Heavy Sites in northern France and who dropped the last Tallboy bombs of the war there on 25 April 1945. After the Germans moved most of their General Staff there, the Allies knew that southern Bavaria would provide rich pickings and just as Dr Wernher von Braun had been identified as human treasure that must not fall into the hands of the Soviets, so the British and Americans were desperate to capture both personnel – particularly those with specialised knowledge of their new Soviet adversary – and German intelligence records.

The hundreds of wooden crates weighed 16 tons and were flown to the United Kingdom where, around the middle of June 1945, they arrived at the headquarters of the US Eighth Air Force in the requisitioned Wycombe Abbey School where they completely filled a wooden barracks in the parkland grounds. Since Medmenham had been an Anglo-American unit, it was considered appropriate for the exploitation of the captured material to be carried out as a joint initiative christened Operation DICK TRACY. The analytical intelligence of the police detective in the American comic strip, who was known for his hard-hitting and fast-shooting approach to life, would certainly have been useful, given the challenge they now faced.

Work began the following month when the Americans and British each appointed six photographic interpreters to begin sifting through the war booty. On opening the hundreds of crates, they discovered aerial photographs measuring 12″ × 12″, aerial film, sortie plots, maps, town plans, mosaics, target material, photo maps, recognition and training material, postcards, and photography of military installations and important buildings in many of the key cities of Europe and Great Britain. The personal scrapbooks and photography of German officers, recording their hobbies and sporting activities were also found, but within twenty-four hours had all been 'mislaid'. Work on Operation Dick Tracy progressed at a great pace – with over 200 officers alone working on the project in August and September 1945 – but by the following month, the decision had come to close 'Pinetree' and return most of the American personnel to the United States. This resulted in the material being once again packed and transported, this time to Medmenham along with fifteen American officers and photographers who continued to work with the British on compiling a provisional catalogue.

At the beginning of the project the Air Ministry decided to copy all the photography, which amounted to around a million frames, but as time went on it was realised that this was going to be a costly affair. The situation was exacerbated by the War Department in Washington D.C. who insisted that the records – which had been found by them in the American Zone of Occupation – should be shipped to the United States after provisional cataloguing. With 31 March 1946 set as the deadline, four special cameras were purchased, each costing £30,000, to reduce the original 12″ × 12″ Luftwaffe prints to the Allies' standard 9″ × 9″ format, but even with this equipment the Exploitation Team had to be selective about what areas should be copied. This meant all their efforts were directed towards copying selected photography of Russia, Poland, the Balkans, Bulgaria, Czechoslovakia, Hungary, Yugoslavia, Turkey, Albania, Greece, Finland and parts of eastern Germany. Copies of all the sortie plots were made so that the Air Ministry would be in a position to order photography of the unselected material in the future from the Americans. And so, while the political frosts of 1945 gave way to the generations-long Cold War, the Luftwaffe aerial photography was destined to become one more military secret. The British decided to concentrate on the photography of the eastern side of the Iron Curtain – of the USSR and Soviet-dominated Europe – on the basis that better-quality photography would be created by the British and American units that were sharing responsibility for an aerial survey of Western Europe and parts of North Africa, a survey christened Operation CASEY JONES by the Americans.

Mindful of the ever-changing political landscape, work had begun on planning this enormous aerial survey that would begin in earnest after victory in Europe had been achieved. Since March 1945, 540 Squadron had been relocated from RAF Benson to Coulommiers, east of Paris – where Sidney Cotton had been based at the outset of the war – to begin work over France. The operation was carried out in the knowledge that the fog of war would lift progressively and that, as the sovereignty of countries was reclaimed and new governing powers emerged, the opportunity to overfly many in such a way would disappear. Unlike after the First World War, the importance of aerial survey for intelligence purposes – from which accurate mapping could be created and useful information extracted – was fully appreciated now and was given the highest priority. This meant that before demobilisation, from June 1945 and throughout 1946, many battle-hardened bomber crews found themselves held back to fly the top-secret missions. Using squadrons of B-17 Flying Fortress and Avro Lancaster bombers, which had been converted to accommodate aerial cameras, many disgruntled crews flew 'tedious in the extreme' sorties. On the British missions each aircraft was typically allocated a block of land forty by thirty miles that was divided into ten runs, each three miles apart. On reaching the designated block, success depended on the ability of the pilot to maintain a steady course, altitude and speed and on the bomb aimer's map-reading and dexterity with the camera.

At the same time as some aircrews flew survey missions over Europe, other RAF bomber crews flew missions over England, Scotland, Wales and Northern Ireland as part of Operation REVUE. Mainly carried out by the photographic-reconnaissance squadrons, it led to the creation of a vast library of photography that provided near-complete 3D coverage of the country, with urban areas photographed in greater detail and oblique photographs of most settlements. Coordinated from the headquarters of 106 Group at RAF Benson, detachments of Spitfires, Mosquitos and Avro Ansons – which had been used as training aircraft during the war – were sent around the country and the wartime reconnaissance pilots now systematically photographed their own nation.

Smog was to prove such a problem over industrial conurbations, such as the Staffordshire Potteries, that the photography often had to be scheduled for Wakes' Weeks when factory production would be brought to a standstill and both the air pollution and most of the local population would go on holiday, leaving behind soulless ghost towns. The survey was completed in the early post-war years and prints were mainly distributed to the Ministry of Town and Country Planning to assist post-war reconstruction and to the Ordnance Survey who used photographic mosaics created by Medmenham to revise their maps. On completion, the comprehensive National Air Photograph Survey superseded the patchy wartime photography taken over the country by the Allies during training exercises, and for the evaluation of camouflage schemes and for bomb-damage assessment. It also meant that no importance was ever attached to the photography taken by the Luftwaffe over the United Kingdom that was included with the

Operation Dick Tracy material and which was never copied before the originals were despatched to the United States.

When most of the American personnel left Medmenham in the summer of 1945, and either returned to the United States or travelled on to the Pacific Theatre, they were presented with a booklet featuring an oblique aerial photograph of Danesfield, surrounded by Nissen huts – titled 'Medmenham USA' – that begins with the introduction:

This is a sentimental dossier. It is a record of all the people, all the work, all the things that made RAF Medmenham so memorable. It is a personal history of all the Americans who lived and worked here with their British Allies. All of us in years to come will seek to recall the flavour of our unique experience. This dossier is your source of comparative cover for future reminiscing.

With photographs and information about each Medmenham section, it contained a photographic record of the living conditions in the huts, a mess party with revellers in Halloween costumes, church marriages to British women, baseball, rowing on the Thames, pubs in the surrounding area, guards on the Medmenham gate, and their victory parade on VE Day. The images strongly convey the impact of the American men and women on Medmenham and, by the same token, the British impact on them. With the document recording the full name, school, societies, date of arrival at the ACIU, sections worked in, peacetime occupation and home address, the Americans are shown to be an equally diverse group of people.

'Medmenham USA' was edited by Captain Francis Xavier Atencio (known as X Atencio) who brought a Disney influence with him to Medmenham, which explained why visitors to the third-phase airfield interpreters were welcomed by Dumbo the Elephant, resplendent in C Section cloak, on the office door. Later he recalled the day when Walt Disney first greeted him with a robust 'Hiya, X', which both thrilled the young artist and ensured that his initial X became his nickname around the studio. X Atencio worked on *Fantasia*, the Mickey Mouse film, and *Dumbo* before arriving at Medmenham on 22 August 1943, where he worked as a second-phase interpreter in Z Section and later in C Section. Listed in 'Medmenham USA' as a 'motion picture cartoon animator' from Hollywood, California, he returned to the Walt Disney Company and continued work as an animator on Academy Award-winning animations and live-action films, including *The Parent Trap* and *Mary Poppins*. Then Walt Disney – who had an ability to see talent in people who didn't even see it in themselves – got X Atencio writing dialogue and music, that resulted in his writing the script for *Pirates of the Caribbean* and its theme song 'Yo Ho (A Pirate's Life for Me)'. In 1996 he was recognised as a 'Disney Legend' for his speciality in animation and imagineering.

The British similarly recorded their history by arranging for histories to be written of each of the sections and lodged with the Air Historical Branch – whose historians produce historical narratives for the RAF – and by arranging for photographers to tour the RAF station with a glass-plate camera. From these a volume titled *The Chalk House with the Tudor Chimneys* was created, complete with the Unit badge, which dated back to 1940 when Captain Gerald Lacoste, a Royal Artillery Officer, designed an informal badge for the Wembley-based Photographic Interpretation Unit with the motto of 'Nil Lyncea Latebit', which translates roughly as 'Nothing can be hidden from the Lynx'. The final design, which featured the Canadian lynx, was developed by Sir John Heaton-Armstrong at the College of Arms, approved by King George VI in June 1944, and presented to the Unit by the Air Officer Commanding 106 Group, Air Commodore D. J. Waghorn, on 6 August 1944. In the foreword to the sixty-six-page volume, it was recorded that:

> This is a photographic record of the Allied Central Interpretation Unit, as it existed at Royal Air Force Medmenham, near Marlow, Buckinghamshire, in the closing months of the war. It is an attempt to show in pictures that comprehensive organisation which was built up during the war to deal with intelligence from air photographs . . .

Within days of completing the copying and provisional cataloguing of the Operation Dick Tracy material, truckloads more captured German photographic-intelligence records began arriving at Medmenham, courtesy of Field Marshal Montgomery and the British 21st Army Group. Although the Russians had reportedly salvaged material from barges on the river Rhine used by the Germans to store intelligence material – which had been set on fire in an attempt to destroy their contents – the British still managed to capture the records of 1,600 sorties, over 10,000 annotated mosaics and thousands of maps, postcards and target material.

With the destruction of tightly bound rolls of aerial film, photographic prints and paper records by fire having proved more difficult than anticipated, the scorched nature of the records doubtless added a sense of drama as the interpreters began sifting through the collection. The initial work on the captured material – known as Operation PATRON – began at Medmenham but, before the exploitation of the intelligence could begin in earnest, the Air Ministry turned RAF Station Medmenham into the Signals Headquarters of the Royal Air Force in preparation for carrying out electronic intelligence against the Soviet Union. This required the Central Interpretation Unit (renamed in August 1945) to relocate by May 1947 to RAF Nuneham Park, the stately Palladian villa at Nuneham Courtenay in Oxfordshire beside the river Thames, which for most of the war had been a satellite station to Medmenham. The landscaped grounds by Capability Brown were now home to a network of Nissen and wooden huts that accommodated the RAF personnel. And while the German photography and Anglo-American personnel working on it would transfer too, the scale of the Print Library meant that for the time being, while its fate was being considered, it would remain at Medmenham.

Somewhat bizarrely, the Patron collection included several large mahogany cases containing 15,000 photographic prints showing the complete pictorial story of the 1936 German expedition into Borneo and Africa, but since they had no intelligence value they were gifted to the Royal Geographical Society.

Around this time, the code name TURBAN was adopted for the Anglo-American project on the captured German photography and the material itself became known as GX photography – after old RAF designations for clandestine photographic sorties – regardless of which collection it was held within. This was to avoid the confusion associated with the countless smaller collections that had been found in places as geographically diverse as Vienna and Oslo and that were code-named Monthly, Filter, Tenant, Lattice, Hutch and Orwell. It applied equally to GX photography obtained by the RAF 'from two gentlemen of a European country' in March 1954 that provided coverage of more targets in the Soviet Union.

Since the majority of the items had been found in the American Zone of Occupation, or by the Americans elsewhere, most of the original photography, or copies thereof, was shipped to the United States after sorting, cataloguing and copying at Nuneham Park. As part of Operation Turban, a reciprocal film-exchange programme was agreed that allowed both governments to make copies of any GX photography they wanted, something that was also happening on Operation Casey Jones and would continue throughout the Cold War on other projects. Great pains were taken to separate the highly flammable nitrate aerial film, stored at RAF Benson, from the Print Libraries at Medmenham and Nuneham Park, and having another copy on the other side of the Atlantic Ocean helped reduce again the risk of the important intelligence ever being lost.

While much of the GX photography survives, few records exist about Operation Turban, a fact put down in an official report to 'those originally responsible clearly preferring the telephone to the pen'. The sheer magnitude of the project to sort, catalogue and plot the photography was underscored by an official estimate that initial errors of judgement alone wasted five hundred people-years of effort. This lack of a written history was clearly further exacerbated by the mass demobilisation of personnel throughout the project, the speed with which the work had to be done, and by the large number of disgruntled people still awaiting demobilisation. When they left, their departure broke so many of the personal contacts with the past that it weakened further the continuity of any oral history.

But however scant the official history might be, the long-term legacy of GX photography was the intelligence it provided for target dossiers on important industrial and military complexes in the western Soviet Union and its strengthening of the partnership between British and American Air Intelligence that had begun during the Second World War.

The value of this partnership was only enhanced by American inter-service rivalry which encouraged the embryonic intelligence unit of the newly created United States Air Force – an independent third service – to look instinctively to the RAF for support rather than to the US Army or Navy. This helped create a merger of Anglo-American air-intelligence effort against the Soviet Union, a cooperation which concentrated on target intelligence and which would benefit enormously from the Turban project. The GX photography provided intelligence on targets in the Soviet Union on a scale unimaginable before the advent of satellite reconnaissance in the mid-1960s and remained a key intelligence asset for more than two decades. And while the British and Americans made a small number of albeit dramatic clandestine flights over the Soviet Union in the 1950s, the coverage obtained was comparatively small and specific to particular targets.

Just as Ultra was a post-war secret until Frederick Winterbotham sensationally revealed its existence in his 1974 book *The Ultra Secret*, the GX photography captured from the Germans remained secret throughout most of the Cold War. Eventually declassified by the Defense Intelligence Agency and Ministry of Defence in the United States and Britain respectively, the GX photography of the United Kingdom (which the British never had any need to copy) may now be accessed at the National Archives and Records Administration (NARA) near Washington D.C. at College Park in Maryland. But apart from one or two iconic photographs, including one featuring a Heinkel He 111 flying over east London on 7 September 1940 – with the West India Docks and the Isle of Dogs below – knowledge of these unique wartime photographs of Britain remains limited. And while the American collection contains unique aerial photography that they captured from the Japanese – code-named JX – the British collection also contains unique photography never copied by the Americans, including a large collection of mosaics covering large parts of Eastern Europe.

GX aerial photography was a key intelligence resource at the outset of the Cold War and, while the advent of the U-2, the SR-71 and satellite imagery would increasingly diminish its importance to air-intelligence agencies, the importance of historical photography for other reasons was demonstrated through a series of later CIA reports. To analyse the photography created using the U-2, the National Photographic Intelligence Center was created in 1955, with photographic interpreters Robert Poirier and Dino Brugioni as founding members. In the late 1970s their re-analysis, using new imaging techniques, of Second World War aerial photography of Auschwitz and other concentration camps taken by the Allies allowed them to forensically analyse history. Using a variety of density-slicing and enlargement techniques they were able to see Holocaust victims, who had arrived at Auschwitz in the infamous enclosed railway wagons, being marched to their deaths in the gas chambers. Poirier would also use the GX collection to investigate the Katyn Massacre, the mass murder by the Soviet secret police in 1940 of Polish nationals whose mass graves the Germans discovered and announced to the world in 1943. From the GX photography it was possible to identify disturbed earth and to establish an understanding of Soviet and later German activity in the area.

In the immediate post-war era, both the United States and the Soviet Union expanded their rocket-research programmes, thanks to the human intelligence provided by captured German rocket scientists and engineers, and their technical drawings. With more and more American Strategic Air Command bomber bases surrounding the USSR, Stalin was encouraged to continue his country's nuclear-weapons programme, making use of long-range bombers and missiles. The seriousness of the Soviet intent was evidenced by their dismantling and transporting the V2-rocket assembly line from Nordhausen to Russia along with German personnel which meant that the Soviet Union would become the second nation after the United States to successfully develop nuclear energy and conduct atomic-bomb tests.

As both superpowers developed aircraft capable of delivering nuclear payloads, the discovery by Anglo-American intelligence in the early 1950s of the rocket-test facility at Kapustin Yar, east of Stalingrad in southern Russia, was something they could not afford to ignore. But just as Peenemunde had been chosen for its remoteness, which naturally enhanced security, the Soviet facility was so far inside the country that photographic-reconnaissance missions would be problematic. Its southern location in the northern hemisphere took advantage of the Earth's rotation, since linear velocity is greatest towards the equator and means that rockets can be launched more easily and eastwards to maximise the Earth's natural gravitational thrust. With this natural phenomenon also prompting the Americans to establish their facility at Cape Canaveral, in the southern state of Florida, the evidence that both superpowers were now accelerating their headlong battle was clear.

In the early Cold War, the brilliant Hungarian-American mathematician John von Neumann, a key player in the wartime Manhattan Project, headed the top-secret Von Neumann Intercontinental Ballistic Missile Committee. Although the technical development problems were great and potentially insurmountable, its purpose was to decide on the feasibility of building a missile large enough to carry a thermonuclear warhead. With a fondness for humorous acronyms, von Neumann is credited with formulating the equilibrium strategy of Mutually Assured Destruction – typically referred to as MAD – which provided a doctrine of military and national security policy in which the full-scale use of weapons of mass destruction by two opposing sides would effectively result in the complete, utter and irrevocable annihilation of both the attacker and the defender.

On this basis it was assumed that neither side would launch a first strike, because the other side would either launch on warning or would have sufficient second-strike capability to ensure that unacceptable losses would be suffered by both parties. The pay-off from the MAD doctrine was that although there would be tension and fear for survival, a stable global peace would nevertheless prevail.

The MAD doctrine was immortalised in the 1964 Stanley Kubrick film *Dr. Strangelove or: How I Learned to Stop Worrying and Love the Bomb*, which satirises the nuclear scare by following the crew of a Boeing B-52 bomber as they try to deliver their nuclear payload. With Peter Sellers playing three roles, one of which is the wheelchair-bound Dr Strangelove who was based on an amalgamation of Cold War strategists including von Neumann, the film follows the consequences of an unhinged United States Air Force general ordering a first-strike attack on the Soviet Union. Sellers also appears as the President of the United States who receives advice from his Joint Chiefs of Staff, and as RAF Group Captain Lionel Mandrake, a role that allowed him to draw on his wartime experience of mimicking senior officers while touring with RAF Gang Shows.

But with the Strategic Air Command in real life under Presidential instructions not to penetrate Soviet airspace, and following a successful top-secret mission with the British the year before, the RAF were approached by the Americans to reconnoitre Kapustin Yar in the summer of 1953. This earlier mission had been led by Squadron Leader John Crampton, who raced his Maserati sports car at Goodwood and had flown Whitley and Halifax heavy bombers during the strategic air campaign against Germany in the Second World War. In July 1951 the six-foot six-inch Old Harrovian was summoned by the Commander-in-Chief of Bomber Command and told that, under conditions of utmost secrecy, he was to command a Special Duty Flight. It would be equipped with the North American RB-45C four-engine jet reconnaissance aeroplane, and the RAF crews involved would proceed immediately to the United States. The Strategic Air Command was responsible for America's land-based strategic bomber aircraft and missile sites, and as a consequence for strategic reconnaissance, and they provided the RAF personnel with sixty days of intensive training.

Returning to Britain in December 1951, Crampton and his crews found themselves based at RAF Sculthorpe in rural Norfolk, alongside an American RB-45C detachment that was carrying out detailed aerial photography of the Rhine Valley all the way from the North Sea to Switzerland. With this information, NATO aimed to create accurate reference information for programming the guidance systems on the ballistic missiles that would shortly be deployed for the defence of Western Europe.

In late February 1952 Crampton was summoned to Bomber Command headquarters in High Wycombe to be told about Operation JU-JITSU. With a momentary feeling of apprehension as charts were unrolled to reveal separate tracks from RAF Sculthorpe to the Baltic States, the Moscow area and central southern Russia, he was nevertheless relieved to know finally what was expected of him and his men.

The operation involved three Tornados flying simultaneously over the Soviet Union – stripped of their USAF markings and repainted with Royal Air Force roundels and colours – to gather electronic intelligence. Timing would be critical since the intelligence agencies would be listening for the Soviet air-defence reaction to the deep penetration of their airspace. Once a target had been located and identified, they were to take 35mm photographs of the aircraft's radar display. Throughout the operation they were to fly without navigation lights and were to maintain radio silence, although they would have an OMG – Oh My God – frequency for any truly desperate emergency.

Before the date of their live sorties was fixed, Crampton took his crew on a gentle probe over the Soviet Zone of Eastern Germany for an hour or so in March 1952, to monitor Soviet radio and radar activity. With nothing of consequence noted and with Prime Ministerial approval, late in the afternoon of 17 April 1952 the three Royal Air Force RB-45Cs departed from Sculthorpe and headed across the North

Sea towards Denmark. After mid-air refuelling by aerial tankers, each doused their lights and breached Soviet airspace.

Crampton had the long-haul sortie, south-east across Russia, and with his navigator feeding him the courses for the 126 air intelligence targets they flew across the wilderness of eastern Europe. And despite signals-intelligence confirmation that the Soviet air-defence system had been rudely awakened by the overflights, and that MiG fighters had been scrambled to intercept them, all three Tornados returned safely to Norfolk, in the case of Crampton after more than ten demanding hours of flying.

Prime Ministerial approval for Operation Ju-Jitsu had been provided by an elderly Winston Churchill who at the age of seventy-six had been reappointed for a second term following the General Election of October 1951. However great the political risk associated with the RAF overflights might have been, Churchill evidently considered the radar-mapping of key Soviet military targets to have overriding priority, presumably on the basis that it would provide Anglo-American air intelligence with essential information for any future war, and that it helped bolster the special relationship with the United States which he valued so highly.

By the summer of 1953, though, when the level of interest in Kapustin Yar had been raised, a predictable consequence of Operation Ju-Jitsu was that overflight missions had become even more dangerous. The Soviet air-defence system had been shown to be seriously wanting and a shake-up was inevitable. There was understandable concern in the Soviet leadership that intelligence had been collected that put their country at risk in the event of war, and that although this time the bombers had been carrying cameras, radar and listening devices, they could easily have been carrying atomic warheads.

Although capturing GX photography was undoubtedly an intelligence coup, the vast scale of the Soviet Union meant that dangerous and politically sensitive penetrations of Soviet airspace were necessary to obtain photographs of places – for which there was no GX coverage – and to allow the comparative analysis where they had. The decision by Stalin, after the first fraught weeks of Operation Barbarossa, to relocate as much Russian heavy industry as possible to the Urals, beyond the reach of the Luftwaffe, meant that in the Cold War overflights were still considered vital to gain intelligence on places in the vast Soviet

Union that had not already been covered, and to provide comparative coverage.

Just as Sidney Cotton had pioneered a new approach to photographic reconnaissance at the beginning of the Second World War, reducing the risk of being shot down and the associated political ramifications meant new approaches to photographic reconnaissance were required at the start of the Cold War.

One of the leading advocates of the need for higher-altitude reconnaissance aircraft was Richard Sully Leghorn. The highly decorated US Air Force colonel had commanded the 30th Photographic Reconnaissance Squadron during the Second World War and had taken aerial-reconnaissance photographs over the Normandy beaches in preparation for the D-Day invasion. On returning to civilian life and his job with Kodak, he maintained his interest in photographic intelligence and wrote papers advocating the need for reconnaissance of a potential enemy before the outbreak of actual hostilities, rather than combat reconnaissance during wartime. Recalled to active duty to plan and photograph the atomic-bomb tests at Bikini Atoll, what he witnessed helped focus his vision that effective reconnaissance would determine the course of world history in the Cold War.

Leghorn had the opportunity to put his ideas into effect when he was recalled for active duty on the outbreak of the Korean War in 1950. His vision was for the United States to develop high-altitude aircraft with high-resolution cameras, and since the MiG-17 – the best Soviet interceptor at the time – struggled to reach 45,000 feet he reasoned that an aircraft that could exceed 60,000 feet would be safe from Soviet interception. Recognising that the quickest way to produce a high-altitude aircraft was to modify an existing one, he began looking for the highest-flying aircraft in the Free World, a search that soon led him to the British twin-engine Canberra bomber, built by the English Electric Company. With its speed of 469 knots and service ceiling of 48,000 feet the Canberra was a natural choice, something that had led the British to develop a photographic-reconnaissance version which began flying in March 1950. To make the Canberra fly even higher, Leghorn worked with English Electric who designed a new configuration with very long high-lift wings, new Rolls-Royce engines, a solitary pilot, and a new airframe that Leghorn calculated would allow the Canberra to reach 63,000 feet and up to 67,000 feet as the diminishing fuel supply lightened the aircraft. With such fine-tunings he estimated that the Canberra could penetrate the Soviet Union and China for a radius of 800 miles from bases around their borders, and that upwards of 85 per cent of the intelligence targets in those countries could be photographed. When Leghorn's concept was ultimately rejected, in frustration he transferred to the Pentagon in early 1952, where he became responsible for planning the US Air Force's reconnaissance needs for the next decade.

Anglo-American cooperation continued with Project ROBIN, which involved the RAF flying a stripped-down Canberra bomber to Hanscom Air Force Base, near Boston, Massachusetts. The Canberra was then fitted with the 'Boston Camera' that had been developed by the Boston University Optical Research Laboratory in 1951 and had an optical system, designed by the Harvard astronomer James Baker, which used a mirror system to pack an enormous 240-inch focal length camera into a 10-foot-long cylinder. It was suspended in the bomb bay. The camera took long-range oblique photography that meant, when the Canberra flew near the border with the Soviet zones of occupation, that it could photograph installations 'over the fence'. The camera was reported to be so powerful that during a test flight over the English Channel – off the coast of Dover – clear shots of St Paul's Cathedral 75 miles away were taken. And in America during tests the camera was so powerful – the legendary CIA PI and historian Dino Bruggioni would later observe – that it photographed people in Central Park, New York, from 72 miles away in a Convair B-36 strategic bomber.

After permission was first given, in June 1953, photographs were taken of targets in the Soviet zone of Germany. This was followed by photography of Czechoslovakia and the Soviet zone of Austria, before the 1955 Austrian State Treaty re-established the country as a sovereign state. The method employed involved flying the Canberra at around 42,000 feet on a course 10 miles inside the British or American zones of occupation. To guard against long condensation trails, which could reveal the activity too clearly, a second aircraft flew alongside the Robin aircraft to monitor the situation.

Ultimately Project Robin sorties were declassified and photographs were released by the Ministry of Defence in 2004, not only of East Germany, Austria, Croatia, Hungary, Slovakia and Slovenia but also of Cyprus, Eygpt, France, Israel, Lebanon, Palestine and Syria in the mid-1950s. According to some oral histories and secondary sources, the Robin camera allowed the Canberra to be used on a daring daytime overflight of the Soviet Union to photograph the Kapustin Yar test facility, north of the Caspian Sea and east of Stalingrad. Just as Crampton had made his epic run the previous year, this overflight is recorded in a declassified Central Intelligence Agency report as having been carried out in the first half of 1953 by the RAF. This particular mission, in which Soviet fighters are recorded to have damaged the Canberra, appears never to have been confirmed by the British.

In the early 1950s, with interest in high-altitude photographic reconnaissance growing, several United States Air Force agencies investigated what aircraft were suitable for such missions. And while the Martin Aircraft Company were awarded a contract to examine the potential of the English Electric Canberra – which they built under licence as the Martin B-57 Canberra – the Lockheed Aircraft Corporation became aware of the competition and proposed that an aircraft should be built that could reach altitudes of between 65,000 and 70,000 feet. In early 1954 the brilliant aeronautical engineer Clarence 'Kelly' Johnson began design work on what would become the Lockheed U-2. As leader of the design team behind the P-38 Lightning, which had been the principal platform for American photographic reconnaissance during the Second World War, he would become the first team leader of the top-secret Lockheed Skunk Works – who developed aircraft as 'black' projects – and was later responsible for the Lockheed SR-71 Blackbird strategic reconnaissance aircraft.

Concerned with the lack of intelligence data coming from behind the Iron Curtain, President Eisenhower approved the U-2 Project and by August 1955 it made its maiden flight. The year 1955 was to prove a year of reflection for President Eisenhower as he contemplated the horror of an atomic war, assessed the technology available for intelligence collection and speculated about how ideas could be advanced that would ensure better relations with the Soviet Union.

At the 1955 Geneva Summit, convened to reduce Cold War tensions, President Eisenhower proposed the concept of 'Open Skies'. This unexpected suggestion presented to a world anxious for peace and more fearful of war than at any previous time in history a practical means for armament control. But though many greeted the proposal as a gesture of sincerity and a step towards peace many others doubted that such a scheme was politically or practically viable. In Eisenhower's own country, plagued by McCarthysim and strong anti-communist sentiment, influential commentators even told their readers that the President would never have made the offer if there was any chance of its being accepted.

The scepticism that existed on all sides was seen by Douglas Kendall as a tragic setback in the quest for peace that reflected a general failure to appreciate that photographic reconnaissance could be an effective means of policing the world. Although aerial photography had been used for propaganda purposes during the Second World War, he decried the lack of genuine understanding of what photographic intelligence had achieved during that war, and regretted that, given the dearth of non-technical literature on the subject, there was such limited public understanding about what it could achieve in the Cold War. In his unpublished memoirs he cites his wartime experience at Medmenham as a useful precedent, where the inter-Allied Unit combined its efforts to hunt for the Nazi V weapons and was a powerful exemplar of what photographic intelligence offered.

When one commentator made the scathing remark that the United States would need to develop X-ray cameras since the Soviets would invariably hide top-secret facilities underground, Kendall was reminded of Medmenham's successes during Operation Crossbow in locating underground facilities, from the manufacturing sites to the Heavy Sites, and that by a process of elimination they had been able to clarify the purpose of each site. Furthermore, in the case of the V2 rockets, careful study of enemy communication systems had allowed them to track the movement of the missiles from the factory to their deployment. The very fact that an underground facility can almost always be spotted, and that putting something underground at great cost would immediately be suspect, means it would be perfectly reasonable for the United Nations to request the right to inspect a particular factory. In Kendall's opinion, the far greater challenge lay in spotting illicit arms production in, for instance, a steel plant that was still sufficiently active in production to mask its illicit activities. And although a further criticism against the photographic-intelligence method of policing is that a small firing installation such as might be used to launch an intercontinental rocket with an atomic warhead could easily be hidden in the vast Soviet Union, a key lesson from

Operation Crossbow, in the case of the V2 rocket, had been acceptance of this fact. For despite the construction of the Heavy Sites, which were comparable to modern underground missile silos, the V2 rockets were actually fired from the Netherlands from any flat area. Cold War missiles were equally mobile.

Photographic intelligence was a major source of information in hundreds of different ways. Rather than decry where it failed or was incomplete, it is arguably more important to consider where Operation Crossbow would have been without it. From the outset of the war, German propaganda had created a climate of uncertainty and fear that secret weapons against which there could be no defence were held by them. Doubts about the veracity of the Oslo Report could easily have meant that without photographic intelligence it would have been mere rumour that secret weapons were being developed at Peenemunde. It would certainly not have been possible to begin developing effective countermeasures, including the attack on Peenemunde itself. And, given the divergence of opinion among the Allies' scientists, the exact nature of the weapons could easily have stayed unclear.

Secret agents undoubtedly revealed the existence of the Bois Carré sites in northern France, but ground sources are notoriously unreliable in giving exact information and geographical positions. Obtaining precise information from within enemy-occupied territory typically multiplies problems tenfold and it is only too easy for the same information to be reported differently by different sources. The Medmenham interpreters successfully and quickly located all ninety-six launch sites through skilled photographic interpretation, and were able to provide the detailed information required to successfully attack and destroy them.

Following the attacks on the Bois Carré sites, photographic reconnaissance was essential for damage assessment, a function that no other source could have performed effectively. With the flying bomb, and a process of masterful analysis of the facts by Kendall, the Allies knew substantially what they faced as early as November 1943. The interpreters knew that the enemy intended to use this weapon against them, and by supplying the necessary information to the operational forces were able to delay the deployment of this weapon until June 1944, after the successful Normandy landings.

During Operation Crossbow some 3,000-plus photographic

reconnaissance sorties were flown – in the battle against the flying bomb – producing over 1.2 million aerial photographs, all of which had to be examined in detail. With losses on photographic-reconnaissance missions running on average at 2 per cent throughout the war, many pilots and aircrew lost their lives, but since the majority of missions were over northern France, rather than deep penetration into Germany, losses were likely to have been less than the sixty men they statistically should have been. To keep pace with the investigation – and with what was being discovered on a daily basis – 3,450 separate reports were issued to interested parties ranging from the Prime Minister's office to the Pentagon. While in general around twenty officers were assigned to the operation, the size of the PI effort depended greatly on the emphasis of the investigation and the state of the weather. A spell of particularly clear weather or an important task meant that as many as 80 officers from the 500-plus based at Medmenham would be deployed onto the operation when required, from across the British Army, Royal Air Force, Women's Auxiliary Air Force and the US Army Air Force. From the photographic intelligence the enemy's intentions were clearly seen well ahead of the time they were put into practice, in spite of the best efforts of the Germans to conceal their activity and purposes. Operation Crossbow had first focused on whether the Germans were developing a long-range rocket. In April 1943 the enemy's intention had been established and by June that year it had been revealed to be 38 feet long. With the rocket not deployed in anger until September 1944, the Allies had significant advance warning of this new weapon and were not taken by surprise when it duly arrived. The Japanese did not have the same advantage of knowing about the atomic bomb, which came as both an operational and intelligence surprise to them. Combined with German rocket technology, this example of secret science cast a long shadow over world peace from the outset of the Cold War.

Although the Axis powers failed to develop photographic intelligence to anything like the level achieved by the Allies, Operation Crossbow was not a complete success for Allied intelligence and underscored many shortcomings. For while the interpreters were able to provide detailed information about the size of the rocket, division among the scientists meant that establishing its tonnage and performance proved inconclusive. Photographic intelligence cannot tell you everything and highlights the importance of those other intelligence sources that contributed to the successful outcome of the operation. The human intelligence contained within the Oslo Report or supplied by the Réseau AGIR and the rocket parts that were smuggled out of Poland were to prove essential, since although ground intelligence was frequently wrong or contained only half-truths, without its direction the interpreters were often blind. The interpreters could provide the answers, correctly and in great detail, but it was the other sources of intelligence that frequently provided the inspiration. Successful intelligence is always a matter of checking one source against another. The examination of aerial photography allowed the reports from different intelligence sources to be checked for their reliability and graded according to their probability.

In the view of Kendall, the Eisenhower plan might have been complex and would undoubtedly have required a major effort to operate effectively. But it would not have been any more complex than Operation Crossbow had been and would not have required an unrealistic amount of resources when it was considered that increased global stability could be better ensured.

After the war Douglas Kendall emigrated to Canada where he continued his career in commercial aerial survey. But when the United Nations Atomic Energy Commission was founded by the United Nations General Assembly in 1946 to deal with the problems raised by the discovery of atomic energy, and with the Soviet Union opposed to the policing methods suggested by the Americans, he was approached by the US representative on the Commission. Former US Ambassador to the United Kingdom throughout most of the war, John Winant was well acquainted with Kendall, and was anxious for him to be called as an expert witness. Kendall found himself on the way to the Rockefeller Center in New York City, where he was greeted by the Canadian delegation. At the time, the key proposal was to control the supply of radioactive material at the point of mining. It was considered that if all the uranium could be stored under control, it would be possible, theoretically, to prevent anyone from collecting enough to produce a bomb. And although the Soviet delegation were not willing to discuss a political solution, they were persuaded to attend a session that focused exclusively on technology. It was at this meeting that Kendall spoke.

The plan devised by Kendall was to control and verify the mining of all radioactive materials via photographic reconnaissance that would be created by a United Nations survey unit that would photograph the globe in its entirety. As incredible as it doubtless sounded to the audience, Kendall considered it technically possible. But he recognised that the United Nations would need the authority to fly everywhere in the world. He calculated that fifty light aircraft would be required, and the first part of the programme would see the creation of a world photo-map at a scale of 1/40,000. Thereafter the reconnaissance unit would be in contact with the intelligence agencies of all United Nations members and from their reports it would rephotograph suspected areas. Comparative analysis would then be undertaken and, if suspicious activity was suspected, the United Nations would have the right to immediate inspection on the ground. When the Soviet delegates took the position that the scheme would not work, and used as an example the failure of photographic intelligence to detect the German V-weapon programme, Kendall had a lucky break and was able to quote chapter and verse on the success of Operation Crossbow. But with Kendall's visionary plan and President Eisenhower's later proposal for an Open Skies programme destined to be rejected by the Soviet Union, golden opportunities were lost.

Also in 1946, the Medmenham Club was established to maintain the camaraderie enjoyed by the photographic interpreters of all three services who served in the Allied Central Interpretation Unit at RAF Medmenham during the Second World War. With the Club's first president, Hamshaw Thomas, they set what would become an annual tradition of revisiting Medmenham for afternoon tea in the Officers' Mess, when they also had the chance to look around Danesfield House and wander around the grounds.

When in 1948 it became apparent that the costs of restoring the property would be more than the place was worth, the house and estate were bought by the Air Ministry and Medmenham became the permanent home of RAF Signals Intelligence. During this time the RAF residents established their own Hell-Fire Club – inspired by the earlier antics of Sir Francis Dashwood at Medmenham Abbey – whose members, on one occasion, are recorded to have piled up all the tables and chairs in the grand hall and left a footprint on the ceiling.

During the 1970 reunion at Danesfield House, which then served as possibly the most lavish Officers' Mess in the country – complete with its own boathouse on the river Thames – it became apparent to club members that some of the contemporary residents, whose world was dominated by electronic signals intelligence, had no conception of what had taken place there during the war. While standing on the veranda overlooking the Thames Valley, one of the architect members suggested that a plaque on the wall, commemorating the place's occupation by the Allied Central Interpretation Unit, might therefore be appropriate. With so many talented architects, artists and sculptors among its membership the task of carving the Portland stone, which was slightly polished to produce a golden-brown effect that would complement the Medmenham stone, fell to Geoffrey Deeley. The one-time Squadron Leader in charge of the model-makers had returned to the Polytechnic School of Art in Regent Street, where he was then Head of the Sculpture Department.

The plaque was formally unveiled in the West Tower archway of Danesfield House in September 1972 by Lord Shackleton – who had served as an Intelligence Officer with Coastal Command during the war – and over seventy members of the Club were there, alongside representatives of the Joint Air-Reconnaissance Intelligence Centre (JARIC), the direct descendant of Medmenham. JARIC at that time was based at RAF Brampton in Cambridgeshire.

The presence of younger interpreters at a Medmenham Club function, then as now, would have provided a timely reminder of the ongoing intelligence challenges that were being faced. By this time the interpreters had been rechristened 'Imagery Analysts' on the basis that they analysed not only monochrome photography – which was nevertheless still the mainstay of their work – but also a variety of complementary imagery obtained by sensor systems on a variety of aerial and satellite platforms. At the same ceremony the Medmenham Trophy, awarded to the photographic interpreter judged to have made the most outstanding contribution in the field during the year, was also presented.

Also present at Medmenham that day was Constance Babington Smith, who in 1958 had published *Evidence In Camera* – available in the United States under the title *Air Spy* – which introduced the story of photographic intelligence during the Second World War to a mass audience for the first time, particularly after it was serialised by the *Sunday Times*. These publications followed a dramatically titled 1957 article written for *Life* magazine in the United States – 'How Photographic Detectives Solved Secret Weapon Mystery' – based on her recollections and the research she had conducted on the V weapons. These writings provided inspiration for the 1965 Metro-Goldwyn-Mayer film *Operation Crossbow* that recreated much of the excitement of the time but, as Hollywood often does, took considerable artistic licence. Constance was portrayed by English screen siren Sylvia Syms, and Sophia Loren appeared in a cameo role – for the benefit of the box office – but the vital role of Douglas Kendall, portrayed by Richard Todd, ended up largely on the cutting-room floor in the cause of dramatic effect. Although Kendall had no influence over the film's production, he was given the opportunity of spending a day on the film set watching Sophia Loren in action, a process he later recounted as being fascinating. This was followed with an invitation to the film premiere at the Radio City Music Hall in midtown Manhattan, New York City, with all the associated lights and glitter, which proved a fascinating distraction from his aerial survey work.

While the British were comprehensively surveying their own country after the war – through Operation Revue – in order that they should have modern mapping that could assist post-war planning and reconstruction and were working with the Americans on Operation Casey Jones to create total photographic coverage of western Europe and North Africa, they were also putting into action plans that had been developed in the darkest days of the war by Whitehall officials with a sense of imperial responsibility. In 1946 the Colonial Office established a central aerial-survey and mapping organisation for British colonies and protectorates. By 1957 it was known as the Directorate of Overseas Surveys (DOS) – most of its customers were increasingly independent Commonwealth nations. The work of DOS involved taking aerial photography of countries in their entirety in Australasia and the Far East, the Indian subcontinent, the Middle East, Africa, the Caribbean, South America, British Antarctic Territory and islands in the Pacific, Atlantic and Indian Oceans and the Mediterranean Sea.

The exploits of the surveyors on the ground who travelled across fifty-five countries of the world in the course of DOS history – battling through blizzards in the high mountains of Lesotho and dust storms in the deserts of Sudan – often drew comparison with characters in a John Buchan novel. More than 1.5 million aerial photographs were created over the course of its existence; initially taken by the RAF, most were created by commercial aerial survey companies, including a reconstituted Aircraft Operating Company.

From a sprawling government building in the south-west London suburb of Tolworth – using a Wild A5 Autograph Machine (Instrument No. 81) which had been smuggled into Britain in 1943 via Gibraltar for use at RAF Nuneham Park – DOS cartography, which underpinned the work of governments, commerce and industry across the world, was created. Over its thirty-eight-year history, and before satellite imagery and computers revolutionised cartography, and the passing of the Cold War – which had further encouraged the British to assist their former colonies rather than see the Soviet Union do so – this unconventional part of the British Civil Service mapped over two million square miles, an area roughly twice the size of western Europe. As a tailpiece to their October 1965 newsletter, Deputy Director John Wright took pride in reporting to employees around the world that DOS now had 'almost the ultimate accolade' after being mentioned in Ian Fleming's novel *The Man with the Golden Gun*. Since one of their 1:50,000 Overseas Survey maps of the Caribbean had helped James Bond to successfully reconnoitre the lair of the Bond villain Scaramanga, he observed that all DOS required now to feel that it had finally arrived was to be 'mentioned in a song by the Beatles'.

As a pioneer of aerial survey before the war in southern Africa, and making use of his wartime service experience, Kendall worked for the Hunting Group who resurrected the fortunes of British aerial survey after the war and were responsible for many of the DOS surveys. Recognising a business opportunity, Kendall emigrated to Canada for the Hunting Group and ran the Photographic Survey Corporation in the 1950s. With the Canadian government setting itself the goal of mapping the country in its entirety, and with its air force fighting in the Korean War, survey companies embarked on the immense task in the poorly mapped and geographically vast country. With a ready source of war-surplus aircraft, photographic equipment and experienced personnel, Kendall chose de Havilland Mosquitos and Boeing B-17 Flying Fortress bombers fitted with Williamson cameras to carry out the high-altitude photography.

Over more than a decade, surveys were flown that allowed the whole of Canada to be ground-mapped and aeronautical charts to be produced for the first time. But although photogrammetric equipment improved, and jet planes with pressurised cabins flown from paved runways superseded the wartime aircraft, the real pioneering work had been done. And while the fate of the photography created in combat during the Second World War lay in the balance back at RAF Medmenham, the Canadians had the foresight and vision to create their own National Air Photograph Library that still serves as a comprehensive historical record. In recognition of Kendall's dedication to the growth of Canadian business by forming more than thirty different companies, not only in the field of aerial survey but also in aviation electronics and associated fields, he was awarded the Order of Canada by his adopted country.

Epilogue

*'Will the Hon. Gentleman give an assurance that
photographs will not be allowed to rot . . .'*
Mr Edward Keeling MP, 10 April 1946

After Victory in Europe (VE) Day on 8 May 1945, Constance Babington Smith was seconded to the United States Army Air Force and continued to undertake PI work of the Pacific Theatre. While she was working at the Pentagon, her part in the hunt for the V weapons was given to the press and, with her photogenic looks doubtless helping what was in any case a strong story, dramatic headlines including 'Connie Saves New York' followed. In 1945 she received the MBE and in 1946 her achievements resulted in her being awarded the Legion of Merit – the first time that the US honour had been awarded to a British woman – with her citation recording that she was: 'the outstanding Allied authority on the interpretation of photographs of aircraft, she provided the Eighth Air Force with extremely vital intelligence for the strategic bombing and destruction of the German aircraft industry and contributed materially to the success of the USAF's strategic mission to Europe'.

When it was discovered that the Germans were developing a long-range rocket, one of the greatest fears that British Scientific Intelligence had was the possibility it could be used to deliver a nuclear payload. An understandable sense of trepidation meant that as early as 1943 the existence of the American-led Manhattan Project was revealed to Douglas Kendall by Dr R. V. Jones at the request of the Prime Minister's Office. In a move that helped the Allies to identify whether nuclear weapons would feature in the war in Europe, and under conditions of strictest secrecy, Winston Churchill needed photographic intelligence to confirm whether or not the Germans had atomic weaponry. Kendall's search proved negative and the Allies enjoyed success in developing a secret weapon that was to prove far more decisive and destructive than the German V-weapon programme. The world's first atomic weapon – code-named Little Boy – was dropped by the Americans on the Japanese city of Hiroshima by the Boeing B-29 Superfortress *Enola Gay* on 6 August 1945. When this was followed three days later with the detonation of the second atomic bomb – code-named Fat Man – over the city of Nagasaki, the futility of the Japanese position resulted in their surrender on 15 August 1945 and the signing of the Japanese instrument of surrender aboard the American battleship USS *Missouri* on 2 September 1945. This brought the Second World War to a close – a war in which the successful deployment of long-range rockets and nuclear weaponry revolutionised warfare for ever.

Within days of Victory over Japan (VJ) Day, a press conference was held at RAF Benson to celebrate the achievements of photographic reconnaissance during the war. In September 1943 the censors had allowed *Flight* magazine, following a visit to Benson, to write an article in very general terms. It revealed nothing about the station's modus operandi, and not very much beyond the somewhat obvious fact that Spitfires and Mosquitos could carry cameras. But with the war now over, the secrecy quickly evaporated and details of how the PRU had evolved since 1939 – when the first Spitfires were fitted with cameras and extra fuel tanks – were revealed at Benson to the gathered pressmen for the first time. With the range of aerial cameras designed and deployed in the 'camera war' on display, the latest Griffon-engined Spitfire Mark 19 – which had a top speed of 460 mph, a pressurised cabin, a ceiling of over 43,000 feet and a range of 1,500 miles – was enthusiastically revealed by the pilots and ground crew.

On the boundary of the airfield, the journalists were shown Fyfield Manor, home to the Joint Photographic Intelligence Committee, chaired by Group Captain Peter Riddell, which prioritised the myriad requests for aerial photography from across the Allies and maintained essential communications with the Joint Intelligence Committee. After the Operations and Briefing Rooms on the RAF station, they were shown the Old Mansion in the nearby village of Ewelme, which had been commandeered for accommodation and first-phase interpretation, and in the neighbouring orchard were shown the wooden huts that housed the photographic equipment used to quickly process the aerial films collected from returning PR pilots and for the mass production of photographic prints that were held in the Medmenham Print Library. In the intelligence room at the Old Mansion they saw an enormous map of western Europe on the wall, which had provided the backdrop for pilots to complete their pilots' traces, recording where they had flown on their missions, and for their debriefings by first-phase interpreters.

The publication in 1958 of *Evidence In Camera* by Constance Babington Smith was the first publicly accessible narrative of Allied photographic intelligence during the Second World War, and greatly increased public awareness. While researching her book, Constance had interviewed most of the principal characters on both sides of the Atlantic, keeping copiously detailed notes on more than a hundred interviews and meetings for future reference. She made typically pithy observations on the characters, from her first impressions on meeting them to their personal mannerisms and changes in their physical appearance from the time when she had worked with many of them. She met everyone from the pioneering Sidney Cotton – interviewed in his palatial suite at the Dorchester Hotel – to the colleagues she worked alongside at Paduoc House in Wembley, RAF Medmenham and later the Pentagon. While some, including Sarah Churchill, daughter of the wartime Prime Minister, were interviewed in the splendour of the Dorchester, others were interviewed at the English Speaking Union, the Aero Club, or in the Oxbridge colleges many of them returned to after the war. She would meet Frederick Winterbotham – Head of Air Intelligence at MI6 – on his Devonshire farm, Douglas Kendall for lunch at L'Escargot in Greek Street, Soho, and Group Captain Peter Stewart at his fashionable Mayfair flat in Hertford Street – noting that while he looked much older, 'his short stoutish figure and strutting walk seem exactly the same'.

Travelling to the United States for research in October 1956, she met American pioneers of photographic intelligence, many of whom still worked for the Pentagon. But in the case of Arthur Charles 'Art' Lundahl she visited Washington D.C. and the Central Intelligence Agency. At the beginning of the war Lundahl had been teaching photogrammetry in the Geology Department at the University of Chicago, and after PI training at the Anacostia Naval Base, Washington D.C., had served in the Pacific Theatre. Lundahl had been recruited from the US Navy to work for the CIA in 1953 by senior official Richard Bissell Jr, who from 1954 was responsible for deploying the Lockheed U-2 reconnaissance aircraft, which had undergone exhaustive testing at the top-secret Area 51 military base in Nevada. At the CIA, Lundahl would be responsible for the National Photographic Interpretation Center (NPIC) which, as fellow imagery analyst Dino Brugioni explained in his book *Eyeball to Eyeball*, used photographic interpretation, data processing, photogrammetry, graphic arts, communications, collateral research and technical analysis to extract intelligence from the aerial photography created using the U-2, other airborne surveillance platforms and satellite imagery. The work of these pioneers led to the rapid expansion of Imagery Intelligence (IMINT) during the Cold War, as military commanders and their political masters came to increasingly appreciate the power it offered due to intelligence questions it could answer.

Also at the CIA, Constance met its Deputy Director General Charles Cabell who reminisced about being sent to England by Brigadier General George Goddard in 1940 on a fact-finding mission, and recalled having met Major Harold Hemming at the Aircraft Operating Company in Wembley where he became convinced of the value of PI. Cabell was an enthusiastic and powerful supporter of the Lockheed U-2 programme, and in June that year had briefed West German Chancellor Konrad Adenauer, who agreed to overflights of the Soviet Union being flown from Wiesbaden Air Base, near Frankfurt. After the failure of the Bay of Pigs invasion of Cuba, Cabell would be forced to resign by President John F. Kennedy on 31 January 1962. But later that year, the value of the U-2 was dramatically highlighted after Soviet nuclear missiles were photographed in Cuba on 14 October 1962. In the thirteen days that followed, the Kennedy administration and Soviet Premier Nikita Khrushchev faced one of the major confrontations of the Cold War. Aerial photography would play its part throughout. At an emergency session of the Security Council on 25 October 1962, in one of the most dramatic moments in UN history, Adlai Stevenson II, the US Ambassador to the United Nations, used U-2 photographs to prove the existence of the missiles minutes after the Soviet ambassador Valerian Zorin had questioned their very existence.

In the suburban town of Chevy Chase, Maryland, Constance met Brigadier General George William Goddard. Born in London, England, the naturalised American citizen had worked as a PI since the First World War and was widely regarded and respected as the father of American photographic intelligence. A legendary innovator – who corresponded with his British counterpart Frederick 'Daddy' Laws on the development of aerial photography throughout the interwar years – Goddard developed techniques in night, high-altitude, stereoscopic and colour aerial photography. After the bombing of Pearl Harbor, Goddard was attached to the Navy before returning to the Air Force where he helped ensure that photographic intelligence had the best aircraft and photographic equipment.

In the village of Gurnard, on the Isle of Wight, Constance records visiting Claude Wavell, the wartime PI who specialised in enemy radar and whose work alongside R. V. Jones and his colleagues in Scientific Intelligence proved critical in comprehensively deceiving enemy radar, which helped determine so much of the success of Operation Overlord. She found a grey-haired, stout, elderly man with a bristly grey moustache, a shortish neck and a heavy double chin. The early riser and non-smoker now doted on his workshop, with its lathes and drills, that was crammed with neatly arranged rows of jars and boxes containing bits and pieces and gadgets. His pre-war experiences in aerial survey around the world, alongside his love of mathematics and precision instruments, had proven vital to the outcome of so much during the war. Wavell was one of many who worked for the Aircraft Operating Company, the British aerial-survey company that worked throughout the world – and which prospered in Britain thanks to its Aerofilms subsidiary – whose employees' skill and experience would prove so vital to the successful development of photographic intelligence.

In the summer of 1956 she visited many of the places featured in the story and described Paduoc House as still a shabby little Wembley factory with a pillbox fortification outside the front entrance, opposite a row of jerry-built Mock Tudor villas that looked as though they were made from cardboard. At Paduoc House, Constance had passed the interpretation course after the feared assessment by Peter Riddell and, after lunching with Constance numerous times throughout 1956, he accompanied her on a visit to Medmenham, Phyllis Court and the Benson area in July that year.

Riddell had been one of a handful of photographic interpreters working with Bomber Command before the war, and alongside his colleagues from the private sector he had rapidly developed photographic intelligence in a markedly different way from the Germans, a way that would ultimately prove successful during the war.

In Danesfield House, Constance found all the rooms surprisingly ornate, having only known the walls as boarded-up surfaces during the war. Travelling on to Henley-on-Thames and Phyllis Court, she found that the whole stretch of river still had the feel of a playground for the nouveau riche. At RAF Benson they found the Officers' Mess, where the photographic-reconnaissance pilots lived and where Riddell recalled that Walley the barman would play snooker with the pilots late into the night, and Sylvia the Catering Corporal who, although hopeless at catering, was a great character.

At the Shillingford Bridge Hotel, on the banks of the Thames, Peter Riddell explained that although the hotel had been taken over as a mess for 106 Group and was popular – as were many local pubs – with the PR pilots, the White Hart Public House in the village of Nettlebed was in a league of its own. Frequented by the elite photographic-reconnaissance pilots, if they could not be found at Benson it was always the first place people checked. Mrs Clements the landlady – Clemmie – was still there in 1956 when Constance visited, and happily reminisced about the pilots she clearly adored, and proved a font of knowledge on how to contact them. On the walls Constance records there being framed paintings of the reconnaissance pilots by the society artist Olive Constance Snell. She also noted that every year on the second Saturday in March a reunion was held. While an annexe to the pub known as 'the cottage' was used for after-hours drinking, the bar itself was the scene of the most famous parties at which the Scottish nobleman and PR pilot Lord Malcolm Douglas-Hamilton often played his bagpipes.

When aerial photography began to be created in significant volume during the First World War the military mind was faced with a dilemma and, while the glass plate may have been replaced by the hard drive, the question of whether imagery should be preserved or destroyed remains to the present day. By the time of the First World War armistice, the British had manually exposed millions of glass-plate negatives over the Western Front in Belgium and France, items that were dutifully transported back to the United Kingdom after the war. In the 1920s they were held by the RAF School of Photography at Farnborough, Hampshire, then under the direction of the father of British photographic reconnaissance Frederick 'Daddy' Laws. Some 130,000 of them would be gifted – in their original wooden boxes – to the newly created Imperial War Museum, Lambeth, who christened the hoard the Box Collection.

In 1927, in a decision hastened by the deterioration of the negatives and a need for space at Farnborough, those glass plates not considered historically important were destined for burial in a large hole in the ground – much to the consternation of O.G.S. Crawford, the Archaeology Officer at the Ordnance Survey who had become a passionate advocate of aerial photography for archaeological prospecting after his wartime experiences.

With interest in photographic reconnaissance minimal within the interwar RAF, frustration resulted in Laws leaving for a career in commercial aerial survey, where other veterans including Francis Wills had continued to develop technology and techniques that would ultimately prove critical to the success of Allied photographic intelligence during the Second World War. Despite the German air attack on Paduoc House in October 1940, the Aerofilms collection survived, unlike the limited amounts of Ordnance Survey aerial photography that perished during the 1940 Southampton Blitz. In spite of the presence of Crawford, the Ordnance Survey had proven to be as indifferent as the RAF during the interwar years.

The question of what would happen to the aerial photography created during the Second World War obsessed Crawford, who at the height of the Battle of Britain was even recorded to have badgered 'Daddy' Laws to consider a proper system of post-war photograph preservation. When the time came – and with the captured GX photography and post-war survey operations including Revue and Casey Jones exacerbating the situation – the Air Ministry faced the problem of what to do with their enormous and rapidly growing collection of aerial photography. Some within the military believed that aerial photography was outdated from an operational perspective shortly after it was taken, and that because of long-term storage costs only the 'latest and best' should be preserved, while others felt that everything should be preserved on the basis that over time it develops an increasing importance and allows for the comparative analysis of places in their historical context. After all, it was the Medmenham Print Library that allowed information to be gathered from the chronological viewing of places over time, information that proved so critical to photographic intelligence during the Second World War, never more so than during Operation Crossbow.

This clash between those who preferred to keep only the latest and best and those who would rather keep everything began after the First World War and continued after the Second World War when service personnel struggled to juggle storage space with other demands. It meant, for example, that with Project Casey Jones providing complete coverage of western Europe it could be claimed that there was little intelligence value in retaining millions of the wartime photographs held in the Medmenham Print Library. This coincided with calls from many of the wartime interpreters who had been drawn from the ranks of peacetime academe and who now returned there, for access to the photography which would be of incalculable value to their own fields. Geographers were arguably the strongest advocates of this, followed by archaeologists and, to a lesser extent, historians. They would now find themselves pitting their arguments against their old masters.

In April 1946 the Member of Parliament for Twickenham, Mr Edward Keeling, asked a crucial question in the House of Commons. The Royal Geographical Society had recently exhibited RAF aerial photography – to demonstrate its application for civilian scientific use – and he wanted to know from the Secretary of State for Air whether the photographs taken during the war would be preserved in good condition until those of value could be selected and placed in a National Library similar to the one which had been established in Canada. To allay his concerns, Keeling was assured that the photographs 'will not be allowed to rot' while arrangements for their future were under review.

But this failed to convince David Linton, the recently appointed Professor of Geography at Sheffield University, who, making use of his wartime experience as an interpreter, had been writing articles analysing aerial photography and its value to geographers. Having made extensive use of the resource since reporting for duty at Paduoc House, he decried the lack of any high-level policy decision, knew that the demobilisation of RAF personnel meant the manpower of the sections concerned would be at a care-and-maintenance level, and correctly foresaw that an impasse would be the result. At the 1946 Conference of the Heads of University Departments of Geography, during a special session on aerial photography, he outlined his vision that university geographers should have copies of the photography in order that they could train students in photographic interpretation, illustrate features of regional and world geography, and undertake local and special studies.

Discussions within the Air Ministry at the time favoured the creation of a civilian-controlled library and recognised the Medmenham Print Library as an important national asset that could also bring them, and the RAF, positive publicity. The fact that service personnel were then being used to meet civilian requests for photography – from academics, local authorities, learned societies and the public at large – was considered a waste of resource. Arguments against the library included a concern that the public thirst for aerial photography was increasing and for reasons of security it ought to be diminished rather than encouraged. It was also recognised that even though requests for photography were judged on their individual merits, enemy agents could be very convincing, and there was a certain risk in supplying anything.

In January 1949 the National Library of Air Photographs Committee was created by the Air Ministry to provide the Air Council with recommendations on the fate of Medmenham Print Library and advice on whether they should create a National Library of Air Photography. By this time, Squadron Leader Claude Wavell was officer in charge of the Print Library and he showed fellow committee members the Library in operation and the methods used. By this time over 20 million aerial photographs of foreign countries and the United Kingdom were held in the Library and it was growing at a rate of approximately 1.5 million aerial photographs per year.

With the 1946 warning from Winston Churchill – that from Stettin in the Baltic to Trieste in the Adriatic an 'iron curtain' had descended across the Continent – doubtless ringing in people's ears, issues of security were paramount at this early point in the Cold War. When assessing the security matters that had to be resolved before a publicly accessible library of aerial photography could be created, the

Air Photographs Committee aimed to strike a balance between the differing viewpoints.

The risks associated with the creation of an aerial-photography library were explicitly outlined to the committee by Air Commodore Ian Brodie, the Deputy Director of Intelligence at the Air Ministry responsible for security, in a communiqué dated January 1949. On the basis that British Intelligence would be delighted if it could obtain aerial photography of another country comparable to that created by the RAF during Operation Revue, he observed that the creation of a publicly accessible aerial-photograph library in Moscow would be an intelligence godsend. While accepting the fact that it would be relatively easy to withhold aerial photographs of vital areas, the existence of an otherwise complete coverage of the United Kingdom would nevertheless provide an enemy with a library of comparative cover. Since, during a war, defences and industry are typically redeployed throughout a country, photographic coverage flown during war by an enemy already in possession of peacetime photography would allow them, through skilled photographic interpretation, to reveal quickly the deployment of wartime measures, their development and capabilities. While the ideal security procedure – as was the case in Soviet Russia – would be to prohibit all unauthorised aerial photography, British unwillingness to adopt such a stance meant that the next best thing was to make it as difficult as possible for the Russians and other enemies to obtain copies of aerial photographs from any new National Library.

Growing security concerns about the publication of aerial photographs led to a D Notice being issued to the press who, along with the Ordnance Survey, mapping agencies and aircraft companies, agreed to control the publication of aerial photography. This meant affording a degree of protection to RAF and naval airfields; anti-aircraft defences; coastal defences; military storage depots; Royal Ordnance and Royal Navy factories; HM dockyards; radio and radar installations; Admiralty, War Office, Air Ministry and Ministry of Supply Research and Development establishments, including atomic establishments; industrial facilities of importance to national defence, including aircraft and aero-engine factories, oil refineries and oil storage plants; and airfields adjoining aircraft factories.

With key installations carefully monitored by the Key Points

Intelligence Directorate, which had originated in the Intelligence Section of the Camouflage Branch of the Air Ministry during the war, aerial photography of Key Points was strictly controlled throughout the Cold War. Since the Ordnance Survey had to ensure that Key Points were removed from maps, the Air Photographs Committee was keenly aware that although the proposed National Library would be publicly accessible, in order not to compromise national security there would be a great many gaps in coverage. Notwithstanding the fact that Air Commodore Brodie was unaware of any enemy attempts to obtain RAF photography, he was mindful that the Russians and their agents were likely to be the best customers of any proposed Library so every effort should be made to ensure that gaps in coverage were not widely known, since wherever there was a gap, the Russians would correctly suspect that there was something to hide.

Although the Ordnance Survey might have ensured that Key Points were removed from cartography, there was to be a period while the creation of the National Library was being considered when aerial photography of Key Points became readily available in a way that it perhaps wouldn't be again until the advent of Google Earth. When the final report of the Davidson Committee – established by the government to consider the effectiveness of the Ordnance Survey – was issued in November 1938, it recommended the formation of an air-survey unit to allow more rapid updating of mapping, much as the Aircraft Operating Company was then mapping parts of the world. Practical difficulties and the outbreak of war meant this did not happen, but when wartime experiments using RAF aerial photography revealed that mosaics could be cost-effectively and relatively quickly created, the Air Ministry agreed to undertake this work once the war with Germany was over. The aim of the mosaics was to remedy the lack of up-to-date large-scale mapping that was badly required for post-war planning and reconstruction.

With the Ministry of Town and Country Planning advising on what areas should be covered, the first series of mosaics was produced, and in a Treasury-encouraged attempt to recoup costs the Air Photo Mosaics were published and sold to the public. To promote their existence, a British Pathé newsreel in 1949 filmed an Avro Anson leaving RAF Benson on a sortie to photograph London. Accompanied by suitably patriotic music, the jaunty narrator explained to cinema audiences how aerial surveyors were now recording the changed face of Britain, that Londoners could now buy a photograph of their district in which they could see how their garden was growing, and that if the public took to the concept, the whole of Britain would be mapped in the same way. But when it became apparent that the mosaics were prohibitively expensive, and many more were freely supplied to the copyright libraries than were sold, the Treasury had second thoughts. When the following year the mosaics were 'republished with slight amendments' the keen-eyed would have noticed that in the republished editions certain installations had been obscured by false fields or painted clouds, whilst in the copyright libraries the undoctored mosaics had been hastily withdrawn. Ultimately, increased Cold War security concerns resulted in the mosaics' inevitable withdrawal from sale in the early 1950s and in access to the doctored copies in the libraries being withdrawn. The relative unpopularity of the mosaics and the ready availability of regular mapping at a more realistic price meant – with the possible exception of Soviet agents – their withdrawal from circulation passed off without regret.

One of the biggest issues identified by the committee that could easily plague the development of the resource was cataloguing. All the Medmenham photography was searched using cover traces – the linen sheets that were laid over corresponding map sheets to identify photography covering a particular location – a system which had proven itself effectively during the war. To find photographs of an area or pinpoint involved identifying the location, in the following order: on a Base Map Index, then on the cover trace which in turn led to the box of prints, inside which was a copy of the sortie plot from which the exact photograph references could be found. To provide an effective service in the proposed library, it had been visualised that both the photography and cover traces should be transferred there from Medmenham. But since this would require military requests for imagery to be passed through a library staffed by civilians, it was deemed that a duplicate copy of the cover trace should be created that could be retained by the RAF, otherwise the civilian library itself would need to be expanded to include a department staffed by military personnel.

The fragility of the cover traces, and the difficulty of duplicating them, led to an ingenious proposal by Claude Wavell to create a punched-card-index system. He proposed replacing cover traces altogether with 20 million punched cards that would contain geographical and subject information. His proposal was examined by Air Ministry experts who calculated that Wavell's system would not only take 50 per cent less effort to create than cover traces but would speed searches too. They recognised that while geographical information was of primary importance when searching for aerial photography, and the compilation of an effective subject index was not without its challenges, the ability to search by both place and subject increased immeasurably the extent to which the photography could be useful. Although the cost of duplicating the information would be considerable, and the Air Ministry were anxious to dispose of as much photography from the Medmenham Print Library as possible that was deemed not to have any immediate intelligence value, they were adamant that they should retain the capacity to undertake their own searches. This was to prove a major stumbling block.

It was proposed that the new Library should be created within a two-mile radius of Kingsway – a major road running through central London – in view of its proximity to Fleet Street, from where a large demand for the photography was expected, and to Adastral House, home to the Air Ministry since 1919. Whilst the Ministry of Works had been advised about storage requirements, consideration had also been given to the number of staff, which should not exceed eighteen, and to the pay grade of the civilian staff who would be employed by the Air Ministry. Since Library staff would be dealing with the general public, something that would require an 'amount of tact and discretion', it was recommended that personnel should be selected with care and be of 'good type' as a fair degree of the success of the public service and security aspects would depend on this. From the outset it was decided that the scheme would have a greater likelihood of success if it was presented to the Treasury and Ministry of Works as something that would generate profit.

By 1957 the National Library of Air Photography had failed to materialise. This fact – alongside evidence that suggested post-war aerial photography taken by the RAF was deteriorating in storage – compelled Alan Walton to appeal for the interest and support of Major Hugh Fraser, the prominent Conservative MP for Stafford and Stone who would be the target of a botched IRA assassination attempt in 1975. A lecturer in geography at the nearby University College of North Staffordshire, Walton was then actively involved in promoting the wider use of aerial photography in geographical studies and was committed to the creation of a National Library that would provide 'bona fide' researchers with easy and cheap access to this still relatively modern photography. After the first attempt in 1949 by David Linton to promote a similar scheme foundered on Security Service obstructions, post-war confusion and the apathy of some government departments, Walton and other university geographers considered the time was now ripe to pursue the matter again. With a commitment from the politician to raise the matter, a questionnaire was circulated around British university departments of geography into their use of aerial photography in the hope that it would help generate support for this new scheme.

In its reply of September 1957 to a letter from Major Fraser, the Air Ministry was at pains to explain that they had not lost sight of the 1946 proposal for a National Library of Aerial Photography. It was unfortunately – despite considerable discussion and thought – issues of security and cost that had led them to consider the scheme impracticable. A demonstration of their commitment to the universities and learned societies of Britain was conveyed by the fact that in 1956 alone over 15,000 prints had been supplied, and that aside from their gradual deterioration, the best possible care was being taken of all the original negatives, and the fate of the Library was being looked at once again.

Since the aerial photography in the Medmenham Print Library covered everywhere in Britain, as well as places overseas, it raised a series of legal and political issues which prompted the Air Ministry to convene an interdepartmental meeting in January 1958. Outside government, the Imperial War Museum, who held all the surviving aerial photography taken during the First World War – plus copies of the aerial photography supplied by Medmenham to the Ministry of Information during the Second World War were invited on the basis they were considered a likely new home for the photography. In advance of the meeting, the Director-General visited Claude Wavell at Medmenham, with his curator of photography, to familiarise himself with the collection so that they could speak with greater confidence in the meeting. Also invited were the Ministry of Education, who could advise on the potential demand from schools and training colleges, and the University Grants Committee who were expected to provide an objective assessment on the use of aerial photography in British universities and the feasibility of running the Library from one. The greatest importance, however, was placed on the views of the Public Record Office – precursor to The National Archives – the Colonial Office, Commonwealth Relations Office and the Foreign Office, who were all asked to provide their written observations before the meeting.

As word of the meeting began to spread among the academic community, the possibility that the aerial photography concerned could be destroyed caused alarm and prompted a number of letters to the committee. This included one from geography professor Andrew O'Dell on behalf of the University of Aberdeen, which had purchased the Tillycorthie Estate in rural Aberdeenshire for their Department of Agriculture. And whilst the land on the estate was actively used they now found themselves in the position of owning a mansion house without a purpose. With Dr R. V. Jones also then ensconced at Aberdeen, as Professor of Natural Philosophy, the University College of North Staffordshire was not the only academic institution with an interest in the collection. The only stumbling block that O'Dell foresaw was obtaining the miles of shelving that would be required to house the collection. With support from the second Lord Tweedsmuir, a former Rector of the university, O'Dell made an impassioned plea ahead of the meeting in January 1958 that all the geographers would unite with him in hoping that this magnificent collection, a treasure trove which had been understandably locked away, could now be preserved and at least portions of it made accessible to the public.

In the meeting there was general agreement about the need for a central library system to be retained in a civilian government department where the master negatives and cataloguing would be preserved now and in the future – preferably the Public Records Office who themselves considered that as the statutory body concerned with the preservation of public records they had 'plain responsibility' and were uneasy about such records being housed by a university.

In July 1955 the conference at Medmenham was attended by members of the RAF, Air Ministry and a panel of senior university geographers. The report of the meeting was encouraging for a first move, and it seemed as though matters were finally progressing, only for the initial impetus to wane as the months passed. The views of the university representatives at the conference were that the whole of the collection was of value – for scientific, historical and teaching purposes – that it should be preserved in its entirety and that it should become accessible to the public, both British and foreign. They recognised that since the photography was taken at different times of the year, in many different conditions of light and shade, it would provide invaluable information to scientists interested in geomorphology, archaeology and the geographical aspects of land utilisation. The absolute ideal, in their view, was for the government to erect and maintain a building, with facilities for the inspection of the photography by accredited members of the public. Even if success was a remote possibility, the academics felt strongly that they should press for such an option since it was most unlikely that such an important collection would ever be brought together again.

With steadily decreasing interest in the Medmenham Library from the Air Ministry the academics clearly felt compelled to make suggestions, apart from the unlikely 'ideal' one, that would provide enough options for it to remain intact. The possibility of rash decision-making, of the authorities preferring the space occupied by the Library over the photography itself, was feared to be a real possibility by those academics with wartime military experience. Their first proposal was a somewhat radical form of public/private partnership involving an existing aerial-survey firm – such as Hunting Aero Surveys – who, it was suggested, could provide land on which the government should build suitable accommodation. After the transfer of photography it was proposed that responsibility for maintaining the collection, and for the provision of viewing facilities for members of the public, would transfer to the company. The second possibility was presented as a short-term option that would require the transfer of all the steel shelving from Medmenham and would see the storage of the Library in a large disused brickworks in the village of Madeley, three miles from the University College of North Staffordshire. Should it subsequently be regarded as impracticable to maintain the entire collection, it was contested that this option would at least allow the RAF to get rid of the Library quickly from Medmenham. Not surprisingly, the academics did not consider it their business to consider issues of national security.

The Air Ministry was not in such a favoured position.

In September 1960, at a meeting of the Air Council, the possibility of the Air Ministry establishing a National Library of Aerial Photography was decisively rejected. The aerial photographs covering England and Wales required by the Ministry of Housing and Local Government were transferred to them, those of Scotland went to the Department of Health in Edinburgh, and those of Northern Ireland to the Ministry of Finance in Belfast. It raised the question of what would happen to the photographs of overseas countries that were surplus to Air Ministry requirements, many of which brought a myriad of political and security difficulties with them.

With the Air Ministry now keen to clarify whether the offer of accommodation by the University College of North Staffordshire still stood, things began to move apace. The Air Ministry had to finalise what it was legally and politically possible to do with the photographs, to agree the best disposal policy when photography had no continued intelligence value, and could be released, and what would happen when it could not.

At a meeting of the University Senate on 23 November 1961, permission to accept the collection was given, although from the outset reservations were raised about its vast scale, the difficulties they would face accommodating it and the possibility that it could become a financial burden. This reticence was understandable, particularly since no financial assistance was ever forthcoming from either the Air Ministry – who remained the owners of the photography – or the Public Record Office who had to ensure that these public records were correctly preserved and publicly accessible.

When the decision finally came after years of 'very lengthy negotiations', work segregating the Medmenham Print Library began in October 1961. Fittingly christened Project SEGMED, the SEGregation of the MEDmenham collection would take until August 1963, and involved the SEGMED team transporting prints to JARIC for retention and to the academics in Staffordshire, while those no longer required were destined for Battersea Power Station in central London. The team were a mixture of regular and National Service airmen, and although a Flight Lieutenant at JARIC was nominally the Project Officer, Corporal Anthony Plumb was the NCO in charge throughout.

For the first few months the team were regularly inspected to see how things were progressing, but they were soon left to their own devices. While on the one hand this was complimentary and implied that a satisfactory job was being done, on the other it did underscore a general lack of interest in the whole process. Since Project SEGMED was being handled by a detachment of JARIC personnel based at RAF Medmenham, then occupied by Signals Command, they found themselves immune from station duties, much to the annoyance of the Station Warrant Officer, who wanted their names added to the guard-duty roster. With this only serving to enhance further the popularity of the posting, it made Corporal Plumb a tad suspicious that delaying tactics were being employed by the team to drag the whole process out.

Even so, the mammoth task of deciding what photography was destined for where, required the examination of each box of prints, plus the associated plots, to determine where each sortie covered. Photography which could not be released to the University College included those countries that were neutral during the war and which the Foreign Office could not approach without embarrassment – including Portugal, Spain and Switzerland – plus photography of Commonwealth and colonial territories, specific countries and the Warsaw Pact or Soviet-occupied countries. These sorties were transferred to JARIC and would not be released until 2004, when a four-year project began to declassify large parts of the JARIC Film Vault into the public domain. When the photography arrived at Brampton, things were approaching a crisis as the wartime Film Library from RAF Benson had already been combined with the aerial films created during the enormous post-war aerial survey programmes.

In December 1960 the Film Library had been transported from RAF Benson to RAF Feltwell in Norfolk, where it was sorted out for dispersal. This included transferring films to the Defence Surveyors in Feltham, west London, who were responsible for the production of military maps and charts, and could answer all manner of geographical, hydrographical, oceanographic and geological questions. And though many films were unwanted and were recycled for their silver content, the remainder were transferred to JARIC where they were destined for the so-called Belle Isle Film Vault, which by August 1962 is recorded as having held a staggering 400,000 rolls of film.

As the Belle Isle vault was used to store dangerous nitrate-based film it featured a specially reinforced concrete roof, which, rather disconcertingly for the occupants of the RAF married quarters surrounding it, was regularly hosed down in hot weather by junior airmen. The volatility of nitrate-based film was experienced by Corporal Plumb during a visit to Feltwell when a hole was made in the side of a revetment into which a roll of film was stuffed. After the centre core of the film was pulled out to form a two-foot-long cone, an airman lit the end with a match very carefully and jumped well clear as the film ignited and produced a violent rocket-like flame. This vindicated the RAF decision to copy the most important nitrate-based films onto acetate-based film in the 1960s and 1970s, which means that millions of Second World War aerial photographs released from JARIC in the twenty-first century are now held on perfectly duplicated safety film.

At Medmenham, the material destined for destruction included both RAF and American photography and fell into three main categories. Oblique photographs of attacks on enemy targets, including some particularly dramatic attacks on shipping, vertical photographs where no sortie plot existed and the identification of where the photography covered was impossible, and thirdly, large amounts of low-level oblique photography taken by the Americans in the closing days of the war, which graphically recorded the impact of the war on Europe. It included photography of POW and concentration camps often taken while their occupants were being liberated by the Allies, the mass destruction of German cities, and, since nothing attracts the attention of photographic intelligence more, photography of airfields and of aircraft in flight. However interesting the photographs might have been, they were classified secret and, on the assumption that the Americans would retain copies, Medmenham was under instructions to carefully destroy them. This was necessary to avoid repeating an incident that occurred during the 1957 relocation of JARIC (UK) from Nuneham Park to RAF Brampton when aerial photography was embarrassingly scattered over the English countryside from the back of an RAF vehicle, all of which had to be surreptitiously recovered.

Since many of the prints were classified secret, Corporal Plumb enquired about the security of the disposal method being proposed. The furnaces of the coal-fired Battersea Power Station on the south bank of the Thames were frequently used by the government as a discreet and foolproof means of destroying used banknotes and classified documents. Or so they thought. It transpired that HM Treasury had once sent a large quantity of used banknotes for destruction, and presumably much to the later surprise of the furnace stokers some of the more tightly bound banknotes survived the process. With this lesson learnt, the SEGMED team had to fill paper sacks loosely with the photography, which they then transported from Medmenham to central London for destruction.

Situated roughly halfway between Manchester and Birmingham, the industrial city of Stoke-on-Trent, Staffordshire, was a global power-house in the pottery and ceramics industry in the late 1940s. Famed for pottery manufacturing since the Industrial Revolution, the sale of pottery back-stamped by the likes of Minton, Royal Doulton, Spode and Wedgwood made the Potteries famous throughout the world, and financed the development of a unique urban landscape characterised by hundreds of bottle kilns. Before the advent of the post-war Clean Air Acts, chronically poor living conditions were experienced by the working class who lived alongside the city's Pot Banks, its iron, steel and coal industry. Despite the creation of Stoke-on-Trent early in the twentieth century from six independent towns – Burslem, Tunstall, Hanley, Stoke-upon-Trent, Fenton and Longton – these communities retained their unique sense of identity long into the century. Whilst this had inspired Hanley-born Arnold Bennett to write his novel *Anna of the Five Towns*, the educational plight of these Potteries towns inspired a group of pioneer academics under the leadership of Lord Alexander Lindsay, Master of Balliol College, Oxford.

After service in the First World War, Lindsay had returned from France determined to develop his academic work in the field of adult education, notably with the Workers Educational Association. During the October 1938 Oxford by-election, immediately after Chamberlain had signed the Munich Treaty, the Scottish Presbyterian had stood on an anti-appeasement platform against a young Quentin Hogg who stood as the government – and, by default, the appeasement – candidate. Despite cross-party support for the socialist Lindsay that included dissident Conservatives and future Prime Ministers Winston Churchill, Harold Macmillan and a young Edward Heath, who received criticism for accusing Chamberlain of 'turning all four cheeks to Hitler at once' during a debate in the Oxford Union, Lindsay and his anti-appeasement campaigners who united around the rallying call that 'a vote for Hogg is a vote for Hitler' were unsuccessful. But this venture into politics was to prove significant and came to influence post-war political giants from across the political spectrum.

When the anticipated war came, Lindsay was appointed chairman of the Joint Recruiting Board that allocated work of national importance other than military service to conscientious objectors. Accepting a peerage in 1945, he would be heavily involved in the reformation of German universities before retiring from Balliol at the age of seventy when he became founding Principal of the University College of North Staffordshire in 1949. His vision for Britain's first post-war University College was for an academic community with an open and democratic ethos that broke down barriers between teachers, students and the local community.

For the academics attending job interviews with Lindsay in 1949 – and with many of them coming from the relative serenity of an Oxbridge college – all five senses would have been stimulated on their arrival at the city's main railway station. Invariably they would be greeted with 'Stoke Smoke', courtesy of the Pot Banks, and a constant blast of light, noise and smell from the enormous blast furnaces of the Shelton Bar steelworks. The place epitomised the industrial city. Travelling westwards along terraced streets and through anonymous interwar suburbia, they would have arrived in the immediately adjoining but doggedly independent town of Newcastle-under-Lyme, which forms another part of this polycentric conurbation.

On a hill three miles west of the market town, the University College now occupied Keele Hall estate after the Sneyd family – faced with crippling death duties – sold their ancestral home of more than 400 years. Situated among beautifully cultivated woodland plantations, lakes and gardens, Keele Hall had been requisitioned by the War Department in 1939, and along with the dozens of Nissen huts that appeared in the parkland grounds was used by the Allied armies throughout the war, becoming afterwards a transit camp for refugees. On their arrival in 1949, the academics discovered – courtesy of its previous occupants – that half of Keele Hall's roof was covered in galvanised iron thanks to a fire that had damaged part of the top floor, and its courtyard was covered with debris. With this following their experience of Stoke Smoke and the customary Keele drizzle, two early, glum, weather-sodden visitors to the University College were not exactly encouraged when, on meeting Lord Lindsay for their interview, they were told that the place had to be seen with 'the eye of imagination'.

One of these academics was Stanley Henry Beaver, an enthusiastic and impatient man of sharp intellect, who was appointed the foundation Professor of Geography. Early in his career he forged a working partnership with Lawrence Dudley Stamp, a fellow geographer at the London School of Economics, and together their names and textbooks would become well-known among geography students. Whilst Stamp focused on the compilation of the Land Utilisation Survey of Britain, which used an army of mostly school-age volunteers who recorded the use and frequent misuse of land in their neighbourhood and helped create the country's first comprehensive land-use survey since the Domesday Book. Beaver concentrated on studying change in Britain's industrial areas, particularly the coal mining, iron and steel industries and how technological change had a spatial and social impact. On the outbreak of war, Beaver had worked on the production of intelligence handbooks for the Naval Intelligence Division of the Admiralty before he was invited to become a Research Officer with the newly created Ministry of Town and Country Planning, where his mastery of economic geography was used to investigate what mineral resources were available and where. This work would prove essential for both post-war reconstruction and New Town construction, and involved the extensive use of aerial photography to supplement his field work.

With the Staffordshire Potteries on its doorstep, the University College was to prove a perfect base for Beaver, who studied the economic geography of an English region that would be transformed in his lifetime. This transformation was epitomised shortly after his death by the 1986 Stoke-on-Trent Garden Festival for which 180 acres of highly contaminated and mine-shafted derelict land, once occupied by the Shelton Bar steelworks, was reclaimed in the hope that it would lead to the regeneration of the city.

When the first students arrived at the University College in 1950, they lived and studied in Keele Hall and the 'temporary' wartime buildings, many of which would become familiar to generations of learners. With strict regulations on the wearing of academic gowns, contact between members of the opposite sex and a strictly enforced curfew of 11 p.m. every night, Lord Lindsay's educational experiment in North Staffordshire began. The presence of an Oxbridge-inspired High Table in the refectory might have seemed a little grand, given that the campus resembled a building site and was frequently engulfed in a sea of mud that made wellington boots indispensable. But these conditions and the austerity of 1950s Britain failed to diminish the strong collegiate spirit and the University College flourished. Becoming the University of Keele in 1962, by the time the aerial-photography collection arrived on campus from the brickworks, the institution had grown dramatically in scale but alongside wider social change the camaraderie, considered so important by the Keele pioneers, had changed irreparably. With 1968 in particular a year of unprecedented revolution, a year in which students demonstrated against the war in Vietnam, Enoch Powell made his controversial 'rivers of blood' speech, the civil rights movement gathered pace and the women's liberation movement raised questions about gender roles. At Keele this student revolution included sit-downs at High Table, a march on the vice-chancellor's house, and, for the more radical, the occupation of the University Registry for two days and nights – although the protesters are recorded as having been careful enough to leave it in excellent order without any damage or interference with the paperwork.

In 1969 Keele students gained notoriety in the national press when, among other incidents, a petrol bomb was thrown into the architecture department and windows in Keele Hall and the library were smashed. This only helped to strengthen further the opinion among many locals that the university was 'the Kremlin on the hill' and a hotbed of left-wing activism. All these incidents, however, were trumped by an incident in June 1970 when the police were called to the campus after twenty or so students stripped off to sunbathe in the nude on the campus lawn. After photographs found their way to the *Daily Mail* and *News of the World* – the action of an entrepreneurial student – the reputation of the university and its seemingly radical students fell to its lowest ebb. Under the circumstances the university was perhaps something of an unusual choice for a once top-secret collection of aerial photography.

From the 43,120 sorties inspected at Medmenham, Project SEGMED resulted in 28,076 sorties being transferred to Keele and another 3,083 to JARIC whilst 11,970 were destined for the furnaces of Battersea Power Station. Even though 5.5 million photographs in 40,000 boxes were now safely in their care, the university faced a myriad of security and diplomatic problems that meant for the time being access to the collection was restricted under the Official Secrets Act. This was still the case when the collection moved to a purpose-built facility – complete with electrically operated mobile shelving – on the university's parkland campus in the summer of 1966, courtesy of finance from the University Grants Committee. But although the vetting process carried out at Medmenham reduced considerably the size of the collection it still contained – based on modern-day geography – photography of thirty-three different European sovereign states and dependent territories. With a concern that Britain must not provide third parties with photography that could endanger its Cold War allies, the university was under strict instructions from the Foreign Office that permission for access must first be given by each of the countries concerned. In February 1967 this was forthcoming from the Belgian, Danish, Dutch, Italian, Luxembourg and West German governments, and although more countries followed, in the early 1980s the French, Greek and Norwegian governments still required clearance for visitors, via the Ministry of Defence, one month in advance of a visit.

After the collection arrived at Keele, Professor Beaver appointed Alan Walton, the geography lecturer who incited Major Hugh Fraser MP to question the fate of the collection in 1957, as its founding Director.

In the early years at Keele, and providing a researcher had sufficient stamina to navigate the diplomatic hurdles, easily the greatest challenge was awaiting them on their arrival. Much to the frustration of Beaver and Walton, the Air Ministry failed to supply the cover traces – painstakingly compiled at Medmenham by centuries' worth of WAAF labour – that made it possible to locate within minutes specific items among the millions of photographs. This made the arrival of the 40,000 boxes that were methodically arranged as they had been in the Medmenham Print Library a bitter-sweet victory after all the years of lobbying. At first there was no effective means by which to search for photography taken on a particular date, by a particular squadron, relating to a particular subject and most critically, given they were geographers, of particular places. Undeterred and with a determination to make the collection progressively searchable, geographically at the very least, Walton set about the mammoth task of developing a catalogue using the information contained on the sortie plots which, thankfully, had been supplied by the Air Ministry. Modifying a cataloguing system used by the Royal Air Force, the Keele system linked each target photographed to the corresponding ten minutes of latitude and longitude. Although in deteriorating health, in 1975 Alan Walton contributed a chapter in a Festschrift on the retirement of Beaver, and was able to report that the process had been completed.

This meant that a search of the ten-minute square cards provided the researcher with a list of the sorties that had photographed in a particular geographical square, and meant they could now trawl through the corresponding sortie plots and discover whether their area of interest had been photographed. But for well-photographed areas, such as those in the industrial heart of the Rhineland, this process could take hours. By viewing the sortie plots, a researcher could determine whether the particular part of the ten-minute square had actually been covered by a sortie. The often colourful plots – compiled by annotating large military maps with the footprint of each photograph – were trimmed and pasted onto large sheets of paper, alongside all the other targets photographed during that sortie. Next to the geographical information, the sortie plots contain data on which squadron or group had flown the sortie: flying date, time, altitude, the focal length of the camera and the quality of the photography taken.

Walton had recorded that by the mid-1970s the principal users of the collection were archaeologists and, increasingly, modern-history academics whose usage had increased as first-hand memory of the war diminished and studies of the wartime period grew in popularity. And although city archivists, air attachés of embassies of European countries – often a euphemism for intelligence gatherers – legal advisers, publishers and explosive-ordnance-disposal units used the collection, the number of geographers is recorded as small, and the number who successfully located coverage even smaller. This was on the basis that geographical areas were only photographed during the war if there was a military reason to photograph them, something which only helped to further diminish interest in the collection amongst geographers. And although the photographic-reconnaissance squadrons might have recorded the coastline of Europe between 1940 and 1945 in greater detail and frequency than any coastline in history, interest in the resource for the study of coastal landforms was seemingly limited. Walton's enthusiasm clearly remained undiminished, with his comparison of the resource to that of a 'Domesday of Europe' that not only recorded the continent at war but the European cultural landscape before its transformation by post-war urbanisation, industrialisation, and improvement programmes such as the *remembrement* in France, where land holdings were re-ordered, consolidated and enlarged. But with the retirement of Beaver followed by the premature death of Alan Walton in 1976, and despite their Herculean efforts and vision, the future of the collection looked increasingly uncertain.

Following the death of her husband Sheila Walton became curator, but the costs of running the collection without any endowment or funding from its owners, the Ministry of Defence, or its official guardians, the Public Record Office, meant the costs to the university were proving increasingly onerous. In 1979, by when the earliest parts of the collection were forty years old – and with concern growing about its physical condition – matters were brought to a head when the university's vice-chancellor wrote to the Lord Chancellor's Department. When the Keeper of the Public Records replied on the Lord Chancellor's behalf, he reminded the university of its responsibilities under the Public Records Act to actively conserve the records in its care before observing – seemingly without irony but doubtless to discourage any appeal for funding – that the Public Record Office couldn't afford to deal with its own conservation needs.

Matters were not left there and prompted the creation of an advisory committee with representatives of the Public Record Office and Air Historical Branch – the historical archive and records service of the Royal Air Force – which appraised the options available to them. With the physical conservation of the collection of increasing concern, advice was sought from Kodak on the creation of a photographic copy of the collection, but by the early 1980s things had reached a polite impasse, and with no sign of financial assistance forthcoming the university was left largely to its own devices and faced the perennial funding problem. Then something unexpected happened that dramatically changed the fortunes of the collection.

In 1984 a large defensive anti-tank ditch containing thousands of explosive projectiles, bombs and ammunition was discovered under a recently constructed housing project in West Berlin. For some time interest in the collection had been growing from bomb-disposal experts, and this discovery and the resultant press coverage propelled matters forward. In Germany the wartime legacy of unexploded ordnance was so great that the West German states established *Kampfmittelbeseitigungsdienstes* – War Ordnance Disposal Services – immediately after the Second World War. Their job was and still is to locate and render harmless unexploded bombs, mines and other munitions, to ensure that before construction work begins on an affected site or when a suspected threat is discovered, it is made safe and destroyed, and the risk of death and injury from inadvertent discovery and accidental detonation is minimised. In the early post-war years, with the war a recent memory, the Germans depended greatly on both eyewitness testimony and the 'Bomb Books', kept by the civilian authorities during the war, that recorded the approximate location of unexploded bombs.

In Britain the government had collected information in a similar way, using the police, air-raid wardens and military personnel to record where, when and what type of bombs had fallen during an air raid. This was passed on to the Ministry of Home Security Bomb Census Organisation, who plotted the positions of German bombs, doodlebugs and rockets onto maps. But with the scale of Allied attacks on Germany dwarfing the Luftwaffe attacks on Britain, and as the frequently inaccurate eyewitness testimony became increasingly more so with the passage of time, alternative methods were clearly required.

Bomb disposal by its very nature is not without risk. In 1990 alone a dozen people would lose their lives in Germany. This risk had prompted bomb-disposal expert Manfred Schubert, from Hamburg, to visit Keele in the late 1970s where he was overwhelmed by what he found. Onto Hamburg alone the Allied air forces had dropped more than 100,000 bombs, of which 10 per cent are considered never to have exploded. With copies of 500 photographs of his city, he returned and put them to practical use. Just as the Damage Assessment interpreters at Medmenham had analysed photography before and after bombing raids to assess their success in destroying the target, Schubert repeated the exercise. With the smaller craters of those bombs that failed to explode distinct from the large white craters of those that did, he pinpointed the location of unexploded bombs in his city, many of which now lay underneath post-war reconstruction. From these few hundred photographs, taken from a source containing millions, Schubert saw the enormous potential and, burning with impatience, he began planning a project to make copies of the wartime photography covering West Germany that would help his fellow citizens locate the deadly legacy.

After the discovery of the anti-tank ditch, Anglo-German relations appear to have been ruffled when the West Berliners were infuriated at being charged for the reproduction of the photographs covering their city. With accusations from the city's housing minister that Britain was making a business from their sorry predicament and a defence from the university that it had to cover its costs, matters were brought to a head.

With the Foreign Office anxious for resolution, the use of the photographs for such humanitarian ends was encouraged in the hope it would help cement Anglo-German relations. After protracted negotiations between NATO, the Joint Services Liaison Organisation in Bonn, the Public Record Office, Keele University and the Federal Republic of Germany, a deal was finally reached. At a ceremony in West Germany, in February 1985, the tripartite 'Hamburg Agreement' was signed and a cheque for £300,000 was handed over by the Burgermeister of Hamburg to the university's vice-chancellor. This money was ring-fenced to pay for the creation of a microfilm copy of the entire 5.5 million photographs in the collection, an enormous five-year-long project that began almost immediately on the university's new, American-inspired science park. In turn the West Germans were allowed to borrow the original photographs of Germany, once they had been microfilmed, for copying by each of the West German states.

In November 1985 Manfred Schubert returned to Keele and took personal delivery of the first batch of 200,000 aerial photographs. Formally handed over at a ceremony attended by the former Secretary-General of NATO, Dr Joseph Luns, a British major general, the university vice-chancellor and a representative of the Federal Republic of Germany – and with the proceedings given a splash of colour by the Newcastle-under-Lyme town crier Mr Frank Shufflebotham – the future of the collection looked more positive than it had at any point in its post-war history.

From a central clearing-house in West Germany the photography was forwarded to each of the West German states, who copied the photography created by their Second World War adversaries, which transformed the approach taken by the bomb-disposal services to locating unexploded ordnance. Although the German photographic interpreters in the Second World War had largely ignored stereoscopic interpretation, a new generation of West Germans now depended on it and many of the budding photographic interpreters who worked in the miniature versions of Medmenham that were evolving in West Germany relished the opportunity to return the original photographs to the collection at Keele.

The decade-long project in Germany to copy and catalogue the photography was a process that has saved lives and continues to do so. In many of the German states the detection rate improved spectacularly as a result and of more than 700 unexploded bombs found in Nordrhein-Westfalen in 1992 and 1993, more than a third owed their detection to the aerial photography, while in Baden-Würtemberg one single photograph helped unearth 52 bombs.

Perhaps imbued with a new spirit of optimism, the university made an unsuccessful offer to take on responsibility for the Central Register for Air Photography for England – CRAPE – after the demise of the Air Photograph Unit at the Department of the Environment. The Central Registers of Aerial Photographs had been established by the government to act as central clearing-houses and held large volumes of RAF aerial photography. While the photography of England was ultimately destined for English Heritage, and the photography of Wales and CRAPW was destined for the Welsh Office in Cardiff, the Central Register of Air Photography for Scotland – rather disconcertingly referred to by its acronym CRAPS – and part of the Scottish Office in Edinburgh, would become part of the Royal Commission on the Ancient and Historical Monuments of Scotland. Keele was to prove more successful at acquiring 150,000 aerial photographs from the Mediterranean Allied Photo-Reconnaissance Wing – MAPRW – which had been gifted to the British School at Rome after the war, where its Director, John Ward-Perkins, used them for archaeological prospection.

The aerial photography taken by the San Severo-based MAPRW was mostly gifted to the Italian government at the end of the war and is now among the most important collections in the Aerofototeca Nazionale in Rome. Similar gifts were made to the government of the Netherlands, where photography is now held by Wageningen University and the Dutch mapping agency, the Kadaster. Much of the Royal Canadian Air Force photography of the war in Europe was taken back to Canada, where 300,000 aerial photographs are now held by Wilfrid Laurier University in Waterloo, southern Ontario. The Americans amassed a vast collection of Second World War aerial photography taken by the Allies, the Germans and the Japanese at the Defense Intelligence Agency before it was declassified. It is now accessible at the United States National Archives near Washington D. C. in College Park, Maryland.

The British GX photography held by JARIC – once such a Cold War secret – was finally declassified and joined the Allied wartime photography of western Europe at Keele. But although the collection might have been growing – alongside the decade-long Hamburg Project which so preoccupied the staff – the perennial problems of conservation and cataloguing continued to plague its development in the 1980s and throughout the 1990s.

One of the consequences of the Hamburg Project was the disappearance of its biggest customer base, the West Germans, who now had a complete licensed copy of the photography covering their country. With the retirement of Sheila Walton in 1996, Marilyn Beech who had joined the collection as a photographic technician in 1984 became responsible single-handed for a collection of more than 6.5 million aerial photographs.

With the Hamburg Project coinciding with the final years of the Cold War, thoughts inevitably turned to the potential of supplying aerial photography then still retained by JARIC to the former German Democratic Republic. Whilst some photography of East Berlin was released by JARIC at the time of German reunification in 1990, the process to declassify and release all the photography would not happen until 2004, when a four-year programme to declassify the JARIC Film Vault began.

This process coincided with the university launching the evidenceincamera.co.uk website which aimed to make the sortie plots from the Medmenham Print Library, which had been digitised using early digital-camera technology, accessible online. The project had been funded by the university using part of the money secured from the university management's secretive, speedy and controversial sale of the Turner Collection in 1999 to a book dealer for one million pounds. The collection chronicled the history and development of mathematics from the late fifteenth century onwards, and included extensively annotated volumes from the library of Sir Isaac Newton. Built up over fifty years by civil servant Charles Turner – who was said to have often gone without life's luxuries, including winter coats, in order to buy the books – the collection had been gifted to the fledgling university in 1968 on the basis that it wouldn't otherwise have the opportunity or good fortune to acquire such a special collection.

With prolonged negative press coverage that the sale price had been too low – and with no guarantees that the collection would remain intact or within the country – the controversy was further fuelled when the book dealer applied for export licences and began selling individual books. With rumours that one of the buyers was Paul Allen, the American investor, philanthropist and co-founder of the Microsoft Corporation – and although the sale was legally permissible – the whole affair outraged many in the academic community who saw it as the unacceptable commercial face of a modern scholarly institution driven by market forces. With this background, there was a political imperative for the university to reinvest the money in the special collections and archives in its care, and this decision was to prove decisive for the aerial photography collections.

When the evidenceincamera website was launched in January 2004 and its treasures were revealed during an interview with John Humphrys on the BBC Radio 4 *Today* programme, film crews and journalists from around the world descended on the university. And while feature articles in *Le Monde* and *Der Spiegel* resulted in web-traffic from around the world overwhelming the website for weeks the widespread interest provided confirmation of the collection's importance and prompted the National Archives to consider the entire JARIC Film Library worthy of permanent preservation. The problem then arose of where the vast library – nicknamed the 'tin mine' by generations of JARIC personnel – was going to be housed. While miles of shelving were allocated to the collection in the university library, by 2006 all the available space had been filled with more Second World War imagery and miles more shelving was required for what remained at RAF Brampton. When this coincided with the university having second thoughts about the collection, which although no longer such a financial burden occupied a vast amount of space, and failed to complement the institution's research or teaching agendas, discussions began with the National Archives about the collection's future. Whilst discussions continued for years, work began on identifying wartime photography for German bomb-disposal experts, and increasingly for customers in the Netherlands and Austria who, following the Germans' lead, were starting to use aerial photography to find unexploded ordnance. With a sense of déjà vu the problem of locating photography from within the vast JARIC Film Library was to be the same that had faced Beaver and Walton when the Medmenham Print Library was transferred to the university in the 1960s.

On arrival at JARIC, all new photography and associated sortie plotting was microfilmed and after the latitude and longitude of every photograph was painstakingly calculated and entered into a computer database the original sortie plots were destroyed. When photography of a particular place was required, the database was interrogated, the microfilm cassettes holding photography were viewed, and once individual frames were identified the film could be retrieved from the tin mine. By 2006 the result was 14,000 microfilms containing upwards of 20 million images, many of which proved to be unique since the policy of only keeping the 'latest and best' meant that millions of original frames had been routinely destroyed throughout the Cold War.

In 2006 the problem facing Keele was that neither the database holding the geographical data nor even a database listing the sorties contained on the microfilm was declassified. Locating Second World War photography from within the JARIC holdings was going to involve cataloguing and barcoding all the original rolls of film that had been delivered courtesy of the Royal Air Force, cataloguing and barcoding the entire microfilm collection, and consulting intelligence records created by the interpreters at Medmenham and held by the National Archives in Kew, west London. As textual records, the interpretation reports – which provide detailed interpretation of the aerial photography held in the collection – sit on shelves divorced from the aerial photography they describe.

By 2008 negotiations with the National Archives had reached the point where a new home had been found for the collection, and the staff relocated from the rooms they then occupied in the attic of Keele Hall to Scotland. Whilst the collection was transferred from the university and JARIC to a records-management facility in central Scotland (RCAHMS), in the process it became the National Collection of Aerial Photography (NCAP).

In 2004 when the evidenceincamera website was launched, Google acquired the pioneering software-development company Keyhole Inc , who specialised in geospatial-data visualisation. Initially created with venture capital from sources including IN-Q-TEL, the firm that invests in high-tech companies to keep the Central Intelligence Agency, and other intelligence agencies, equipped with the latest information technology. Their company name of Keyhole Inc paid homage to the CORONA satellite programme, used by the Central Intelligence Agency for photographic surveillance of the Soviet Union and the People's Republic of China and which assumed a greater importance after the 1960 Gary Powers U-2 incident.

Keyhole technology resulted in the release of Google Earth in June 2005, which alongside other virtual globes has continued to drive public interest, awareness and familiarity with geospatial technology. When the NCAP website was launched in November 2009, the predictable comparison to it being a Google Earth of the Second World War was drawn by many journalists. But this digital technology revolution and the end of the Cold War meant that the vast photographic processing facilities at JARIC, and the photographers who staffed them, were surplus to requirements. The same factors meant that the last remaining Canberra PR9 aircraft of No. 39 (1PRU) Squadron – an aircraft that replaced the Mosquito and served throughout the Cold War – was retired by the RAF in 2006, having been increasingly replaced with satellites and Unmanned Aerial Vehicles.

The development of digital technology has transformed the public understanding of aerial photography and has provided a means by which people can access something that historically has only been accessible to those within the intelligence community. It has given people both an expectation and a thirst for more. In addition to Second World War coverage around the world, the aerial film declassified by JARIC reflects key aspects of British government foreign policy throughout the Cold War and covers elements as diverse as the war in Korea, various African communist uprisings, atom-bomb testing, the Suez Crisis and peace-keeping operations around the world. Among the films that will be declassified in the years to come, before all the imagery is born digital, is the photography taken during the disorder after the Cold War, during both Gulf Wars and during the humanitarian disaster of Yugoslavia.

Acknowledgements

I am fortunate and grateful for the help and support of colleagues, friends and family, without whom this book would never have been possible. I am particularly indebted to Wing Commander Mike Mockford OBE (Ret'd) and Major Chris Halsall (Ret'd), their wives, Shirley and Christine. Their friendship, and willingness to share their encyclopedic knowledge of PI has been invaluable during the writing of this book, and throughout the project to declassify JARIC imagery to NCAP, which now holds over 20 million images.

The disbanding of JARIC in July 2012, means there is no longer a single, definable unit responsible for Imagery Intelligence analogous to Medmenham during the Second World War. But whilst the PI may have been replaced with the IA, and the photographers and model makers with computer technology, the cadre of RAF, Army and Royal Navy personnel at the Defence Geospatial Intelligence Fusion Centre, continue to keep a watchful eye. Whilst the Five Eyes Intelligence Community – UK, USA, Canada, Australia and New Zealand – ensures this inter-service unit remains international in its outlook. The Medmenham Club, and its Collection, provide a vital link between veterans, active-service imagery analysts and PI history. I am grateful to the Trustees and Chairman of the Medmenham Collection, Group Captain David Hollin (Ret'd), for permission to use the imagery featured in this book, and the many Club members who have helped me, in particular my friend Major Geoffrey Stone (Ret'd), Squadron Leader Anthony Plumb (Ret'd), and the late Ian Daglish.

I am indebted to Trevor Dolby at Preface Publishing, for his patience and support and for the editorial prowess of Kate Johnson. I am thankful for the copy editing of Nick Austin, proof reading of Alison Rae, and to Katherine Murphy. I would like to thank my mother for the enormous amounts of genealogical research she undertook and Rebecca Bailey for her speedy reading of the manuscript and her many insightful suggestions. The topographical drawings featured in this book were prepared thanks to Michelle Lile.

I am grateful for the support and dedication of the NCAP team: Andreas Buchholz; David Buice; Ruta Gauld; Sam Martin and Kevin McLaren. I would particularly like to thank Alan Potts, our digital imaging specialist, for his work enhancing all the imagery featured in this book. Much of the Operation Crossbow photography at NCAP was declassified by JARIC between 2004 and 2008, thanks particularly to the efforts of Squadron Leader Tim Burt (Ret'd), Andy Moor and team.

Many enjoyable hours were spent at The National Archives, whose staff continue to provide an outstanding service that facilitates access to an extraordinary range of primary sources. I am also indebted to the RAF Air Historical Branch, in particular Flight Lieutenant Mary Hudson (Ret'd), Dr. Sebastian Cox, Group Captain Steve Lloyd (Ret'd), Lee Barton. I am grateful to Dr Hans-Georg Carls, Wolfgang Muller and their colleagues at Luftbilddatenbank GmbH, Tom McAneer at NARA, Mr Dave Lefurgey and Helen Burton at Keele University.

I am thankful to Sarah Davies, Vice President of Factual Programming for Discovery Networks International, who encouraged me to write this book. In 2011, whilst with the BBC, she secured a broadcast commission for the documentary Operation Crossbow – which benefited from a research collaboration between NCAP and the Medmenham Collection – and thanks to its Director, Tim Dunn, and the production team, many were introduced to the world of the wartime PI for the first time.

Allan Williams
31 March 2013

Notes and Bibliography

Three main sources of primary information were consulted during research for this book, the Medmenham Collection (MDM), the National Collection of Aerial Photography (NCAP) and The National Archives (TNA). References to MDM sources commence with 'MDM', and are followed by class and item number (e.g. MDM DFG/5701). Many of the sorties detailed in this book are held by NCAP and are being progressively catalogued and made accessible on the NCAP website – http://ncap.org.uk – where information on all the collections and services offered can be found. References to TNA records commence with 'TNA', and are followed by the department code, series number, and piece number (e.g. TNA AIR 34/703).

The following abbreviations are used throughout the notes:

ACIU	Allied Central Interpretation Unit
ADI	Assistant Director Intelligence
AFHRA	Air Force Historical Research Agency (United States)
AHB	Air Historical Branch
AIR	Air Ministry
AOC	Aircraft Operating Company
BS	Interpretation Reports, written by B Section, on secret weapon deployment sites
CBS	Constance Babington Smith
CIU	Central Interpretation Unit
ESU	English Speaking Union (Dartmouth House, Charles Street, Mayfair, London)

INF	Ministry of Information
IWM	Imperial War Museum
JARIC	Joint Air Reconnaissance Intelligence Centre
MDM	Medmenham Collection
NCAP	National Collection of Aerial Photography
ORB	Operations Record Book
PDU	Photographic Development Unit
PI	Photographic Interpreter / Photographic Interpretation / Photographic Intelligence
PIU	Photographic Interpretation Unit
RAF	Royal Air Force
TNA	The National Archives
V1	Vergeltungswaffe 1 – the Fieseler Fi 103 flying bomb
V2	Vergeltungswaffe 2 – the A4 long-range rocket

Prologue

Bibliography

Abrams, L. (1991) Our Secret Little War, Bethesda: International Geographic Information Foundation, pp. 52-53.

Babington Smith, C. (1958) Evidence In Camera: The Story of Photographic Intelligence in World War II, London: Chatto and Windus, pp. 200-201.

Buckinghamshire County Council, Bombs Over Bucks – 70th Anniversary 1940 – 1945, [Online], Available: http://www.buckscc.gov.uk/bcc/archives/ea_Blitz.page [26 March 2013].

Golley, J. (1987) Whittle: The True Story, Shrewsbury: Airlife Publishing, pp. 212-213.

Hatton, R. G. (2006) Recollections of a schoolboy during the 40s, [Online], Available: http://www.bbc.co.uk/history/ww2peopleswar/stories/61/a8775561.shtml [26 March 2013].

Raleigh, W. (1922) The War in the Air: Being the Story of the Part Played in the Great War, Royal Air Force, Volume 1, Oxford: Clarendon Press, p. 446.

Seaton, P. (2009) A Sixpenny Romance: Celebrating a Century of Value at Woolworths, London: 3D and 6D Pictures Limited, pp. 86-87.

Wallington, N. (1981) Firemen at War: The Work of London's Fire-fighters in the Second World War, Huddersfield: Jeremy Mills Publishing, p. 134.

Chapter 1
Wizard War

Primary Sources

MDM DFG/5667: CBS interview notes, Sidney Cotton, Room 440, Dorchester Hotel, Park Lane, London, 15 April 1957.

MDM DFG/5668: CBS interview notes, Sidney Cotton, Dorchester Hotel, Park Lane, London, 18 April 1957.

MDM DFG/5669: CBS interview notes, Sidney Cotton, Dorchester Hotel, Park Lane, London, 23 April 1957.

MDM DFG/5798: CBS interview notes, Frederick Winterbotham, Wotton Farm, Kingsbridge, Devon, 10 May 1956.

MDM: Unpublished memoirs, A War of Intelligence, by Douglas Neville Kendall.

MDM MDM/056. AIR, AHB, (Undated) RAF Narrative – Photographic Reconnaissance by the Royal Air Force in the war of 1939-1945 – Volume 1.

TNA AIR 40/1176: "X" Series, flight records, January – May 1940.

TNA AIR 40/2572: The Oslo Report, scientific intelligence sent covertly to the British Naval Attaché, Oslo, November 1939; notes by Dr. R. V. Jones.

Bibliography

Baker, R. (1969) Aviator Extraordinary, London: Chatto & Windus, pp. 103-146.

Churchill, W. (1948) The Second World War, Volume I, The Gathering Storm, London: Cassell.

Churchill, W. (1949), The Second World War, Volume II, Their Finest Hour, London: Cassell, pp. 337-352.

Downing, T. (2010) Churchill's War Lab: Code-Breakers, Boffins and Innovators – The Mavericks Churchill Led to Victory, London: Little Brown, pp. 149-150, 156.

Jones, R. V. (1978) Most Secret War, London: Book Club Associates, pp. 3, 58, 63-64, 275.

Lindemann, F. A. (1934) 'Science and Air Bombing' in The Times, 8 August, p. 11.

Winterbotham, F. (1969) Secret and Personal, London: William Kimber, p. 36.

Chapter 2
The Phoney War, Special Flight and Aerofilms

Primary Sources

MDM DFG/3221: Notes taken by CBS during visit to Beresford Avenue, Wembley, London, June 1956.

MDM DFG/5653: CBS interview notes, Alan Butler, Hyde Park Gardens, Westminster, London, 6 June 1956.

MDM DFG/5663: CBS interview notes, lunch with David Brachi (and Mrs. Brachi) at the University of Hull, Kingston upon Hull, Yorkshire, 25 July 1956.

MDM DFG/5668: CBS interview notes, Sidney Cotton, Dorchester Hotel, Park Lane, London, 18 April 1957.

MDM DFG/5672: CBS interview notes, Glyn Daniel, lunch at The Plough, Fen Ditton, Cambridgeshire, 14 April 1956.

MDM DFG/5685: CBS interview notes, Walter Heath, 1 May 1956.

MDM DFG/5699: CBS interview notes, Hugh Hamshaw Thomas, 16 April 1956.

MDM DFG/5704: CBS interview notes, Frederick Victor Laws, ESU, London, 21 June 1956.

MDM DFG/5709: CBS interview notes, Paul Lamboit (and Mrs. Lamboit), dinner at the Cowdray Club, Cavendish Square, London, 23 August 1956.

MDM DFG/5756: CBS interview notes, Tom Muir Warden, tea at the ESU, London, 31 August [year unknown].

MDM DFG/5757: CBS interview notes, Hugh McPhail, lunch at the Whyte Lion Inn, Hartley Wintney, Hampshire, 7 May 1956.

MDM DFG/5759: CBS interview notes, Teddy Pippett, at Pippett's house in Radlett, Hertfordshire, 1 June 1956.

MDM DFG/5764: CBS interview notes, Peter Riddell, lunch at The Ritz, Piccadilly, London, 13 April 1956.

MDM DFG/5765: CBS interview notes, Peter Riddell, lunch at the Mandeville Hotel, Marylebone, London, 16 May 1956.

MDM DFG/5769: CBS interview notes, Peter Riddell, ESU, London, 7 February 1957.

MDM DFG/5782: CBS interview notes, Humphrey Spender, lunch at the Rice Bowl, 1 June 1956.

MDM DFG/5783: CBS interview notes, Stephen Spender and Rose Macaulay, lunch at the Arts Theatre Club, Frith Street, Soho, London, 11 May 1956.

MDM DFG/5784: CBS interview notes, Harry Stringer, tea at Wimpole Street, London, 6 September 1956.

MDM DFG/5788: CBS interview notes, Hugh Hamshaw Thomas, Medmenham Club reunion, Danesfield House, Buckinghamshire, 11 August 1956.

MDM DFG/5791: CBS interview notes, Geoffrey Tuttle, at Monckswood, his home near the village of Wembury, and RAF Mount Batten, Plymouth, Devon, 9 and 10 May 1956.

MDM DFG/5797: CBS interview notes, Francis Wills, 20 April 1956.

MDM DFG/5798: CBS interview notes, Frederick Winterbotham, Wotton Farm, Kingsbridge, Devon, 10 May 1956.

MDM MDM/056. AIR, AHB, (Undated) RAF Narrative – Photographic Reconnaissance by the Royal Air Force in the war of 1939-1945 – Volume 1, pp. 13, 79, 81-85, 106-107, 113-116, 137-138.

MDM: Unpublished memoirs, A War of Intelligence, by Douglas Neville Kendall.

NCAP GB/551/NCAP/8/17/3/1: Photostat copy of unpublished typescript, dated 1964, titled 'Memories of my Association with the Hunting Family and Group' by Francis Wills.

TNA AIR 2/4540: The AOC, the use of its services and facilities by the Air Ministry, 1940-1945.

TNA AIR 20/5749: Notes on the initial organisation of the PDU by Group Captain Frederick Winterbotham.

TNA AIR 29/33: No. 1 Camouflage Unit ORB, October 1939 – July 1944.

TNA AIR 29/434: PIU ORB, September 1940 – April 1941.

TNA AIR 34/83: The AOC / PIU / CIU, their internal organisation, procedures and personnel.

TNA AVIA 2/669: AIR, Imperial Air Survey Organisation, proposals by E. F. Turner and Sons and Major H. Hemming.

TNA FO 371/91781: Foreign Office, Eastern Department (Saudi

Arabia), reports on the activities of yacht owner, Mr Sidney
Cotton.

TNA FO 371/104879: Foreign Office, Eastern Department (Saudi
Arabia), reports on the activities of Mr Sidney Cotton.

TNA PREM 8/1007: Prime Minister's Office, the supply of arms to
Hyderabad and the activities of Mr. Sidney Cotton.

TNA WO 181/32: War Office, Relation of air survey to economic
development and defence, report by Major H. Hemming, 1938.

Bibliography

Babington Smith, C. (1958) Evidence In Camera: The Story of
Photographic Intelligence in World War II, London: Chatto &
Windus, pp. 42, 48-49, 115-116.

Baker, R. (1969) Aviator Extraordinary, London: Chatto & Windus,
pp. 61, 94, 156-157, 168-169, 173.

British Broadcasting Corporation (c. 1946) Children's Hour – How It's
Done: Aerial Photography, [Online], Available: http://www.bbc.
co.uk/archive/aerialjourneys/5301.shtml [26 March 2013].

Cobham, A. (1978) A Time to Fly, London: Shepheard-Walwyn,
pp. 45-48

Daniel, G. (1986) Some Small Harvest: The Memoirs of Glyn Daniel,
London: Thames and Hudson, p. 108.

Flight (1919) 'New Companies Registered', in Flight, 29 May, p. 722.

Flight (1920) 'A Case at Bow Street', in Flight, 22 July, p. 826.

Flight (1927) 'The Bournemouth Whitsun Meeting', in Flight, 9 June,
p. 368.

Flight (1929) 'Air Surveying and its Development: Aircraft Operating
Company's Pioneering Work', in Flight, 11 July, pp. 663-665.

Goodden, H. (2007) Camouflage and Art: Design for Deception in
World War 2, London: Unicorn Press, pp.24-25.

Hansard HC Deb 20 March 1940, vol. 358, cc2000-1: Statement by
the Secretary of State for Air (Sir Kingsley Wood) on the attacks
carried out by the RAF, the previous day, on the island of Sylt.

Powys-Lybbe, U. (1983) The Eye of Intelligence, London: William
Kimber, pp. 22-24, 29-30, 58.

Lake, A. (1999) Flying Units of the RAF: The Ancestry, formation and
disbandment of all flying units from 1912, Shrewsbury: Airlife
Publishing, p. 42.

Laws, F. V. (1942) 'Air Photography In War', in Flight, August, pp. 229-232.

Leigh, C. (1987) The Aerofilms Book of Britain's Railways from the Air, London: Ian Allen Publishing, pp. 6-7.

Martin, A. J. and Lapworth, P. B. (1994) 'Norman Leslie Falcon, 29 May 1904 – 31 May 1996', in Biographical Memoirs of Fellows of the Royal Society, vol. 44, November, pp. 160-174.

Spender, S. (1940) The Backward Son, London: The Hogarth Press.

Sutherland, J. (2004) Stephen Spender: The Authorised Biography, New York: The Viking Press, p. 44.

Winchester, C. and Wills, F. (1928) Aerial Photography: A Comprehensive Survey of its Practice and Development, London: Chapman and Hall.

Chapter 3
The Chalk House

Primary Sources

IWM Interview, Shirley Komrower, Catalogue reference: 22107.

IWM Interview, Pamela Bulmer, Catalogue reference: 22112.

MDM DFG/5663: CBS interview notes, lunch with David Brachi (and Mrs. Brachi) at the University of Hull, Kingston upon Hull, Yorkshire, 25 July 1956.

MDM DFG/5665: CBS interview notes, Sarah Churchill, Dorchester Hotel, Park Lane, London, 29 March 1957.

MDM DFG/5672: CBS interview notes, Glyn Daniel, lunch at The Plough, Fen Ditton, Cambridgeshire, 14 April 1956.

MDM DFG/5673: CBS interview notes, Harry Dawe, 20 April 1956.

MDM DFG/5690: CBS interview notes, Eve Holiday, dinner at the ESU, London, 30 May 1956.

MDM DFG/5691: CBS interview notes, Eve Holiday, drinks at the ESU, London, 9 August 1956.

MDM DFG/5704: CBS interview notes, Frederick Victor Laws, ESU, London, 21 June 1956.

MDM DFG/5785: CBS interview notes, Peter Stewart in his Hertford Street apartment, Mayfair, and lunch at L'Escargot, Greek Street, Soho, London, 26 July 1956.

MDM: Unpublished memoirs, A War of Intelligence, by Douglas Neville Kendall.

TNA AIR 20/4267: Miscellaneous papers accumulated by the AHB relating to the PDU and CIU.

TNA AIR 29/227: CIU ORB, July 1941 – December 1943.

TNA AIR 34/83: The AOC / PIU / CIU, their internal organisation, procedures and personnel.

TNA AIR 34/84: Historical record of each section at the ACIU, prepared for the AHB.

TNA AIR 34/703: 'The Chalk House with the Tudor Chimneys', a photographic record of the ACIU in the closing months of the Second World War.

TNA AIR 40/1169: RAF Station Medmenham, Intelligence Reports and Papers, 1 January 1941 – 3 August 1943.

TNA AIR 40/2574: AIR, Directorate of Intelligence, Intelligence Reports and Papers, 1 January 1943 – 31 December 1944.

TNA HO 45/25105: Home Office, Defence Regulation 18B detainee: RUTLAND, Frederick Joseph; pre-war agent for Japanese secret service.

TNA INF 1/293: INF, Home Intelligence Division, Home Intelligence Reports, January 1943 – December 1944.

TNA KV 2/333: The Security Service, Japanese Agents and Suspected Agents, Frederick Joseph Rutland.

TNA KV 2/337: The Security Service, Japanese Agents and Suspected Agents, Frederick Joseph Rutland.

Bibliography

Abrams, L. (1991) Our Secret Little War, Bethesda: International Geographic Information Foundation, pp. 52-53.

Buckinghamshire County Council, Bombs Over Bucks – 70th Anniversary 1940 – 1945, [Online], Available: http://www.buckscc.gov.uk/bcc/archives/ea_Blitz.page [26 March 2013].

Churchill, S. (1981) Keep on Dancing: An Autobiography, London: Weidenfeld and Nicolson, pp. 58-68.

David, V. (1951) Advice to my Godchildren, London: Duckworth.

Daniel, G. (1986) Some Small Harvest: The Memoirs of Glyn Daniel, London: Thames and Hudson, pp. 94-143.

Powys-Lybbe, U. (1983) The Eye of Intelligence, London: William Kimber, pp. 17-19, 56-57.

Flight (1940) 'Photographic Reconnaissance: Day and Night Problems solved by the RAF', in Flight, 26 December.

Flight (1955) 'Major H. Hemming Returns to Aviation', in Flight, 26 December, p. 746.

The Times (1930) 'Funerals: Mrs Hornby Lewis', in The Times, 29 December.

Norton-Taylor, R. (2000) 'British flying ace was spy for Japan', in The Guardian, 10 November.

Young, D. (1963) Rutland of Jutland, London: Cassell.

Ministry of Information (1942) Coastal Command, London: HMSO, pp. 8-16.

Chapter 4
Target for Tonight

Primary Sources

MDM DFG/3200: CBS interview notes, Robert Quackenbush, Hotel Washington, 515, 15th Street, Washington D.C., 16 October 1956.

MDM DFG/5665: CBS interview notes, Sarah Churchill, Dorchester Hotel, Park Lane, London, 29 March 1957.

MDM DFG/5674: CBS interview notes, Geoffrey Deeley, ESU, 7 March 1957.

MDM DFG/5701: CBS interview notes, Douglas Kendall, lunch at L'Escargot, Greek Street, Soho, the ESU and the Grosvenor House Hotel, Park Lane, Mayfair, London, 14 November 1956.

MDM DFG/5764: CBS interview notes, Peter Riddell, lunch at The Ritz, Piccadilly, London, 13 April 1956.

MDM DFG/5783: CBS interview notes, Stephen Spender and Rose Macaulay, lunch at the Arts Theatre Club, Frith Street, Soho, London, 11 May 1956.

MDM DFG/5794: CBS interview notes, Claude Wavell at his home, Leindenhurst, Solent View Road, Gurnard, Cowes, Isle of Wight, Hampshire, 24 May 1956.

MDM DFG/5800: CBS interview notes, Ted Wood 'Woody', lunch at Reubens Hotel, Buckingham Palace Road, 7 December 1956.

MDM: Unpublished memoirs, A War of Intelligence, by Douglas Neville Kendall.

TNA AIR 34/83: The AOC / PIU / CIU, their internal organisation, procedures and personnel.

TNA AIR 34/84: Historical record of each section at the ACIU, prepared for the AHB.

TNA AIR 34/703: 'The Chalk House with the Tudor Chimneys', a photographic record of the ACIU in the closing months of the Second World War.

TNA AIR 40/1169: RAF Station Medmenham, Intelligence Reports and Papers, 1 January 1941 – 3 August 1943.

TNA AIR 40/2574: AIR, Directorate of Intelligence, Intelligence Reports and Papers, 1 January 1943 – 31 December 1944.

TNA DEFE 40/2: Papers of Dr. R. V. Jones, ADI (Science) reports on intelligence aspects of the Bruneval raid (capture of Wurzburg apparatus).

TNA INF 1/210: INF, Film Division, Target for Tonight.

Bibliography

Binney, G. (1926) With Seaplane and Sledge in the Arctic, New York: George Doran Company.

Daniel, G. (1986) Some Small Harvest: The Memoirs of Glyn Daniel, London: Thames and Hudson, pp. 94-143.

Flight (1932) 'The Survey of Rio de Janeiro: Aircraft Operating Company, Ltd., Completes Important Contract', in Flight, 20 May, p. 746.

Ford, K. (2010) The Bruneval Raid: Operation Biting 1942, Oxford: Osprey Publishing.

Ministry of Information (1943) Combined Operations: 1940-1942, INF, London: HMSO.

Orr, R. (1998) Musical Chairs: An Autobiography, London: Thames Publishing, pp. 45-52.

Powys-Lybbe, U. (1983) The Eye of Intelligence, London: William Kimber, pp. 57-59, 185-187.

Sims, C. (1958) Camera in the Sky, London: Temple Press, pp. 1-33.

Thirsk, I. (2006) de Havilland Mosquito: An Illustrated History – Volume 2, Manchester: Grecy Publishing, pp. 372-374.

Jones, R. V. (1978) Most Secret War, London: Book Club Associates, pp. 132-134.

Verity, H. (1978) We Landed By Moonlight, London: Ian Allan Limited.

Chapter 5
Peenemunde and Bodyline

Primary Sources

IWM Interview, Suzie Morgan, Catalogue reference: 22106.

IWM Interview, Myra Nora Collyer, Catalogue reference: 23845.

IWM Interview, Sophie Wilson, Catalogue reference: 22102.

IWM Interview, Hazel Scott, Catalogue reference: 22105.

MDM DFG/5655: CBS interview notes, Bernard Babington Smith, Yelford Manor, Hardwick-with-Yelford, Oxfordshire, 19 April 1956.

MDM DFG/5657: CBS interview notes, Bernard Babington Smith, Neale House, Fitzwilliam College, Cambridge, Cambridgeshire, 12 July 1956.

MDM DFG/5674: CBS interview notes, Geoffrey Deeley, ESU, 7 March 1957.

MDM DFG/5699: CBS interview notes, Hugh Hamshaw Thomas, 16 April 1956.

MDM DFG/5701: CBS interview notes, Douglas Kendall, lunch at L'Escargot, Greek Street, Soho, the ESU and the Grosvenor House Hotel, Park Lane, Mayfair, London, 14 November 1956.

MDM DFG/5708: CBS interview notes, David Linton, ESU, 6 July 1956.

MDM DFG/5708: CBS interview notes, David Linton, Sale Hill, Sheffield, 24 July 1956.

MDM DFG/5781: CBS interview notes, Neil Simon, ESU, 19 February 1957.

MDM DFG/5788: CBS interview notes, Hugh Hamshaw Thomas, Medmenham Club reunion, Danesfield House, Buckinghamshire, 11 August 1956.

MDM DFG/5789: CBS interview notes, Hugh Hamshaw Thomas, 23 September 1956.

MDM DFG/5790: CBS interview notes, Hugh Hamshaw Thomas, 3 Millington Road, Cambridge, Cambridgeshire, 3 January 1957.

MDM DFG/5800: CBS interview notes, Ted Wood, lunch at Reubens Hotel, Buckingham Palace Road, London, 7 December 1956.

MDM: Unpublished memoirs, A War of Intelligence, by Douglas Neville Kendall.

TNA AIR 29/227: CIU ORB, July 1941 – December 1943.

TNA AIR 29/228: CIU / ACIU ORB, January 1944 – August 1944.

TNA AIR 29/229: ACIU / CIU ORB: September 1944 – March 1946.

TNA AIR 34/80: ACIU: Operation Crossbow, History of the PI Investigation (February 1943 – April 1945).

TNA AIR 34/84: Historical record of each section at the ACIU, prepared for the AHB.

TNA AIR 34/117: CIU, Interpretation Reports: BS1 – BS70.

TNA AIR 34/118: CIU, Interpretation Reports: BS71 – BS100.

TNA AIR 34/184: CIU, Interpretation Report: K3084 (Peenemunde) with target map.

TNA AIR 34/196: CIU, Interpretation Reports: DS1 – DS40.

TNA AIR 34/609: CIU, Target Folder: Mohne and Eder Dams.

TNA AIR 34/703: 'The Chalk House with the Tudor Chimneys', a photographic record of the ACIU in the closing months of the Second World War.

TNA DEFE 40/5: Papers of Dr. R. V. Jones, ADI (Science), draft report on long range rockets.

TNA DEFE 40/12: Papers of Dr. R. V. Jones, ADI (Science), Miscellaneous papers on the V1 and V2.

KV 2/3523: The Security Service, Personal (PF Series) Files: Communists and suspected communists: Jacob Bronowski (18 October 1939 – 14 January 1953).

Bibliography

Abrams, L. (1991) Our Secret Little War, Bethesda: International Geographic Information Foundation, pp. 28-29, 35, 46.

Babington Smith, B. (1942) 'What's wrong with the broadcasters? The curious case of the "monopolists" who do not control their own programmes', in Harper's Magazine, June.

Babington Smith, C. (1958) Evidence In Camera: The Story of Photographic Intelligence in World War II, London: Chatto and Windus, pp. 200-206.

Bode, V. and Kaiser, G. (2008) Building Hitler's Missiles: Traces of History in Peenemunde, Berlin: Ch. Links Verlag, p. 38.

Bronowski, J. (1973) The Ascent of Man, London: British Broadcasting Corporation.

Churchill, W. (1952) The Second World War, Volume V, Closing The Ring, London: Cassell, pp. 201-202.

Daniel, G. (1986) Some Small Harvest: The Memoirs of Glyn Daniel, London: Thames and Hudson, p. 137.

Golley, J. (1987) Whittle: The True Story, Shrewsbury: Airlife Publishing, pp. 212-213

Harris, A. (1947) Bomber Offensive, London: Collins, pp. 182-185, 196.

Heitman, J. (1991) 'The Peenemunde Rocket Centre' in After The Battle, no. 74, pp. 1-25.

Irving, D. (1964) The Mare's Nest: The German Secret Weapons Campaign and the British Counter-Measures, London: William Kimber, p. 34-40, 50-52, 337.

Jeffery, K. (2010) MI6: The History of the Secret Intelligence Service 1909-1949, London: Bloomsbury, pp. 512-513, 533-535.

Lowe, K., (2007), Inferno: The Devastation of Hamburg, 1943, London: Viking, p. 56.

Margry, K. (2008) 'RAF Target Mapping Centre at Hughenden Manor' in After The Battle, no. 141, pp. 34-37.

Middlebrook, M. (1982) The Peenemunde Raid: The Night of 17/18 August 1943, London: Allen Lane.

Middlebrook, M. and Everitt, C. (2011) The Bomber Command Diaries: An Operational Reference Book 1939-1945, Hersham: Midland Publishing, pp. 422-424.

Powys-Lybbe, U. (1983) The Eye of Intelligence, London: William Kimber, pp. 60-64, 191.

Johnson, B. (1978) The Secret War, London: British Broadcasting Corporation, pp. 123-188

Jones, R. V. (1978) Most Secret War, London: Book Club Associates, pp. 33-34, 291, 322, 332-341, 346, 373.

Ramsay, W. (1988) 'The US Prison at Shepton Mallet', in After The Battle, no. 59, pp. 28-34.

CHAPTER 6
The Bois Carré Sites

Primary Sources

MDM DFG/5701: CBS interview notes, Douglas Kendall, lunch at L'Escargot, Greek Street, Soho, the ESU and the Grosvenor House Hotel, Park Lane, Mayfair, London, 14 November 1956.

MDM DFG/5775: CBS interview notes, Robert Rowell, 9 April 1956.

MDM DFG/5776: CBS interview notes, Robert Rowell, 15 April 1956.

MDM DFG/5781: CBS interview notes, Neil Simon, ESU, 19 February 1957.

MDM DFG/5782: CBS interview notes, Humphrey Spender, lunch at the Rice Bowl, 1 June 1956.

MDM: Unpublished memoirs, A War of Intelligence, by Douglas Neville Kendall.

TNA AIR 27/2013/27: RAF, 541 Squadron ORB, Summary of Events Y (1 November – 30 November 1943).

TNA AIR 34/80: ACIU: Operation Crossbow, History of the PI Investigation (February 1943 – April 1945).

TNA AIR 34/118: CIU, Interpretation Reports: BS71 – BS100.

TNA DEFE 40/10: Papers of Dr. R. V. Jones, ADI (Science), miscellaneous technical intelligence reports.

TNA DEFE 40/12: Papers of Dr. R. V. Jones, ADI (Science), miscellaneous papers on the V1 and V2.

Bibliography

Campbell, C. (2012) Target London: Under Attack from the V-Weapons During WWII, London: Little Brown, pp. 122-123, 441.

Fourcade, M. M. (1973), Noah's Ark: The Secret Underground, St Leonards: Allen & Unwin.

Jeffery, K. (2010) MI6: The History of the Secret Intelligence Service 1909-1949, London: Bloomsbury, p. 535.

Martelli, G. (1960) Agent Extraordinary: The Story of Michel Hollard, London: Collins.

McKay, C. G. (1993) From Information to Intrigue: Studies in Secret Service based on the Swedish Experience, 1939-1945, London: Frank Cass, pp. 106-110.

Jones, R. V. (1978) Most Secret War, London: Book Club Associates, pp. 349-355, 360, 373-375.

Tennant, P. (1993) Obituary: Captain Henry Denham, in The Independent, 23 July.

CHAPTER 7
The V1 Flying Bomb

Primary Sources

MDM DFG/5701: CBS interview notes, Douglas Kendall, lunch at L'Escargot, Greek Street, Soho, the ESU and the Grosvenor House Hotel, Park Lane, Mayfair, London, 14 November 1956.

MDM DFG/5753: CBS interview notes, John Merifield, ESU, 27 July 1956.

MDM DFG/5754: CBS interview notes, John Merifield, dinner at Oxshott, Surrey, 19 February 1957.

MDM DFG/5775: CBS interview notes, Robert Rowell, 9 April 1956.

MDM DFG/5781: CBS interview notes, Neil Simon, ESU, 19 February 1957.

MDM: Unpublished memoirs, A War of Intelligence, by Douglas Neville Kendall.

TNA AIR 27/2007/28: RAF, 540 Squadron ORB, Summary of Events Y (1 November – 30 November 1943).

TNA AIR 34/80: ACIU: Operation Crossbow, History of the PI Investigation (February 1943 – April 1945).

TNA AIR 34/84: Historical record of each section at the ACIU, prepared for the AHB.

TNA AIR 34/119: CIU, Interpretation Reports: BS101 – BS140.

TNA AIR 34/120: CIU, Interpretation Reports: BS141 – BS170.

TNA AIR 34/232: CIU, G Section, Interpretation Report No. G1(R): Special Report on 'Spires'.

TNA INF 1/967: INF, Ministerial correspondence files, V1 and V2 censorship and security measures.

Bibliography

Babington Smith, C. (1958) Evidence In Camera: The Story of Photographic Intelligence in World War II, London: Chatto & Windus, pp. 199-232.

Churchill, S. (1981) Keep on Dancing: An Autobiography, London: Weidenfeld and Nicolson, p. 67.

Golley, J. (1987) Whittle: The True Story, Shrewsbury: Airlife Publishing, pp. 212-213.

Golovine, M. N. (1962) Conflict in Space: A Pattern of War in a New Dimension, London: Temple Press.

Jeffery, K. (2010) MI6: The History of the Secret Intelligence Service 1909-1949, London: Bloomsbury, p. 529.

Jones, R. V. (1978) Most Secret War, London: Book Club Associates, p. 231, 367-341.

Keen, P. F. (1996) Eyes of the Eighth: A Story of The 7th Photographic Reconnaissance Group 1942-1945, Sun City: CAVU Publishers.

Middlebrook, M. and Everitt, C. (2011) The Bomber Command Diaries: An Operational Reference Book 1939-1945, Hersham: Midland Publishing, pp. 455.

Nijboer, D. (2005) Graphic War: The Secret Aviation Drawings and Illustrations of World War II, Ontario: Boston Mills Press, pp. 9, 14-20.

Powys-Lybbe, U. (1983) The Eye of Intelligence, London: William Kimber, p. 202.

United States Army Air Force (1945) With the Eyes of the Seventh Photo Group: Now it can be Told!, United States Army Air Force: Seventh Photo Group Publication.

Flavin, J. E. and Hohner, W. J. (1945) The Story of the 31st Photo Reconnaissance Squadron, Nurnberg: United States Army Air Force.

Chapter 8
No-Ball Targets, Hottot and Belhamelin Sites
Primary Sources

MDM DFG/5676: CBS interview notes, Geoffrey Dimbleby, lunch at the ESU, 6 September 1956.

MDM: Unpublished memoirs, A War of Intelligence, by Douglas Neville Kendall.

TNA AIR 34/80: ACIU: Operation Crossbow, History of the PI Investigation (February 1943 – April 1945).

TNA AIR 34/67: CIU: Operation Crossbow reports from R.E.8., Ministry of Home Security.

TNA AIR 34/84: Historical record of each section at the ACIU, prepared for the AHB.

TNA AIR 34/121: CIU, Interpretation Reports: BS171 – BS200.

TNA AIR 34/126: CIU, Interpretation Reports: BS351 – BS380.

TNA AIR 34/127: CIU, Interpretation Reports: BS381 – BS440.

TNA AIR 34/129: ACIU, Interpretation Reports: BS491 – BS540.

TNA AIR 34/703: 'The Chalk House with the Tudor Chimneys', a photographic record of the ACIU in the closing months of the Second World War.

TNA AIR 37/485: AM, Allied Expeditionary Air Force, Operation Crossbow, Offensive counter-measures.

TNA AIR 40/1674: AM, Directorate of Intelligence, Air Intelligence 3c(1), No-Ball target list.

Bibliography

Cox, S. (1998) 'An Unwanted Child – The Struggle to Establish a British Bombing Survey' in British Bombing Survey Unit, The Strategic Air War Against Germany: 1939-1945, London: Frank Cass, pp. xvii-xxii.

Powys-Lybbe, U. (1983) The Eye of Intelligence, London: William Kimber, pp. 79-82, 204.

Craven, W. F. and Cate, J. L. (1951) The Army Air Forces in World War II, Volume III, Europe: Argument to V-E Day, January 1944 – May 1945, Chicago: University of Chicago Press, pp. 97-100.

Zuckerman, S. (1977) From Apes to Warlords: The autobiography (1904-1946) of Solly Zuckerman, London: Hamish Hamilton.

Chapter 9
Diver Diver Diver

Primary Sources

MDM DFG/5685: CBS interview notes, Walter Heath, 1 May 1956.

MDM: Unpublished memoirs, A War of Intelligence, by Douglas Neville Kendall.

TNA AIR 20/4261: Papers accumulated by the AHB on obituary notices, and their use by enemy-agents plotting the fall of flying bombs.

TNA AIR 20/11399: Papers accumulated by the AHB on the V2, dated 1 January 1943 – 31 December 1943.

TNA AIR 34/80: ACIU: Operation Crossbow, History of the PI Investigation (February 1943 – April 1945).

TNA AIR 34/129: ACIU, Interpretation Reports: BS491 – BS540.

TNA AIR 40/2166: AM, Directorate of Intelligence A.I.2.(g) reports 2174-2283.

TNA INF 1/292: INF, Home Intelligence Division, Home Intelligence weekly reports: January 1940 – December 1944.

Bibliography

Addison, P. and Crang, J. A. (eds) (2010) Listening to Britain: Home Inelligence Reports on Britain's Finest Hour, May to September 1940 , London: Bodley Head, pp. xi-xii.

British Broadcasting Corporation (1945) B.B.C. Year Book 1945, London: British Broadcasting Corporation, p. 141.

Christopher, J. (2004) Balloons at War: Gasbags, Flying Bombs and Cold War Secrets, Stroud: Tempus, pp.142-144.

Churchill, W. (1949) The Second World War, Volume II, Their Finest Hour, London: Cassell, p. 331.

Hansard HC Deb 16 June 1944, vol. 400, cc2301-3: Statement by the Secretary of State for the Home Department (Mr. Herbert Morrison) on air raids (pilotless machines).

Hansard HC Deb 6 July 1944, vol. 401, cc1322-39: Statement by the Prime Minister (Mr. Winston Churchill) on flying bombs.

Hinsley, H., with Thomas, E. E., Ransom, C.F.G. and Knight, R. C. (1984) British Intelligence in the Second World War: Its Influence on Strategy and Operations, Volume Three (Part 1), London: HMSO, p. 429.

Nijboer, D. (2005) Graphic War: The Secret Aviation Drawings and Illustrations of World War II, Ontario: Boston Mills Press, pp. 20.

Ogley, B. (1992) Doodlebugs and Rockets: The Battle of the flying bombs, Westerham: Froglets Publications, pp. 24-31, 38-40, 48-50, 56, 61.

Ramsey, W. (ed) (1990) The Blitz: Then and Now (Volume 3), London: Battle of Britain Prints International, pp.392-393, 403.

Seventh Photographic Reconnaissance Group Association (1986) Journal of the 7th Photo Recon Group, p. xviii.

Thompson, G. P. (1947) Blue Pencil Admiral: The Inside Story of the Press Censorship, London: Sampson Low, Marston & Company.

Chapter 10
Hell's Corner

Primary Sources

TNA AIR 20/4140: Manuscript 'Battle of the flying bomb', by
   Squadron Leader H. E. Bates.

TNA AIR 25/790: AIR, No. 106 (Photographic Reconnaissance) Group
   ORB: April 1944 – May 1945.

TNA AIR 34/80: ACIU: Operation Crossbow, History of the PI
   Investigation (February 1943 – April 1945).

TNA AIR 34/129: ACIU, Interpretation Reports: BS491 – BS540.

TNA AIR 34/130: ACIU, Interpretation Reports: BS541 – BS600.

TNA AIR 34/134: ACIU, Interpretation Reports: BS771 – BS840.

TNA INF 1/292: INF, Home Intelligence Division, Home Intelligence
   weekly reports: January 1940 – December 1944.

Bibliography

Abrams, L. (1991) Our Secret Little War, Bethesda: International
   Geographic Information Foundation, pp. 52-53.

Bates, H. E. and Ogley, B. (ed) (1994) Flying Bombs over England,
   Westerham: Froglets Publications.

Buckinghamshire County Council, Bombs Over Bucks – 70th
   Anniversary 1940 – 1945, [Online], Available: http://www.buckscc.
   gov.uk/bcc/archives/ea_Blitz.page [26 March 2013].

Ogley, B. (1992) Doodlebugs and Rockets: The Battle of the flying
   bombs, Westerham: Froglets Publications, pp. 102-103, 107, 123,
   128, 133.

Chapter 11
The V2 Rocket

Primary Sources

MDM DFG/5678: CBS interview notes, Eric Fuller, lunch at The
   Hungaria restaurant, off Piccadilly, London, 7 September 1956.

MDM DFG/5687: CBS interview notes, Ray Herschel, Elstree,
   Hertfordshire, 20 April 1956.

MDM DFG/5688: CBS interview notes, Ray Herschel, ESU, London,
   6 March 1957.

MDM DFG/5692: CBS interview notes, Gordon Hughes, lunch at the
   Royal Aero Club, 119 Piccadilly, W1, London, 26 April 1956.

MDM DFG/5697: CBS interview notes, Gordon Hughes, lunch and
tea at the ESU, 6 December 1956.

MDM DFG/5704: CBS interview notes, Frederick Victor Laws, ESU,
London, 21 June 1956.

MDM DFG/5766/1: Expedition by CBS and Peter Riddell to RAF
Medmenham, Phyllis Court, and RAF Benson, 19 July 1956.

MDM: Unpublished memoirs, A War of Intelligence, by Douglas
Neville Kendall.

TNA AIR 34/80: ACIU: Operation Crossbow, History of the PI
Investigation (February 1943 – April 1945).

TNA AIR 34/84: Historical record of each section at the ACIU,
prepared for the AHB.

TNA AIR 34/126: CIU, Interpretation Reports: BS351 – BS380.

TNA AIR 34/128: ACIU, Interpretation Reports: BS441 – BS490.

TNA AIR 34/133: ACIU, Interpretation Reports: BS701 – BS770.

TNA AIR 34/134: ACIU, Interpretation Reports: BS771 – BS840.

TNA AIR 34/197: CIU / ACIU, Interpretation Reports: DS42 – DS70.

TNA AIR 34/198: ACIU, Interpretation Reports: DS71 – DS147.

TNA AIR 40/2517: AIR, Directorate of Intelligence, Intelligence Reports
on offensive and defensive measures against V Weapon attacks.

TNA DEFE 40/10: Papers of Dr. R. V. Jones, ADI (Science),
miscellaneous technical intelligence reports.

Bibliography

Beevor, A. (2009) D-DAY: The Battle for Normandy, London: Viking,
pp. 75-76.

Bode, V. Kaiser, G. (2008), Building Hitler's Missiles: Traces of History
in Peenemunde, Berlin: Ch. Links Verlag, pp. 58-59.

Center for Air Force History, (1991), U.S. Army Air Forces in World
War II – Combat Chronology – 1941-1945, Washington D.C.:
Center for Air Force History, p. 427.

Hewlings, R. (2008) 'Sir Howard Colvin: Architectural historian
whose biographical dictionaries laid a foundation for all other
scholars in his field', in The Independent, 1 January.

Hinsley, H., with Thomas, E. E., Ransom, C.F.G. and Knight, R. C.
(1984) British Intelligence in the Second World War: Its Influence
on Strategy and Operations, Volume Three (Part 1), London:
HMSO, pp. 437, 439, 443, 445, 447-448, 445, 453-454.

Irving, D. (1964) The Mare's Nest: The German Secret Weapons Campaign and the British Counter-Measures, London: William Kimber, p. 173.

Johnson, B. (1978) The Secret War, London: British Broadcasting Corporation, pp. 166-169.

Jones, R. V. (1978) Most Secret War, London: Book Club Associates, pp. 431-434, 364-366, 400-408. 432-436, 444-445.

Kemp, R. C., Lewis, C. G., Scott, C. W., and Robbins, C. R. (1925) Aero-Photo Survey and the Mapping of the Forests of the Irrawaddy Delta, Maymo: Office of the Superintendent, Government Printing, Burma.

Longdon, S. (2009) T Force: The Race for Nazi Secrets, 1945, London: Constable.

Mediterranean Allied Photo Reconnaissance Wing (1944) The Mediterranean Allied Photo Reconnaissance Wing: A Pictorial History, San Severo: MAPRW.

Nutting, D. (ed) (2003) Attain By Surprise: Capturing Top Secret Intelligence in World War II, Chichester: David Clover.

Spooner, T. (1987) Warburton's War: The Life of Wing Commander Adrian Warburton, London: Williams Kimber.

The Telegraph (2004) Obituary: 'Wing Commander Rupert Cecil' in The Telegraph, 14 July.

Thompson, H. L. (1956) New Zealanders with the Royal Air Force (Volume 2), Wellington: Department of Internal Affairs (Historical Publications Branch), p334-335.

Tweddie, N. (2003) 'RAF's Wartime Daredevil finally laid to rest' in The Telegraph, 15 May.

Chapter 12
From V1 to V2?

Primary Sources

MDM DFG/3198: CBS interview notes, Geoffrey Platt, lunch at the Champs Elysee, New York, 31 October 1956.

MDM DFG/5701: CBS interview notes, Douglas Kendall, lunch at L'Escargot, Greek Street, Soho, the ESU and the Grosvenor House Hotel, Park Lane, Mayfair, London, 14 November 1956.

MDM: Unpublished memoirs, A War of Intelligence, by Douglas Neville Kendall.

TNA AIR 10/3866: The Strategic Air War against Germany, 1939-1945, the report of the British Bombing Survey Unit.

TNA AIR 34/80: ACIU: Operation Crossbow, History of the PI Investigation (February 1943 – April 1945).

TNA AIR 34/84: Historical record of each section at the ACIU, prepared for the AHB.

TNA AIR 34/85: Operational record of Damage Assessment 'K' Section, CIU.

TNA AIR 34/703: 'The Chalk House with the Tudor Chimneys', a photographic record of the ACIU in the closing months of the Second World War.

TNA AIR 40/1678: AIR, Directorate of Intelligence, Air Intelligence 2(h) investigation of the Heavy Sites in northern France. Report by the Sanders Mission to the chairman of the Crossbow Committee.

Bibliography

Bowman, M. (1999) Mosquito Photo-Reconnaissance units of World War 2, Oxford: Osprey Publishing, p.46-47.

Cox, S. (1998) 'An Unwanted Child – The Struggle to Establish a British Bombing Survey' in British Bombing Survey Unit, The Strategic Air War Against Germany: 1939-1945, London: Frank Cass, pp. xvii-xxii.

Daniels, G. (ed) (1981) Guides and Handbooks Supplementary Series: A Guide to the Reports of the United States Strategic Bombing Survey, London: Royal Historical Society.

Finn, P. (1992) 'Obituary: Wilfred Aurthur Seaby 1910-1991', in Spink Numismatic Circular, vol. 100, p. 8.

Henshall, P. (2002) Hitler's V Weapon Sites, Stroud: Sutton Publishing, p. 33.

Middlebrook, M. and Everitt, C. (2011) The Bomber Command Diaries: An Operational Reference Book 1939-1945, Hersham: Midland Publishing, pp. 524-525.

Pace, E. (1985) 'Geoffrey Platt is dead at 79; Led City Preservation Move' in New York Times, 15 July.

Powys-Lybbe, U. (1983) The Eye of Intelligence, London: William Kimber, pp. 157-158.

Ramsay, W. (ed) (1974) 'The V Weapons' in After The Battle, no. 6, pp. 19-22.

Time Magazine (1945) 'Awesome and Frightful', in Time Magazine, 5 November.

United States Strategic Bombing Survey (1945) Summary Report: European War, Washington D.C.: United States Government Printing Office.

Chapter 13
Big Ben

Primary Sources

IWM: CGE 11428~11714, Combat Film No 11692. Flying Officer Baxter of 602 Squadron, 14/2/1945, 0950, flying Spitfire. Target: Ground Targets.

IWM: CGE 11428~11714, Combat Film No 11535. Flying Officer Maksymowicz of 303 Squadron, 17/1/1945, 1530, flying Spitfire. Target: V2 Sites.

MDM DFG/3185: CBS interview notes, Richard Hughes, breakfast meetings at the Hotel Continental, Cambridge, Massachusetts, 23 and 24 October 1956.

MDM DFG/3202: CBS interview notes, Walt Rostow, Massachusetts Institute of Technology, Cambridge, Massachusetts, 23 and 24 October 1956.

MDM DFG/5701: CBS interview notes, Douglas Kendall, lunch at L'Escargot, Greek Street, Soho, the ESU and the Grosvenor House Hotel, Park Lane, Mayfair, London, 14 November 1956.

MDM DFG/5763: CBS interview notes, George Reynolds, ESU, 19 February 1957.

MDM: Unpublished memoirs, A War of Intelligence, by Douglas Neville Kendall.

TNA AIR 25/790: AIR, No. 106 (Photographic Reconnaissance) Group ORB: April 1944 – May 1945.

TNA AIR 34/80: ACIU: Operation Crossbow, History of the PI Investigation (February 1943 – April 1945).

TNA AIR 34/84: Historical record of each section at the ACIU, prepared for the AHB.

TNA AIR 34/137: ACIU, Interpretation Reports: BS990 – BS1038.

TNA AIR 34/138: ACIU, Interpretation Reports: BS1039 – BS1089.

TNA AIR 34/198: ACIU, Interpretation Reports: DS71-DS147.

TNA AIR 34/626: Miscellaneous papers on German V weapons accumulated by JARIC.

TNA AIR 34/703: 'The Chalk House with the Tudor Chimneys', a photographic record of the ACIU in the closing months of the Second World War.

TNA AIR 37/999: Supreme Headquarters Allied Expeditionary Force (Air) and 2nd Tactical Air Force, Account of Operation Crossbow.

TNA DEFE 40/17: Papers of Dr. R. V. Jones, ADI (Science), information from prisoners of war on German rocket sites.

TNA INF 1/292: INF, Home Intelligence Division, Home Intelligence weekly reports: January 1940 – December 1944.

Bibliography

Bowyer, M. J. F. (1972) 2 Group RAF: A Complete History – 1936-1945, London: Faber Paperbacks, pp.337-345, 363, 411.

Bode, V., and Kaiser, G. (2008) Building Hitler's Missiles: Traces of History in Peenemunde, Berlin: Ch. Links Verlag, pp. 38, 58-78.

Cabell, C. and Thomas, G. (2006) Operation Big Ben: The Anti-V2 Spitfire Missions, Stroud: Spellmount, p. 29.

Cadbury, D. (2005) Space Race, London: BBC Worldwide Limited.

Collins, L. and Lapierre, D. (1991) Is Paris Burning?, New York: Warner Books.

Hodgson, G. (2003) 'Obituary: Walt Rostow', in The Guardian, 17 February.

Ismay, H. L. I. (1960), The Memoirs of General the Lord Ismay, London: Heinemann, p. 360.

King, B. and Kutta, T. (1998) Impact: The History of Germany's V Weapons in World War II, Cambridge: Da Capo Press, p. 230-231.

Kreis J. (ed) (1986) Piercing The Fog: Intelligence and Army Air Forces Operations In World War II, Washington D.C.: Air Force History and Museums Program, pp. 76-77, 194, 203-204.

Ogley, B. (1992) Doodlebugs and Rockets: The Battle of the flying bombs, Westerham: Froglets Publications, pp.128, 148-150, 184-186.

Magry, K. (1998) 'Nordhausen', in After The Battle, no. 101, pp. 2-32, 41-42.

Ramsay, W. (ed) (1974) 'The V Weapons', in After The Battle, no. 6, pp. 2-41.

Simpson, B. (2007) Spitfire Dive-Bombers Versus the V2: Fighter

Command's Battle With Hitler's Mobile Missiles, Barnsley: Pend & Sword Aviation, pp. 167-171.

Smith, P. (2006) Air-Launched Doodblebugs: The Forgotten Campaign, Barnsley: Pen & Sword Aviation, pp.22-25, 172-179.

Smith, J. C. and Richardson, N. (1988) 'Flying Bombs Over The Pennines: The Story of the V1 attack aimed at Manchester on December 24th 1944', privately published.

Powys-Lybbe, U. (1983) The Eye of Intelligence, London: William Kimber, pp. 135-140, 142.

The Telegraph (2006) 'Obituary: Raymond Baxter' in The Telegraph, 16 September.

Jones, R. V. (1978) Most Secret War, London: Book Club Associates, pp. 432-433.

Chapter 14
Open Skies

Primary Sources

AFHRA IRISNUM/01114094: United States Air Force, Strategic Air Command, report on 'Project Casey Jones: Post-hostilities Aerial Mapping; Iceland, Europe and North Africa, June 1945 to December 1946', dated 30 September 1988, by Robert J. Boyd.

MDM DFG/3174: CBS interview notes, Harvey Brown, dinner at O'Donnells Sea Grill and Hotel Statler, Washington D.C., 10 October 1956.

MDM DFG/3174: CBS interview notes, Harvey Brown, dinner at Harvey's Restaurant, Washington D.C., 12 October 1956.

MDM DFG/3174: CBS interview notes, Harvey Brown, dinner at The Broadmoor resort, Colorado Springs, Colorado, 20 October 1956.

MDM DFG/5701: CBS interview notes, Douglas Kendall, lunch at L'Escargot, Greek Street, Soho, the ESU and the Grosvenor House Hotel, Park Lane, Mayfair, London, 14 November 1956.

MDM DFG/5782: CBS interview notes, Humphrey Spender, lunch at the Rice Bowl, 1 June 1956.

MDM: Unpublished memoirs, A War of Intelligence, by Douglas Neville Kendall.

NCAP GB/551/NCAP/4/3: Declassified catalogues of the aerial photographic coverage held by Hansa Luftbild, Tempelhof, Berlin.

TNA AIR 14/4104: Papers GX photography declassified by JARIC.

TNA AIR 19/1106: AIR, Project ROBIN files (1 January 1954 – 31 December 1955).

TNA AIR 20/12042: Papers accumulated by the AHB on the activities of the Medmenham Club.

TNA AIR 25/791: AIR, No. 106 (Photographic Reconnaissance) Group ORB, June 1945 – May 1946.

TNA AIR 29/229: ACIU / CIU ORB, September 1944 – March 1946.

TNA AIR 34/62: ACIU report on 'The German Photo Reconnaissance and Photo Intelligence Service'.

TNA AIR 34/77: ACIU reports on enemy PI.

TNA AIR 34/703: 'The Chalk House with the Tudor Chimneys', a photographic record of the ACIU in the closing months of the Second World War.

TNA AIR 34/749: JARIC declassified report on Project 'Turban'.

TNA AIR 14/4104: Papers accumulated, declassified by JARIC on GX photography.

TNA OD 70/1: Directorate of Overseas Surveys Newsletters 1 – 25.

Bibliography

Aldrich, R. J. (2001) The Hidden Hand: Britain, America and Cold War Secret Intelligence, New York: Overlook Press, pp. 206-207.

Babington Smith, C. (1957) 'How Photographic Detectives Solved Secret Weapon Mystery', in Life, 28 October, pp. 126-146.

Brugioni, D. (1991) Eyeball to Eyeball: The Inside Story of the Cuban Missile Crisis, New York: Random House.

Brugioni, D. (1999), Photo Fakery: The History and Techniques of Photographic Deception and Manipulation, Dulles: Brassey's, pp. 126, 137-139.

Burrows, W. E. (2001) By Any Means Necessary: America's Secret Air War in the Cold War, New York: Farrar, Straus and Giroux, pp. 131-135, 148-153.

Central Intelligence Agency (1979) The Holocaust Revisited: A Retrospective Analysis of the Auschwitz-Birkenau Extermination Complex, Washington D.C.: Central Intelligence Agency.

Fleming, E. A. (2010) 'Mapping A Northern Land: Obtaining The High Altitude Photography Required for the Completion of the Mapping of Canada by the Use of Wartime Aircraft (1951-1963)', in Geomatica, Vol. 64, No. 4, pp. 463-472.

Godziemba-Maliszewski, W. (1995) Katyn: An Interpretation of Aerial Photographs considered with Facts and Documents, Warsaw: Polish Geographical Society.

The Telegraph (2010) 'Obituary: Squadron Leader John Crampton', in The Telegraph, 1 August.

Fleming, I. (1965) The Man With the Golden Gun, London: Jonathon Cape.

Hall, C. H. (2003) 'Early Cold War Overflight Programs: An introduction', in Hall, C. H. and Laurie, C. (eds) Early Cold War Overflights – 1950-1956 – Symposium Proceedings – Volume 1: Memoirs, Washington D.C.: National Reconnaissance Office, Washington D.C., pp. 1-14.

Macdonald, A. (1996) Mapping the World: A History of the Directorate of Overseas Surveys 1946-1948, London: HMSO, pp. xi-xiii, p168.

Pedlow, G. W. and Welzenbach, D. E. (1998) The CIA and the U-2 Program – 1954-1974, Central Intelligence Agency: History Staff Centre for the Study of Intelligence, pp. 5-8, 23-24.

Williams, P. and Denney, J. (2004) How to Be Like Walt: Capturing the Disney Magic Every Day of Your Life, Deerfield Beach: HCI Books, p. 94.

Epilogue

Primary Sources

MDM DFG/3176: CBS interview notes, Charles Cabell, lunch at the CIA, Washington D.C., 11 October 1956.

MDM DFG/3184: CBS interview notes, George Goddard, 3714 Leland Street, Chevy Chase, Maryland, 14 October 1956.

MDM DFG/3191: CBS interview notes, Arthur Charles 'Art' Lundhal, CIA, Washington D.C., 11 October 1956.

MDM DFG/3192: CBS interview notes, Arthur Charles 'Art' Lundhal, Anacostia, Washington D.C., 16 October 1956.

MDM DFG/3221: Notes taken by CBS during visit to Beresford Avenue, Wembley, June 1956.

MDM DFG/5665: CBS interview notes, Sarah Churchill, Dorchester Hotel, Park Lane, London, 29 March 1957.

MDM DFG/5667: CBS interview notes, Sidney Cotton, Room 440, Dorchester Hotel, Park Lane, London, 15 April 1957.

MDM DFG/5668: CBS interview notes, Sidney Cotton, Dorchester Hotel, Park Lane, London, 18 April 1957.

MDM DFG/5669: CBS interview notes, Sidney Cotton, Dorchester Hotel, Park Lane, London, 18 April 1957.

MDM DFG/5701: CBS interview notes, Douglas Kendall, lunch at L'Escargot, Greek Street, Soho, the English Speaking Union, Dartmouth House, Charles Street, and the Grosvenor House Hotel, Park Lane, Mayfair, London, 14 November 1956.

MDM DFG/5764: CBS interview notes, Peter Riddell, lunch at The Ritz, Piccadilly, London, 13 April 1956.

MDM DFG/5766/1: Expedition by CBS and Peter Riddell to RAF Medmenham, Phyllis Court, and RAF Benson, 19 July 1956.

MDM DFG/5785: CBS interview notes, Peter Stewart in his Hertford Street apartment, Mayfair, and lunch at L'Escargot, Greek Street, Soho, London, 26 July 1956.

MDM DFG/5794: CBS interview notes, Claude Wavell at his home, Leindenhurst, Solent View Road, Gurnard, Cowes, Isle of Wight, Hampshire, 24 May 1956.

MDM DFG/5798: CBS interview notes, Frederick Winterbotham, Wotton Farm, Kingsbridge, Devon, 10 May 1956.

NCAP GB/551/NCAP/3/1: Papers on Project SEGMED accumulated by NCAP.

NCAP GB/551/NCAP/3/1/1: Extract of communication from Squadron Leader Anthony Plumb (Ret'd) to Allan Williams, detailing his involvement in Project SEGMED, 20 July 2011.

NCAP/GB/551/NCAP/7/1/1: Correspondence and papers documenting the transfer of photography from RAF Medmenham to the University College of North Staffordshire (now Keele University).

TNA AIR 2/10955: AIR, papers relating to the National Library of Air Photographs Committee, and proposed formation of a National Library of Air Photographs (1 May 1946 – 30 April 1950).

TNA AIR 2/15151: AIR, papers relating to the proposed formation of a National Library of Air Photographs (1 January 1957 – 31 December 1961).

TNA AIR 2/16079: AIR, papers relating to the proposed formation of a National Library of Air Photographs (1 January 1961 – 31 December 1963).

TNA AIR 25/791: AIR, No. 106 (Photographic Reconnaissance) Group ORB, June 1945 – May 1946.

TNA AIR 34/81: AIR, historical record of the CIU / ACIU Print Library, April 1941 – September 1945.

Bibliography

Beaver, S. (1986) 'The Keele Campus and its Environment', in Harrison, C. (ed), Essays on the History of Keele, Keele: University of Keele Press, pp. 155-159.

Board, C. (2004) 'Air Photo Mosaics: A short-term solution to topographic map revision in Great Britain 1944-1951' in Sheetlines: The Journal of the Charles Close Society, vol. 71, December, pp. 24-35.

Campbell, J. (1993) Edward Heath: A Biography, London: Jonathon Cape, p. 35.

Freeman, C. (1984) 'Anger over war pictures', in Daily Mail, 21 May.

Going, C. (2009) 'Déjà vu all over again? A Brief Preservation History of Overseas Service Aerial Photography in the UK', in Stichelbaut, B., Bourgeois, J., Saunders, N., and Chielens, P. (eds), Images of Conflict: Military Aerial Photography and Archaeology, Newcastle Upon Tyne: Cambridge Scholars Publishing, pp. 121-134.

Hansard HC Deb 10 April 1946, vol. 421, cc1913-4: Question to the Secretary of State for Air, from Mr Edward Keeling M.P., on the future preservation of Second World War aerial photography.

Harley, J. B. (1975) Ordnance Survey Maps: A Descriptive Manual, Southampton: Ordnance Survey.

Morley, J. D. (1995) 'Bomb Buster' in The Times Magazine, 11 February, pp. 18-20.

Mountford, J. (1972) Keele: An Historical Critique, London: Routledge, pp. 188-190, 246-25.

Ottaway, R. H. (2010) RAF Benson: A diary of wartime losses, Andover: Speedman Press.

Sawbridge, C. (2000) 'The Royal Air Force and Ewelme' in Chisolm, A. (ed), Glimpses of an Oxfordshire Village: Ewelme 1900-2000, Ewelme: The Ewelme Society.

The Sentinel (1985) 'RAF air photos loan to Germany', in The Sentinel, 1 November.

Swanton, O. (1999) 'Wave Goodbye to the silver', in Guardian Higher Education, 2 February.

Ryder, N. D. (1945) 'Keeping watch on Jerry', in Flight, 13 September 1945, pp. 290-291.

Walton, A. D. (1975) 'Air Photograph interpretation and the Keele University collection' in Phillips, A. D. M., and Turton, B. J. (eds), Environment, Man and Economic Change: Essays presented to S. H. Beaver, London: Longman, pp. 492-496.

Plate section images by kind permission of the Medmenham Collection, Dave Lefurgey, Allan Williams, US National Archives & Records Administration (NARA), RAF Air Historical Branch and Geoffrey Stone.

Maps by Michelle Lile.

Index